Transgenic Plants

METHODS IN MOLECULAR BIOLOGY™

John M. Walker, SERIES EDITOR

300. **Protein Nanotechnology:** *Protocols, Instrumentation, and Applications,* edited by *Tuan Vo-Dinh, 2005*
299. **Amyloid Proteins:** *Methods and Protocols,* edited by *Einar M. Sigurdsson, 2005*
298. **Peptide Synthesis and Application,** edited by *John Howl, 2005*
297. **Forensic DNA Typing Protocols,** edited by *Angel Carracedo, 2005*
296. **Cell Cycle Protocols,** edited by *Tim Humphrey and Gavin Brooks, 2005*
295. **Immunochemical Protocols,** *Third Edition,* edited by *Robert Burns, 2005*
294. **Cell Migration:** *Developmental Methods and Protocols,* edited by *Jun-Lin Guan, 2005*
293. **Laser Capture Microdissection:** *Methods and Protocols,* edited by *Graeme I. Murray and Stephanie Curran, 2005*
292. **DNA Viruses:** *Methods and Protocols,* edited by *Paul M. Lieberman, 2005*
291. **Molecular Toxicology Protocols,** edited by *Phouthone Keohavong and Stephen G. Grant, 2005*
290. **Basic Cell Culture,** *Third Edition,* edited by *Cheryl D. Helgason and Cindy Miller, 2005*
289. **Epidermal Cells,** *Methods and Applications,* edited by *Kursad Turksen, 2004*
288. **Oligonucleotide Synthesis,** *Methods and Applications,* edited by *Piet Herdewijn, 2004*
287. **Epigenetics Protocols,** edited by *Trygve O. Tollefsbol, 2004*
286. **Transgenic Plants:** *Methods and Protocols,* edited by *Leandro Peña, 2004*
285. **Cell Cycle Control and Dysregulation Protocols:** *Cyclins, Cyclin-Dependent Kinases, and Other Factors,* edited by *Antonio Giordano and Gaetano Romano, 2004*
284. **Signal Transduction Protocols,** *Second Edition,* edited by *Robert C. Dickson and Michael D. Mendenhall, 2004*
283. **Bioconjugation Protocols,** edited by *Christof M. Niemeyer, 2004*
282. **Apoptosis Methods and Protocols,** edited by *Hugh J. M. Brady, 2004*
281. **Checkpoint Controls and Cancer, Volume 2:** *Activation and Regulation Protocols,* edited by *Axel H. Schönthal, 2004*
280. **Checkpoint Controls and Cancer, Volume 1:** *Reviews and Model Systems,* edited by *Axel H. Schönthal, 2004*
279. **Nitric Oxide Protocols,** *Second Edition,* edited by *Aviv Hassid, 2004*
278. **Protein NMR Techniques,** *Second Edition,* edited by *A. Kristina Downing, 2004*
277. **Trinucleotide Repeat Protocols,** edited by *Yoshinori Kohwi, 2004*
276. **Capillary Electrophoresis of Proteins and Peptides,** edited by *Mark A. Strege and Avinash L. Lagu, 2004*
275. **Chemoinformatics,** edited by *Jürgen Bajorath, 2004*
274. **Photosynthesis Research Protocols,** edited by *Robert Carpentier, 2004*
273. **Platelets and Megakaryocytes, Volume 2:** *Perspectives and Techniques,* edited by *Jonathan M. Gibbins and Martyn P. Mahaut-Smith, 2004*
272. **Platelets and Megakaryocytes, Volume 1:** *Functional Assays,* edited by *Jonathan M. Gibbins and Martyn P. Mahaut-Smith, 2004*
271. **B Cell Protocols,** edited by *Hua Gu and Klaus Rajewsky, 2004*
270. **Parasite Genomics Protocols,** edited by *Sara E. Melville, 2004*
269. **Vaccina Virus and Poxvirology:** *Methods and Protocols,*edited by *Stuart N. Isaacs, 2004*
268. **Public Health Microbiology:** *Methods and Protocols,* edited by *John F. T. Spencer and Alicia L. Ragout de Spencer, 2004*
267. **Recombinant Gene Expression:** *Reviews and Protocols, Second Edition,* edited by *Paulina Balbas and Argelia Johnson, 2004*
266. **Genomics, Proteomics, and Clinical Bacteriology:** *Methods and Reviews,* edited by *Neil Woodford and Alan Johnson, 2004*
265. **RNA Interference, Editing, and Modification:** *Methods and Protocols,* edited by *Jonatha M. Gott, 2004*
264. **Protein Arrays:** *Methods and Protocols,* edited by *Eric Fung, 2004*
263. **Flow Cytometry,** *Second Edition,* edited by *Teresa S. Hawley and Robert G. Hawley, 2004*
262. **Genetic Recombination Protocols,** edited by *Alan S. Waldman, 2004*
261. **Protein–Protein Interactions:** *Methods and Applications,* edited by *Haian Fu, 2004*
260. **Mobile Genetic Elements:** *Protocols and Genomic Applications,* edited by *Wolfgang J. Miller and Pierre Capy, 2004*
259. **Receptor Signal Transduction Protocols,** *Second Edition,* edited by *Gary B. Willars and R. A. John Challiss, 2004*
258. **Gene Expression Profiling:** *Methods and Protocols,* edited by *Richard A. Shimkets, 2004*
257. **mRNA Processing and Metabolism:** *Methods and Protocols,* edited by *Daniel R. Schoenberg, 2004*

METHODS IN MOLECULAR BIOLOGY™

Transgenic Plants

Methods and Protocols

Edited by

Leandro Peña

Instituto Valenciano de Investigaciones Agrarias,
Valencia, Spain

HUMANA PRESS TOTOWA, NEW JERSEY

Softcover reprint of the hardcover 1st edition 2004
999 Riverview Drive, Suite 208
Totowa, New Jersey 07512

humanapress.com

This publication is printed on acid-free paper. ∞
ANSI Z39.48-1984 (American Standards Institute)

Permanence of Paper for Printed Library Materials.
Cover illustration:Figure 2 from Chapter 11, "Regeneration of Transgenic Cassava From Transformed Embyronic Tissues," by Peng Zheng and Johanna Puonti-Kaerlas

Production Editors: Angela Burkey and Amy Thau.
Cover design by Patricia F. Cleary.

For additional copies, pricing for bulk purchases, and/or information about other Humana titles, contact Humana at the above address or at any of the following numbers: Tel.: 973-256-1699; Fax: 973-256-8341; E-mail: humana@humanapr.com; or visit our Website: www.humanapress.com

10 9 8 7 6 5 4 3 2 1

ISBN 978-1-61737-461-6 e-ISBN 978-1-59259-827-4

Library of Congress Cataloging-in-Publication Data

Transgenic plants : methods and protocols / edited by Leandro Peña.
p. cm. -- (Methods in molecular biology ; v. 286)
Includes bibliographical references (p.).

1. Transgenic plants--Laboratory manuals. I. Peña, Leandro. II. Series.
SB123.57.T728 2004
631.5'233--dc22
2004042471

Preface

The aim of *Transgenic Plants: Methods and Protocols* is to provide a source of information to guide the reader through a wide range of frequently used, broadly applicable, and easily reproducible techniques involved in the generation of transgenic plants. Its step-by-step approach covers a series of methods for genetically transforming plant cells and tissues, and for recovering whole transgenic plants from them. The volume then moves on to the use of selectable and reporter markers, positive selection, marker elimination after recovery of transgenic plants, and the analysis of transgene integration, expression, and localization in the plant genome. Although contributors usually refer to model plants in most chapters, the protocols described herein should be widely applicable to many plant species. The last two sections are devoted to methods of risk assessment and to exploring the current and future applications of transgenic technology in agriculture and its social implications in a case study.

Transgenic Plants: Methods and Protocols is divided into six major sections plus an introduction, comprising 27 chapters. Part I, the Introduction, is a review of the past, present, and perspectives of the transgenic plants, from the discovery of *Agrobacterium tumefaciens* as a feasible transformation vector, to its use as a tool to study gene expression and function, and the current and possible future applications of this technology in agriculture, industry, and medicine. Part II covers the most commonly used transformation systems, including *Agrobacterium tumefaciens*, *Agrobacterium rhizogenes*, particle bombardment, electroporation, floral dip, and chloroplast transformation. Part III covers regeneration of whole transgenic plants by both organogenesis and somatic embryogenesis from different explant cells/tissues and from such diverse plant species as tomato, cassava, conifers, and citrus. Part IV covers the use of selectable and reporter markers, exemplified by the utilization of the *nptII* and *bar* genes for wheat transformation, and by β-glucuronidase (GUS) and green fluorescent protein (GFP) detection and quantification, respectively. Positive selection (for maize transformation) is also described as an alternative to the use of antibiotic and herbicide resistance genes as selectable markers. Also covered in this section is the controlled excision and removal of marker genes from both nuclei and plastids once transgenic shoots have been efficiently generated. Part V treats the study of transgene copy number and organization by quantitative real-time polymerase chain reaction (PCR), and the analysis of transgene expression by Northern and dot-blot hybridizations using nonradioactive probing methods, by reverse transcription (RT)-PCR, and by

RNA *in situ* hybridization. Also described is the use of matrix attachment regions (MARs) flanking the transgenes to obtain predictable and stable expression of the transgenic traits. Fluorescence *in situ* hybridization (FISH) is described as a method to map transgenes physically in specific plant chromosome regions. This section also covers the use of thermal asymmetric interlaced (TAIL)-PCR to amplify (and precisely determine by sequencing) genomic sequences flanking transgene insertions. Part VI covers risk assessment methods for studying *Agrobacterium* persistence in plant tissues and to investigate the possibility of transgene dispersal through pollen. Part VII provides an overview of the current and next generations of transgenic crops based not only in the most recent scientific literature, but also in patent applications. Social implications of the transgenic crops are exemplified by the development and impact of the virus-resistant transgenic papayas in Hawaii, Jamaica, and Venezuela.

Transgenic Plants: Methods and Protocols has been planned, written, and edited with the intention of being useful for those beginners and experienced scientists looking for a laboratory manual covering all aspects of plant genetic transformation. I greatly hope you will find it helpful.

I would like to thank the staff of Humana Press, series editor John Walker, and the authors for all their effort and for being so supportive and patient.

Leandro Peña

Contents

Contributors

GEORGE C. ALLEN • *Department of Crop Science, North Carolina State University, Raleigh, NC*

LORELEI J. BILHAM • *Department of Crop Genetics, John Innes Centre, Norwich, UK*

ROBERT H. BRAUN • *New Zealand Institute for Crop & Food Research, Christchurch, New Zealand*

MAGDALENA CERVERA • *Department of Plant Protection and Biotechnology, Instituto Valenciano de Investigaciones Agrarias, Valencia, Spain*

YUANLIN CHEN • *College of Life Science, South China Agricultural University, Guangzhou, China*

MARY C. CHRISTEY • *New Zealand Institute for Crop & Food Research, Christchurch, New Zealand*

STEVEN J. CLOUGH • *USDA-ARS, Department of Crop Science, University of Illinois, Urbana, IL*

JAIME CUBERO • *Department of Plant Protection, INIA, Madrid, Spain*

IAN S. CURTIS • *Department of Biotechnology, National Institute of Agrobiological Sciences, Tsukuba, Japan*

ABHAYA M. DANDEKAR • *Department of Pomology, University of California, Davis, CA*

HENRY DANIELL • *Department of Molecular Biology and Microbiology, University of Central Florida, Orlando, FL*

MICHAEL R. DAVEY • *Plant Sciences Division, School of Biosciences, University of Nottingham, Loughborough, UK*

ANIL DAY • *School of Biological Sciences, University of Manchester, Manchester, UK*

AMIT DHINGRA • *Department of Molecular Biology and Microbiology, University of Central Florida, Orlando, FL*

JIM M. DUNWELL • *Department of Agricultural Botany, School of Plant Sciences, The University of Reading, Reading, UK*

HIROYASU EBINUMA • *Pulp and Paper Research Laboratory, Nippon Paper Industries Co., Ltd., Tokyo, Japan*

SAORI ENDO • *Pulp and Paper Research Laboratory, Nippon Paper Industries Co., Ltd., Tokyo, Japan*

GUSTAVO FERMÍN • *Facultad de Ciencias, Universidad de Los Andes, Mérida, Venezuela*

HENRY J. FISK • *Department of Pomology, University of California, Davis, CA*
ANNE FRARY • *Department of Molecular Biology and Genetics, Izmir Institute of Technology, Izmir, Turkey*
CAROL GONSALVES • *Pacific Basin Agricultural Research Center, Hilo, HI*
DENNIS GONSALVES • *Pacific Basin Agricultural Research Center, Hilo, HI*
JULIA L. GOODWIN • *Crop Performance and Improvement Department, Rothamsted Research, Hertfordshire, UK*
MATTHEW D. HALFHILL • *Department of Plant Sciences, University of Tennessee, Knoxville, TN*
GENEVIÈVE HANSEN • *Protein Therapeutics, Diversa Corporation, San Diego, CA*
WENDY A. HARWOOD • *Department of Crop Genetics, John Innes Centre, Norwich, UK*
LUIS HERRERA-ESTRELLA • *Departamento de Ingeniería Genética de Plantas, Centro de Investigación y de Estudios Avanzados, Guanajuato, Mexico*
LAURA C. HUDSON • *Department of Plant Sciences, University of Tennessee, Knoxville, TN*
SIRILUCK IAMTHAM • *School of Biological Sciences, University of Manchester, Manchester, UK*
DAVID J. INGHAM • *BASF Plant Sciences, L.L.C., Research Triangle Park, NC*
HUW D. JONES • *Crop Performance and Improvement Department, Rothamsted Research, Hertfordshire, UK*
JOSÉ JUÁREZ • *Department of Plant Protection and Biotechnology, Instituto Valenciano de Investigaciones Agrarias, Valencia, Spain*
JULIE R. KIKKERT • *Department of Horticultural Sciences, New York State Agricultural Experiment Station, Cornell University, Geneva, NY*
KRYSTYNA KLIMASZEWSKA • *Natural Resources Canada, Canadian Forest Service, Quebec, Canada*
VASUMATHI KODE • *School of Biological Sciences, University of Manchester, Manchester, UK*
DAVID LEE • *Department of Applied Economics and Management, Cornell University, Ithaca, NY*
YAO-GUANG LIU • *South China Agricultural University, College of Life Science, Guangzhou, China*
MARÍA M. LÓPEZ • *Department of Plant Protection and Biotechnology, Instituto Valenciano de Investigaciones Agrarias, Valencia, Spain*
PANAGIOTIS MADESIS • *School of Biological Sciences, University of Manchester, Manchester, UK*

MIGUEL MARTÍNEZ-TRUJILLO • *Departamento de Ingeniería Genética de Plantas, Centro de Investigación y de Estudios Avanzados, Guanajuato, Mexico*
ETSUKO MATSUNAGA • *Pulp and Paper Research Laboratory, Nippon Paper Industries Co., Ltd., Tokyo, Japan*
REGINALD J. MILLWOOD • *Department of Plant Sciences, University of Tennessee, Knoxville, TN*
SUBHASH C. MINOCHA • *Department of Plant Biology, University of New Hampshire, Durham, NH*
LUIS NAVARRO • *Department of Plant Protection and Biotechnology, Instituto Valenciano de Investigaciones Agrarias, Valencia, Spain*
ANDREW F. PAGE • *Department of Plant Biology, University of New Hampshire, Durham, NH*
GABRIELA M. PASTORI • *Crop Performance and Improvement Department, Rothamsted Research, Hertfordshire, UK*
LEANDRO PEÑA • *Department of Plant Protection and Biotechnology, Instituto Valenciano de Investigaciones Agrarias, Valencia, Spain*
JOHANNA PUONTI-KAERLAS • *European Patent Office, Munich, Germany*
BRUCE I. REISCH • *Department of Horticultural Sciences, New York State Agricultural Experiment Station, Cornell University, Geneva, NY*
OSCAR N. RUIZ • *Department of Molecular Biology and Microbiology, University of Central Florida, Orlando, FL*
ROBERT G. RUTLEDGE • *Candian Forest Service, Natural Resources Canada, Quebec, Canada*
HAROLDO SALVO-GARRIDO • *Biotechnology Unit, INIA Carillanca, Temuco, Chile*
ARMAND SÉGUIN • *Natural Resources Canada, Canadian Forest Service, Quebec, Canada*
JUNE SIMPSON • *Departamento de Ingeniería Genética de Plantas, Centro de Investigación y de Estudios Avanzados, Guanajuato, Mexico*
JOHN W. SNAPE • *Department of Crop Genetics, John Innes Centre, Norwich, UK*
STEVEN SPIKER • *Department of Genetics, North Carolina State University, Raleigh, NC*
C. NEAL STEWART, JR. • *Department of Plant Sciences, University of Tennessee, Knoxville, TN*
KOICHI SUGITA • *Pulp and Paper Research Laboratory, Nippon Paper Industries Co., Ltd., Tokyo, Japan*

PAULA TENNANT • *Department of Life Sciences, University of the West Indies, Kingston, Jamaica*
WILLIAM F. THOMPSON • *Department of Botany, North Carolina State University, Raleigh, NC*
SILVIA TRAVELLA • *Institute of Plant Biology, University of Zurich, Zurich, Switzerland*
JOYCE VAN ECK • *The Boyce Thompson Institute for Plant Research, Ithaca, NY*
JOSÉ R. VIDAL • *Department of Horticultural Sciences, New York State Agricultural Experiment Station, Cornell University, Geneva, NY*
ALLAN WENCK • *BASF Plant Sciences L.L.C., BASF Corporation, Research Triangle Park, NC .*
KEIKO YAMADA • *Pulp and Paper Research Laboratory, Nippon Paper Industries Co., Ltd., Tokyo, Japan*
PENG ZHANG • *Institute of Plant Sciences, ETH-Zentrum, Zurich, Switzerland*
QUNYU ZHANG • *College of Life Science, South China Agricultural University, Guangzhou, China*

I

INTRODUCTION

1

Transgenic Plants

An Historical Perspective

Luis Herrera-Estrella, June Simpson, and Miguel Martínez-Trujillo

Summary

The development of technologies that allow the introduction and functional expression of foreign genes in plant cells has extended in less than two decades to the production of transgenic plants with improved insect and disease resistance, seeds and fruits with enhanced nutritional qualities, and plants that are better adapted to adverse environmental conditions. Vaccines against serious human diseases and other important products have also been developed using transgenic plants. Many more agronomic and quality traits are currently being engineered in both academic and industrial laboratories, which are limited only by our poor knowledge of plant gene function. The emergence of new functional genomic strategies for the identification and characterization of genes promises to provide a wealth of information with an enormous potential to enhance traditional plant breeding and to genetically engineer plants for specific purposes. This chapter describes some of the highlights in the development of these technologies and some of the major achievements in production and commercialization of transgenic crops. We also discuss some of the biosafety issues related to release of this novel class of plants into the environment.

Key Words: Biosafety regulations; disease and pest resistance; genetic engineering; metabolic engineering; plant protection; transgenic plants.

1. Introduction

To date, the world population stands at more than 6 billion people, and it is expected to reach 9 billion by the year 2050. Food production will need to increase at the same rate or more to satisfy the needs of such an enormous number of people. Plants, the first link in the food chain, obtain energy from sunlight and transform it into compounds that directly or indirectly provide the

From: *Methods in Molecular Biology, vol. 286: Transgenic Plants: Methods and Protocols*
Edited by: L. Peña © Humana Press Inc., Totowa, NJ

food necessary for the survival of other living organisms. Traditionally, plants have been improved through selection during many crop cycles, and to date this has produced high yielding varieties, especially in the case of hybrids, which formed the basis of the green revolution. One challenge that faces traditional agriculture is the fact that normally, only individuals of the same species can be crossbred. If natural resistance to a specific insect or fungus does not exist, then traditional breeders cannot create resistance or introgress this trait. Therefore, it is necessary to search for alternative sources of genes in other species of plants, microbes, or fungi. The genes harbored in other species can now be transferred to different plant species with the appropriate regulatory sequences so as to add a new trait or modify an existing one. Plant genetic engineering has become possible as a result of the work of many researchers during the last two decades.

Twenty years ago, results of the first experiments describing the successful transfer and expression of foreign genes in plant cells were published. Since then, transgenic plants have become an essential tool for studying plant biology and for the development of novel plant varieties that have been cultivated extensively in some regions of the world. Transgenic technology has had a profound impact on the rapid development of plant biology in the past 15 yr by providing the means of producing gene-tagged populations, cell markers to study plant development, and the technology to study gene function. In terms of agricultural production, the impact of transgenic technology has not achieved its full potential because of the controversies that this new technology has generated and the strict regulatory systems that have been adopted by many countries. This chapter briefly discusses some of the highlights that led to the development of transgenic plants; also reviews some of the tools that transgenic technology has provided to study plant biology together with several of the major plant improvements achieved using this technology.

2. Plant Transformation Methods

Different methods have been developed to introduce foreign genes into plants. A common feature is that the transforming DNA has to bypass different membrane barriers; it first has to enter the plant cell by penetrating the plant cell wall and the plasma membrane and then must reach the nucleus and integrate into the resident chromosomes. For the majority of species gene transfer is carried out using explants competent of regeneration to obtain complete, fertile plants. This implies the development of a tissue culture technology that frequently becomes an art. Although gene transfer technology has become routine in working with several plant species, in others the limiting step is not the transformation itself but rather the lack of efficient regeneration protocols.

The most widely used and successful transformation methods are the *Agrobacterium tumefaciens*-mediated DNA transfer and direct transfer through particle bombardment.

2.1. The Agrobacterium *System*

In 1907, Smith and Townsend demonstrated that the Gram-negative soil bacterium *Agrobacterium tumefaciens*, a member of the eubacterial family Rhizobiaceae, is the organism responsible for the elicitation of crown gall tumors in plants; formation of these tumors occurs as a result of bacterial infection, usually at wound sites, on many dicotyledonous and some monocotyledonous plants *(1)*. This discovery had no major repercussion until Armin Braun demonstrated that tumor cells are transformed and that the uncontrolled proliferation of the tumor cells was not dependent on the continuous presence of *Agrobacterium*, implying the existence of a transformation-inducing principle *(2)*. In 1974, Ivo Zaenen, Jeff Schell and Marc Van Montagu *(3)* at the University of Ghent, Belgium, identified a megaplasmid that was present only in the virulent strains of *Agrobacterium* and absent in the avirulent ones, and named it Ti plasmid for tumor-inducing plasmid. Three years later, Eugene Nester, Milton Gordon, and Mary-Dell Chilton *(4)*, at the University of Washington, demonstrated that only some genes of the Ti plasmid were transferred to the chromosomes of the plant cell and were responsible for inducing tumors. The DNA segment transferred to plant cells was named T-DNA and is delimited by left and right borders, which are 25-basepair imperfect, direct repeats. Researchers reasoned that any piece of DNA between these borders could be transferred into the plant cell and randomly integrated into the genome of the plant. Taking into account this consideration, research teams at the University of Ghent, the Monsanto Company, and the University of Washington at St. Louis, Missouri, inserted heterologous genes with the appropriate regulatory regions into the T-DNA region and showed that foreign genes became integrated and functionally expressed in plant cells. Later, disarmed Ti plasmids, which contain a T-DNA lacking genes involved in tumor formation, were used to produce the first transgenic plants *(5–7)*.

In the years since these early experiments using *Nicotiana tabacum* and *Petunia hybrida*, the *Agrobacterium* system has been used to transform a large range of dicotyledonous plant species. Although initially the transformation of cereals was considered impossible, a few years later it was shown that cereals such as maize and rice could also be transformed. More recently fungi have also been transformed using this system *(8)*. A large number of plant species has been transformed with this method *(9)*.

The *Agrobacterium* system has several advantages over other transformation methods and it is considered as the first option to transform plants. These advantages include the following aspects: (a) In a significant percentage of the transformation events, a single copy of the T-DNA is integrated into the chromosomes of the transformed cell ***(10)***. (b) Numerous vector systems are now available containing the T-DNA borders and various reporter and selectable marker genes, allowing researchers to choose the most appropriate combination to insert heterologous genes. (c) It is possible to transfer large fragments of DNA, including bacterial artificial chromosomes ***(11)***. (d) Transformation *in planta*, without the necessity of tissue culture, is possible in some species such as *Arabidopsis thaliana* and *Medicago truncula*t*a* ***(12)***. For a more detailed description of the *Agrobacterium*-mediated gene transfer system, *see* Chapter 2.

2.2. The Biolistic Method

The biolistic method was developed as a necessity to transform plant species originally recalcitrant to transformation by the *Agrobacterium* system including the economically important cereals. This method consists of the delivery of microprojectiles, usually of tungsten or gold, coated with DNA and propelled into the target cells by acceleration. The acceleration can be provided by an explosion of gunpowder or a discharge of high-pressure gases such as helium or CO_2 ***(13,14)***. Molecular analysis of plants transformed biolistically in general reveals a complex pattern of transgene, indicating the integration of multiple copies of the bombarded DNA. However, it has been demonstrated that in most cases, these multiple copies are arranged as a single locus and segregate in a Mendelian pattern ***(15)***. As with *Agrobacterium*, a great number of diverse plant species have been transformed by the biolistic method ***(9)***. Some advantages of the biolistic method are the following: (a) A wide variety of types of explants can be used to undergo bombardment and obtain fertile plants. (b) There is no need for specialized transformation vectors. (c) This is the only reliable method for chloroplast transformation. More detail information of the biolistic method is provided in Chapter 4.

2.3. Other Transformation Methods

The direct transfer of DNA to protoplasts using polyethylene glycol (PEG), calcium phosphate, or electroporation has been shown to be possible in various of plants including maize (*see* Chapters 5 and 8 and **ref. *16***). Low reproducibility and the regeneration of plants were the main problems because these methods are often specific for certain cultivars. The microinjection technique employs immobilized cells into which the DNA is internalized individually. However, the tedious manipulation, need for sophisticated equipment, and difficulty of regeneration of plants have not permitted its wider utilization.

3. The Use of Transgenic Plants to Study Gene Expression and Function

Transgenic plants have been used extensively to study gene expression and function. For this purpose, plants are transformed with chimeric gene constructs in which a reporter gene is under the control of the regulatory sequences of the gene to be analyzed.

Several reporter genes are commonly used in plants, including β-glucuronidase, luciferase, and genes involved in anthocyanin biosynthesis (*see* Chapter 14 and **ref. *17***). More recently, the gene for the green fluorescent protein (GFP) has become an important in vivo reporter in plants. When expressed in plant cells and illuminated with blue light, GFP produces a stable bright green fluorescence that is easily monitored nondestructively (*see* Chapter 15 and **ref. *18***). Thus, it can be used as a means to visualize the fate of transformed cells over time and rapidly test the influence of various factors on gene expression.

These new generations of reporter genes are easily monitored for expression, and allow rapid determination of sequences important in regulating the temporal, spatial, and environmental expression of a gene in great detail. Reporter genes have been instrumental in the analysis of gene expression under a myriad of environmental stimuli, including light, wounding, temperature, growth hormones, and so forth and in different plant tissues. These studies are leading to the unraveling of the complex interactions involved in the responses of plants to these stimuli.

Reducing or increasing the expression of the target gene by sense and antisense or cosuppression strategies can be used to study gene function. Analysis of the phenotype or changes in mRNA or metabolite profiles can provide valuable information to determine gene function. Plant transformation is now also widely used as a tool for insertional mutagenesis, either directly by the T-DNA or by the mobilization of transposons into species in which these elements have not been characterized. This strategy produces a collection of individuals containing transposon or T-DNA insertions throughout the genome. These insertion mutants can then be systematically screened for interesting phenotypes and the affected genes identified and isolated with relative ease. This strategy has been carried out successfully in several plant species to date, including *Arabidopsis*, tomato, and rice ***(19–21)***.

4. Production of Transgenic Plants With Important Applications in Agriculture, Industry, and Medicine

With the development of plant transformation methods, knowledge of the structure and function of certain genes, and the desire to resolve some of the classic problems in traditional agriculture, the race to obtain better plants by genetic engineering began with satisfactory results. Initial strategies considered the introduction of single genes into plants of interest; now, however, strategies involving multiple genes from a single metabolic pathway can be used. The main strategies used to produce improved transgenic plants with commercial or agricultural applications are mentioned in the following subheadings.

4.1. Nutrients and Quality of Seeds and Fruits

Postharvest losses are one of the main constraints in preservation and commercialization of agricultural products. In the case of fruits, it is essential to conserve quality properties during transport and storage to ensure successful marketing. The main problem is fruit softening resulting from the ripening process. Using plant antisense technology it has been possible to delay ripening by decreasing the expression of genes important in this process, such as those involved in cell wall degradation or the biosynthesis of ethylene *(22,23)*. To date, tomatoes have been modified for slower ripening and higher solid content and commercialized by three different companies *(24)*. This strategy has an enormous potential for tropical fruit such as mango and papaya, grown in many developing countries. In such countries, the lucrative export market cannot be exploited because the fruits ripen rapidly and there is a lack of appropriate storage conditions and efficient transport systems to enable them to reach the end consumer *(25)*.

The major sources of proteins for a large portion of the human population are cereal grains and legume seeds. However, a characteristic of these seeds is a deficiency in lysine in cereals and cysteine and methionine in legumes. One obvious solution to this problem would be the consumption of both kinds of seeds in adequate proportions; however, in the case of human nutrition there are cultural traditions and economical factors that prevent this. An alternative is to change the seed protein composition of certain crops. Efforts in this direction include the production of methionine-rich proteins in transgenic tobacco *(26)* and canola seeds, which results in an increase of up to 33% in methionine *(27)*; the expression of a sunflower seed albumin gene in lupins (*Lupinus angustifolius*) causing the methionine content to double *(28)*; or a synthetic gene that encodes a protein with 43% lysine content in tobacco seeds *(29)*. A 100-fold increase in free lysine in soybean and canola was possible by modifying the regulatory properties of enzymes involved in synthesis of this essential amino acid *(30)*. Potato, the most important vegetable food crop, was transformed with a gene from amaranth that encodes a seed-specific nonallergenic protein (AmA1), with a balanced amino acid composition that promises to improve the nutritional value of this food source *(31)*. Recently, the Indian government has authorized cultivation of these transgenic potatoes to help alleviate the serious malnutrition problems in that country.

Vitamin A deficiency is a very important nutritional problem in many countries, especially in Asia, where 124 million children suffer from blindness caused by deficiency of this vitamin. As a potential solution to this problem, a strategy has been developed to produce vitamin A in rice. Rice does not normally produce vitamin A, but the genes encoding the three enzymes for β-carotene (pro-vitamin A) biosynthesis that are absent in rice were specifically expressed in the endosperm of transgenic rice seeds *(32)*. Because rice is an important dietary component in Asia, consumption of the transgenic so-called "golden rice" could help alleviate vitamin A deficiency in this region.

Plant oils have many uses in both food and industrial applications, therefore, the manipulation of fatty acid composition and content is one of the areas of greatest interest in metabolic engineering. One of the goals in manipulating fatty acid composition is to produce healthier vegetable oils. Most vegetable oil used for food applications is partially or fully hydrogenated to produce semisolid spreads, a process that results in the production of *trans*-isomers of unsaturated fatty acids that are normally not present in plants and have been associated with coronary heart disease. Increasing the content of stearic acid in soybean, cotton and *Brassica* oilseeds using antisense, cosuppression, and RNA interference to down-regulate endogenous stearyl-ACP synthase has allowed the production of semisolid margarine without the need for hydrogenation *(33)*. An oxidatively stable liquid oil low in saturated fatty acids has also been produced in soybeans by suppression of the oleoyl desaturase. This oil has been produced commercially and is extremely stable for high-temperature frying applications *(33)*. In the future, it will be necessary to increase the production of oil, including novel types of oils, to satisfy the demand not only for edible oils but also for industrial oils, considering that the nonrenewable fossil oils will be depleted in the future, and vegetable oils are a renewable resource that can be harvested at a rate of several million tons annually.

The phytoene synthase from the bacterium *Erwinia uredovora* has been overexpressed in tomato (*Lycopersicon esculentum*). Fruit-specific expression was achieved by using the tomato polygalacturonase promoter and the levels of phytoene, lycopene, β-carotene, and lutein levels were increased 2.4-, 1.8-, and 2.2-fold, respectively *(34)*. These changes in flux coefficients have revealed a shift in the regulatory step of carotenogenesis, which has important implications for future metabolic engineering strategies. In addition, a high consumption of tomatoes in the human diet could help to reduce the onset of chronic diseases such as coronary heart diseases and certain cancers.

Although coffee has no nutritive function in the human diet, it does have an important social and psychological role as it accompanies various social events and is part of the daily habits of many individuals. The consumption of caffeine, however, can adversely affect sensitive individuals by producing insomnia and increasing blood pressure, among other effects. With this in mind, in plants of *Coffea canephora* the gene encoding theobromine synthase, an enzyme involved in the synthesis of caffeine, was inhibited using RNA interference ***(35)***. The caffeine content of these plants was reduced by up to 70%, indicating that it would be possible to produce naturally "decaffeinated" coffee seeds. This strategy is now being applied to *Coffea arabiga*, which accounts for roughly 70% of the world coffee market.

4.2. Insect and Virus Resistance

The damage to crops caused by insects, viruses, and other pathogens represents one of the most important problems in agricultural production. Therefore, major efforts both in public institutions and the private sector have focused on developing pest- and disease-resistant transgenic plants that have resulted in some of the most successful genetically engineered plant products.

The development of insect-resistant transgenic plants was based on the knowledge of insecticidal proteins. *Bacillus thuringiensis* (*Bt*) is a soil microorganism that produces proteins called δ-endotoxins during sporulation. When δ-endotoxins are ingested by insects, they bind to midgut epithelial cells, causing their osmotic lysis ***(36)***. Many *Bt* strains express different δ-endotoxins, each with their own spectrum of activity, against different types of insects ***(37)***. For example, the Cry1A and Cry1C proteins are specific to lepidopteran larvae such as those of the European corn borer (*Ostrinia nubilalis*), whereas the Cry3A protein is toxic to coleopteran larvae such as those of the Colorado potato beetle (*Leptinotarsa decemlineata*) ***(38)***. δ-Endotoxin genes have been manipulated for expression in plants by generation of truncated versions of the genes, decreasing their GC content and changing codons that are seldom used in plants ***(36)***. Transgenic plants expressing δ-endotoxins from several strains of *Bacillus thuringiensis* (known as *Bt* lines), have to date been generated and effectively used for insect control in various species, including tobacco ***(39)***, tomato ***(40)***, cotton ***(41)***, potato ***(42)***, maize ***(43)***, canola ***(44)***, soybean ***(45)***, and rice ***(46)***. *Bt* maize plants provide excellent protection against European corn borer (*Ostrinia nubilalis*) under insect pressure several hundred-fold higher than natural infestations, and transformed potatoes are resistant to Colorado potato beetle larvae ***(36)***. Although *Bt* maize has been a major biotechnological success, the insects that attack this crop differ depending on the geographical region, and field trials are necessary to evaluate their effectiveness for insect control in different environments. Corn plants resistant to beetle rootworms were also generated by the expression of two novel proteins from *Bacillus thuringiensis* ***(47)***. Elite hybrid *Bt* rice lines have been produced that show resistance to two of the most important lepidopteran rice pests (leaffolder and yellow stem borer) without reduced yield ***(48)***. Transgenic lines of insect-resistant sugarcane, an important crop in developing countries, have also been developed ***(49)***. Several insect-resistant transgenic crop plants including maize, potato, and cotton are currently produced commercially.

The production of *Bt* endotoxins is the most widely used strategy to produce insect-resistant transgenic plants; however, these proteins are not effective against all pests and alternative insecticidal proteins are required to avoid the development of resistance in the target pest. Therefore, it has been necessary to find alternative insecticidal and nematicidal proteins, such as proteinase inhibitors, which are part of the defense system of many plants. Proteinase inhibitors of plant origin effective against certain target insects have been engineered into different plant species, such as canola, potato, alfalfa, lettuce, petunia, and tomato; however, they have not been commercialized ***(38)***. Transgenic *Arabidopsis* plants expressing a gene encoding a proteinase inhibitor have been shown to suppress the growth and egg production of two root nematodes ***(50)***.

Avidin, a glycoprotein that sequesters the vitamin biotin, was expressed in maize at levels that prevent the development of insects that damage grain during storage. This toxicity is caused by biotin deficiency and suggests that avidin could be used as a biopesticide in stored grains ***(51)***. Fertile transgenic tobacco plants with leaves expressing avidin in the vacuole have been produced and shown to halt growth and cause mortality in larvae of two nocturnal lepidopterans, *Helicoverpa armigera* and *Spodoptera litura* ***(52)***.

Many virus-resistant transgenic plants have exploited genes derived from viruses themselves, in a concept referred to as pathogen-derived resistance (PDR). The first example of this strategy was the expression of the *Tobacco mosaic virus* (TMV) coat protein gene in tobacco plants. These transgenic plants were found to have effective resistance against TMV ***(53)***. Using a similar strategy, transgenic yellow squash resistant to *Zucchini yellow mosaic virus* (ZYMV) and *Watermelon mosaic virus* II (WMVII) have been produced ***(54)***.

Papaya ringspot virus (PRSV) transmitted by aphids causes one of the most important diseases in papaya. Several attempts using conventional breeding failed to produce PRSV-resistant papaya varieties. However, the use of plant biotechnology succeeded in producing PRSV-resistant, transgenic plants by expressing the coat protein gene from this virus ***(55)***. The future of the papaya crop in Hawaii and other regions of the world effectively rests on the development of virus-resistant plants, an aim already achieved by transgenic technology ***(56)*** (*see* Chapter 27). Transgenic tomatoes resistant to *Tomato mosaic virus* (ToMV), transgenic potatoes resistant to *Potato virus X* (PVX) and *Potato virus Y* (PVY), and transgenic cucumbers resistant to *Cucumber mosaic virus* (CMV) have all been produced ***(57)***. In China, watermelon has been transformed with the WMV-II coat protein gene showing high resistance to the infection by this virus *(58)*. To date, transgenic, virus-resistant papaya and yellow squash have been commercialized. The protective mechanism conveyed by the coat protein or other viral genes is not yet completely understood, and an RNA-mediated cosuppression model has been proposed *(59)*.

Because the potyvirus genome is initially translated into a polyprotein, the completion of the life cycle of this virus depends on the site-specific processing of this precursor by the action of self-processing viral cysteine proteinases. It has been shown that the expression of a rice cysteine proteinase inhibitor in tobacco induces resistance against two important potyviruses, *Tobacco etch virus* (TEV) and *Potato virus Y* viruses (PVY). This represents an alternative method to control this agriculturally important group of aphid-transmissible plant viruses *(60)*.

4.3. Resistance to Phytopatogenic Fungi and Bacteria

Fungal pathogens cause some of the most devastating diseases of crop plants; therefore much effort has been spent on producing resistant plants. Success in the production of transgenic resistant plants has been limited. Expression in transgenic plants of genes encoding enzymes capable of degrading the major constituents of fungal cell walls (chitin and β-1,3-glucan) have been used as a strategy to control these organisms. Expressing the genes of two of these enzymes in tomato showed a useful level of resistance to a *Fusarium* wilt disease *(54)*.

Production of toxins by phytophatogenic bacteria is an important virulence factor. For instance, *Xanthomonas albilineans* produces a family of toxins that lead to characteristic chlorotic symptoms by blocking chloroplast development. The introduction and expression of the albicidin detoxifying gene (*albD*) from the bacterium *Protoea dispersa* resulted in sugarcane in a significant reduction of disease symptoms and decreased multiplication of the pathogen *(61)*. The bacteria *Pseudomonas syringae* pv *phaseolicola* produces phaseolotoxin, which inhibits ornithine carbamoyltransferase (OCTase), an enzyme involved in the biosynthesis of citrulline. As *P. syringae* also produces citrulline, this bacterium harbors a gene encoding a phaseolotoxin-resistant OCTase, which when introduced into the tobacco genome provides resistance against this pathogen *(62)*.

The identification, characterization, and understanding of the mode of action of plant disease resistance genes will provide new avenues to generate disease resistance plants. Several genes encoding key components of the machinery that recognizes avirulance factors produced by viral, bacterial, and fungal pathogens have been cloned and characterized in the past 10 yr. Although the mechanisms by which these disease resistance genes work is still not completely understood, some of them have been successfully used to engineer disease resistance. For instance, the *Xa21* gene from the wild rice *Oryza longistaminata*, which confers resistance to *Xanthomonas oryzae* pv. *oryzae* (*Xoo*), has been successfully transferred to four elite *Indica* varieties, providing significant improved resistance to this important rice pathogen ***(63)***. If these transgenic lines display resistance to the pathogen under field conditions and maintain desirable qualities, they could be used as breeding material or directly to provide resistant plants ***(64)***. Another strategy used is based on the use of the ferredoxin-like protein gene *ap1* in a japonica rice variety; several transgenic lines showed enhanced resistance to *Xoo* ***(65)***.

4.4. Photosynthesis and Sugar Metabolism

CO_2 fixation in C3 plants is carried out directly by ribulose bisphosphate carboxylase (Rubisco), an enzyme that functions as a carboxylase and oxygenase. The use of O_2 instead of CO_2 in the reaction catalyzed by Rubisco results in a loss of up to 50% of the carbon fixed in a process known as photorespiration. This occurs in economically important crops such as wheat, rice, soybean, and potato. C4 plants, such as maize and sugarcane, are more efficient because they first fix CO_2 using the phosphoenolpyruvate carboxylase enzyme (PEPC) to produce oxaloacetate, which is either reduced to malate or transaminated to produce aspartate. These four-carbon compounds are used to generate higher CO_2 concentrations by decarboxylating enzymes in bundle sheet cells, increasing the carboxylase activity of Rubisco. Although the majority of C4 plants have an anatomy that separates PEPC and Rubisco activities, some submerged aquatic macrophytes carry out C4 metabolism in a single type of cell. Strategies to introduce the C4 metabolism into C3 plants have focused on the overexpression of PEPC and one decarboxylating enzyme, either NADP-ME (EC 1.1.1.40) or pyruvate orthophosphate dikinase (PPDK) (EC 2.7.9.1). This has been possible in potato, tobacco, and rice ***(66,67)***. In transformed rice plants that overexpress PEPC and PPDK, the photosynthetic capacity was increased by 35% and the grain yield by 22% ***(67)***.

A key enzyme in the sucrose synthesis pathway is the sucrose phosphate synthase enzyme (SPS), which produces sucrose phosphate from UDP-glucose and fructose 6-phosphate. This enzyme is regulated either by covalent modification by phosphorylation or allosterically by inorganic phosphate. As a way to modify carbon partitioning and increase the synthesis of sucrose, genes for SPS of one species have been introduced into another. The maize SPS gene was overexpressed in tomato, increasing by sixfold the activity of the SPS enzyme, the amount of sucrose produced was doubled ***(68)***, and the rate of photosynthesis was increased by 20% in CO_2-saturated conditions ***(69)***. The difference in the level of increase of the enzyme activity and that of the sucrose produced reflects the fact that the sucrose synthesis pathway is highly regulated at different levels.

4.5. Transgenic Strategies to Produce Abiotic Stress Tolerance

Drought, salinity and cold-induced dehydratation produce osmotic stress and are some of the most important abiotic factors that decrease agricultural production. One strategy to increase the tolerance to such stresses is the production of osmoprotective compounds (osmolytes), such as sugars, alcohols, amino acids, and quaternary ammonium compounds (glycinebetaine) which raise the osmotic potential of the cell, allowing the influx of water, and stabilize membranes and/or macromolecular structures ***(70)***. Some plants adapted to stress conditions naturally produce these osmolytes; however, many important crop plants do not accumulate sufficient osmoprotective compounds to be stress tolerant. Advances have been made to achieve or increase the production of osmolytes in transgenic plants. Glycinebetaine has been produced in tobacco plants by the expression of a bacterial gene; these plants show an enhanced NaCl stress tolerance ***(71)***. Transgenic *Arabidopsis* lines containing genes encoding enzymes responsible for the synthesis of glycine betaine in their chloroplast genome are more tolerant to salt and cold stress, being capable of growing in 100 m*M* NaCl ***(72)*** and low temperatures ***(73)*** (*see* Chapter 26). Overproduction of trehalose in tobacco increases drought tolerance of both intact plants and detached leaves ***(74)***. More recently, an abscisic acid (ABA) inducible promoter to express a bifunctional enzyme that synthesizes trehalose was used to produce trangenic rice plants that exhibited sustained plant growth, less photooxidative damage, and more favorable mineral balance under salt, drought, and low-temperature stress conditions ***(75)***. Other osmolytes such as mannitol have been overproduced in *Arabidopsis*, in which an enhancement of seed germination under high salt conditions has been observed ***(76)***. Recently, it has been demonstrated that the ectopic expression of the *mtlD* gene of *E. coli*, for the biosynthesis of mannitol in wheat improves tolerance to water stress and salinity ***(77)***.

An alternative strategy to the production of osmoprotective compounds is the overexpression of genes that encode ion transporters, such as the vacuolar Na^+/H^+ antiport, which transports sodium away from the cytosol and compartmentalizes it into the vacuole, maintaining an osmotic balance. Transformed *Arabidopsis* plants overproducing a Na^+/H^+ ion antiport have improved tolerance to salt, and have sustained growth and development in soil watered with up to 200 m*M* NaCl ***(78)***. Transgenic tomato plants overproducing the *Arabidopsis* Na^+/H^+ ion antiport were capable of growing, flowering and producing fruits in the presence of a high salt concentration (200 m*M*), with a very low sodium content in the fruit, preserving the quality of this product ***(79)***. The expression in a salt-sensitive variety of rice of a gene encoding a vacuolar type Na^+/H^+ antiporter from a halophytic plant survived under conditions of 300 m*M* NaCl while conventional plants died ***(80)***.

Transcription factors that regulate the expression of genes involved in cold/drought tolerance, such as DREB1A, have been identified. Ectopic expression of these transcription factors using the 35S CaMV lead to enhanced salt, cold, and drought tolerance but also have some detrimental effects on plant growth and development. The use of the stress inducible *rd29A* promoter to drive expression of DREB1A induces the expression of stress tolerance genes without detrimental effects on plant growth in *Arabidopsis* ***(81)***. The CBF1 (DREB1B) is another transcriptional activator that binds regulatory elements in the promoter region of cold-regulated (*COR*) genes that respond to both low temperature and water deficit. Overexpression of this *Arabidopsis* transcription factor increases tolerance to freezing in *Brassica napus* ***(82)*** and confers water deficit resistance in transgenic tomato plants ***(83)***.

Another important abiotic stress problem is aluminum toxicity and low P availability in acid soils. Acid soils comprise up 40% of the world's arable land (68% of tropical America, 38% of tropical Asia, and 27% of tropical Africa) ***(84)***. Aluminum is the most abundant metal in the Earth's crust and is toxic to many plants at low concentrations in solution, which is a problem in acidic soils ***(85)***. Phosphate, the anionic form in which P is assimilated by living organisms, is extremely reactive and is available for plant uptake only at a narrow range of pH. In acid soils, P forms weakly soluble molecules with aluminum and iron, whereas in alkaline soils it combines efficiently with calcium and magnesium to form sparingly soluble phosphate compounds ***(86)***. Production of organic acids and their exudation in the soil is a strategy used by some plants to combine the aluminum forming chelated species and eliminate their toxicity ***(87)***. Alkaline soil adapted plants (calcicolas) growing in calcareous soils exude three to four times more low-molecular-weight organic acids than nonadapted plants (calcifuges), allowing more capacity to extract phosphate and iron ***(88)***. Tobacco and papaya were transformed with a citrate synthase gene from *Pseudomonas aeruginosa* and the plants secreted five to six times more citrate from their roots than controls, providing tolerance to toxic levels of aluminum that were 10-fold higher than those tolerated by control plants ***(89)***. Tobacco plants overproducing citrate also showed better growth in low phosphorous alkaline soils, demonstrating the efficient use of insoluble forms of P, such as Ca-P ***(90)***. A mitochondrial citrate synthase of *Arabidopsis thaliana* introduced into carrot (*Daucus carota*) resulted in an enhanced capacity of phosphate uptake from insoluble sources of phosphorous ***(91)***. Manipulation of organic acid synthesis in transgenic plants could be used to obtain novel plant varieties better adapted to grow under adverse soil conditions.

Phytoremediation of metal-contaminated soils is a complex problem. The use of plants that accumulate metals to remove and recycle excessive soil metals selectively is a potential practical and more cost-effective technology than soil replacement and other strategies ***(92)***. Expression of the gene encoding a bacterial mercuric ion reductase (MerA) in *Arabidopsis thaliana* provides tolerance to toxic levels of mercury, reducing it to Hg (0), a volatile and nontoxic form of the element ***(93)***. Another strategy is to accumulate heavy metals in the plants themselves and therefore remove them from the soil, such in the following cases: (a) A wheat gene that encodes a phytochelatin synthase (TaPCS1) was introduced into a wild fast growing, high-biomass plant, *Nicotiana glauca*. These transformed plants resulted in increased tolerance to Cd and Pb and accumulated double the normal concentration of the latter metal ***(94)***. (b) In the Indian mustard (*Brassica juncea*) a glutathione synthase or γ-glutamylcysteine synthase were overproduced and the transformed plants accumulated threefold more Cr, Cu, and Pb ***(95)***.

4.6. Plants as Bioreactors

The production of foreign proteins in plants has many advantages, as the proteins can be produced in seeds, naturally protective packages that can be easily stored and transported or in fruits or tubers that can be consumed directly or processed to obtain the product of interest. Larrick and Thomas ***(96)*** have also mentioned several other advantages: (a) The cost on an agricultural scale is low; (b) the use of plants offers reduced capitalization costs relative to fermentation methods; (c) production can be rapidly upscaled; (d) unlike bacteria, plants can produce complex multimeric proteins, such as antibodies that are correctly assembled; and (e) plant proteins are considered to be safer, as plants do not serve as hosts for human pathogens. Biopharmaceutical products are the best choice for production in plants because of the high cost of using other systems, such as mammalian cells. β-Interferon has been produced in turnip and it could potentially be used for hepatitis B and C treatment ***(97)***.

Bovine trypsin is an enzyme widely used commercially to digest or process other proteins including some therapeutic proteins. The biopharmaceutical industry is trying to eliminate animal-derived proteins from manufacturing processes owing to the possible contamination of these products by human pathogens. An obvious solution is to express these proteins in plants. Trypsin at commercial levels has been produced in transgenic maize and is functionally equivalent to native bovine pancreatic trypsin ***(98)***. The availability of this reagent should allow for the replacement of animal-derived trypsin in the processing of pharmaceutical proteins.

The absence of glucocerebrosidase produces Gaucher's disease, a recessively inherited disorder. The enzyme has traditionally been extracted from placentas, at a high cost. Production of glucocerebrosidase in transgenic tobacco strongly supports the future commercial viability for therapy ***(99)***. Production of hirudin, an anticoagulant to treat thrombosis, has been achieved in oilseed rape and is now produced commercially ***(100)***. Somatotropin, a human hormone used in the treatment of hypopituitary dwarfism in children and other disorders, has been produced in bacteria and recently its expression in tobacco chloroplasts at high levels demonstrates the potential of this organelle as a highly efficient vehicle for production of pharmaceutical proteins ***(101)***.

Generation and assembly of functional secretory antibodies in tobacco plants ***(102)*** have significant implications for passive immunotherapy. To date, only four antibodies have been made in plants that are potentially useful in human therapeutics ***(103)***. Only one of these so far has been tested in humans: a chimeric secretory IgG–IgA antibody against a surface antigen of *Streptococcus mutans*, the primary causal agent of tooth decay. This tobacco-produced antibody was applied topically to teeth and found to be effective. The second antibody, a humanized anti-herpes-simplex virus (HSV) antibody made in soybean, was effective in the prevention of vaginal HSV-2 transmission in a mouse model. A third antibody, against carcinoembryonic antigen (CEA), has recently been expressed in rice and wheat. The fourth antibody has been produced in tobacco for the treatment of B-cell lymphoma. Maize has been proposed as an advantageous plant, as the corn seed system allows stable accumulation of high levels of recombinant protein, demonstrated in the case of a secretory immunoglobulin A ***(104)***.

Antigens are capable of activating the mucosal immune system in digestive and respiratory tracts. The partial protection of plant proteins during digestion owing to the presence of cell walls allows the conservation of antigens and maintains their effectiveness to activate the mucosal immune system in the intestine. However, one factor that has limited the development of this technology is the relatively modest levels of accumulation of some antigenic proteins in plant tissues. A hepatitis B antigen was produced in tobacco ***(105)*** and potato ***(106)***, showing in the latter case immunization in mice ***(107)***. A cholera antigen has been produced in tobacco ***(108)*** and potato ***(109)*** and in the latter case demonstrated to generate protective immunity in mice against the cholera holotoxin ***(110)***. Tomatoes have also been used to produce an antigen against the rabies virus ***(111)***. A subunit of the heat-labile enterotoxin of *Escherichia coli* was expressed in potato, and these tubers were shown to induce oral immunization in mice ***(112)***. The expression of a synthetic enterotoxin gene at high levels also protected mice ***(113)***. Human trials progress with antigens of a heat-labile toxin (LT) of *E. coli* produced fourfold increases in levels of serum LT-neutralizing antibodies ***(114)***; similar results were obtained using antigens of the hepatitis B virus and antigens of the Norwalk virus ***(114)***. Recently, transgenic spinach expressing epitopes from the rabies virus was orally delivered to 14 human volunteers and 8 showed significant elevation in rabies-specific antibodies ***(115)***.

Human vaccines derived from plants are important in developing countries, where high production costs, the management of vaccination programs, and conservation of vaccines are a problem. The use of tropical fruits is an interesting alternative option and efforts are being made to express antigens against malaria and rotavirus in transgenic tropical fruit plants.

Alcaligenes eutrophus and other bacteria produce polyhydroxyalkanoates as a carbon reserve in high carbon containing media. These compounds have properties ranging from brittle plastics to rubberlike materials and because of their biodegradability are an attractive source of nonpolluting plastics and elastomers for special uses ***(116)***. Because of the high cost of production by bacterial fermentation, plants have been explored as an alternative to produce biodegradable plastics. Introduction of three bacterial genes in *Arabidopsis* produced plants that accumulate polyhydroxybutyrate (PHB) to up to 14% of the dry weight without growth or fertility problems ***(117)***. This opens the possibility to use plants to produce PHB at a commercial scale. Production of PHB in cotton plants has been carried out to modify the cotton fiber properties, and although the amounts of PHB are small (0.34% fiber weight), there is an enhanced insulation characteristic that may have applications in winter fabrics. The positive changes in fiber qualities demonstrate the potential of this technology ***(118)***.

4.7. Herbicide-Resistant Plants

The control of weeds that compete with cultivated plants for nutrients and space is achieved with herbicides that affect photosynthesis or the biosynthesis of essential compounds. However, cultivated plants often are not resistant to the herbicide and this strategy cannot be used efficiently. Therefore, production of herbicide-tolerant transgenic plants is a good option. Glyphosate is a herbicide that inhibits the biosynthesis of aromatic amino acids by inhibiting the activity of the enzyme 5-enolphyruvylshikimate 3-phosphate synthase (EPSPS). Expression of a mutant allele of the *AroA* locus of *Salmonella typhimurium* produces an EPSPS enzyme insensitive to glyphosate. Expression of this allele in transformed plants produced resistance to glyphosate in tobacco ***(119)*** and tomato ***(120)***, as does overproduction of the EPSPS in petunia ***(121)***. Using these strategies, glyphosate resistance also has been achieved in soybean ***(122)***, canola, cotton, and sugarbeet ***(24)***.

Phosphinotricin (PPT), the active compound of many herbicides, inhibits the enzyme glutamate synthetase (GS). The *bar* gene from *Streptomyces hygroscopicus* encodes a detoxifying enzyme that acetylates the free NH_2 group of PPT. Tobacco, tomato, and potato plants transformed with the bar gene were resistant to commercial formulations of Bialaphos and PPT herbicides ***(123)***. Transgenic plants with resistance to the herbicide glufosinate using these detoxifying enzymes include sugarcane, rice, maize, canola, and cotton ***(24)***. Recently, wheat was transformed with the *bar* gene, resulting in agronomic resistance to the BASTA herbicide ***(124)***. Oilseed rape plants resistant to both glyphosate and ammonium glufosinate have been also produced ***(125)***. Cotton plants resistant to the commonly used herbicide bromoxynil also have been produced and are now on the market ***(126)***. Another gene, derived from *Bacillus subtilis*, encoding a protoporphyrinogen oxidase has been used to transform rice plants to confer resistance to the diphenyl ether oxyfluorfen herbicide ***(127)***.

5. Commercialization and Biosecurity

Slow ripening tomato was the first type of transgenic plant to reach the market in 1994, and currently plants with tolerance to herbicides and pests (viruses and insects) of various species are also produced commercially. The cultivated area of transgenic plants increased from 1.7 million hectares in 1996 to 58 million hectares in 2002, with the United States, Argentina, Canada, and China having 99% of the cultivated area ***(128)***. Soybean, maize, cotton, canola, potato, squash, and papaya are the main transgenic crops cultivated, and herbicide tolerance, insect resistance, and virus resistance are the modified traits under commercial use to date (**Fig. 1**). Herbicide tolerant soybean continued to be the dominant transgenic crop grown commercially in seven countries in 2002 (the United States, Argentina, Canada, Mexico, Romania, Uruguay, and South Africa). Globally, herbicide tolerant soybean occupied 36.5 million hectares, representing 62% of the global transgenic crop area of 58.7 million hectares for all crops. The second most dominant crop was insect resistant maize, which occupied 7.7 million hectares, equivalent to 13% of the global transgenic area and planted in seven countries (the United States, Canada, Argentina, South Africa, Spain, Honduras, and Germany). The third most dominant crop was herbicide tolerant canola, which occupied 3.0 million hectares, equivalent to 5% of the global transgenic area and planted in two countries, Canada and the United States. The other five crops listed all occupy 4% each of the global transgenic crop area and include, in descending order of cultivated area: herbicide tolerant maize on 2.5 million hectares (4%); insect-resistant cotton, on 2.4 million hectares (4%); herbicide tolerant cotton on 2.2 million hectares (4%); insect resistant /herbicide tolerant cotton on 2.2 million hectares (4%), Bt/herbicide tolerant maize on 2.2 million hectares (4%), insect resistant potato (<0.1%), virus tolerant squash (<0.1%), and virus tolerant papaya (<0.1%).

The global adoption rates of the four principal crops (soybean, cotton, canola, and maize) in which transgenic technology is used indicate that in 2002, 51% of the 72 million hectares) of soybean planted globally were transgenic (**Fig. 1**). Of the 34 million hectares of cotton, 20% (6.8 million hectares) were transgenic. The area planted of transgenic canola was 12%, or 3.0 million hect-

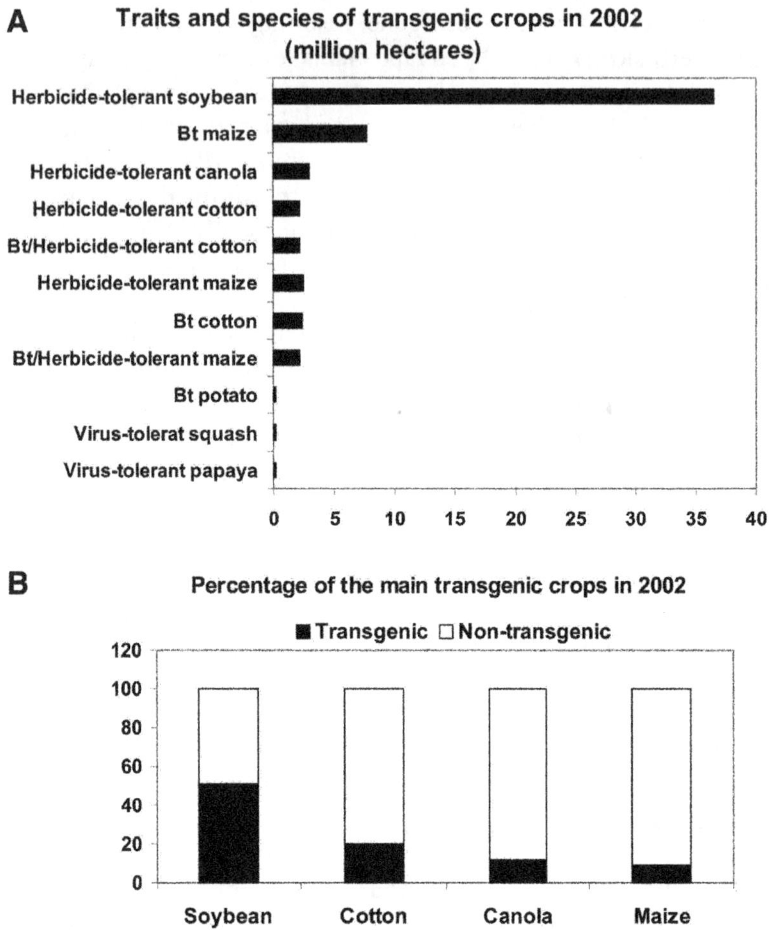

Fig. 1. Main cultivated transgenic plants. **(A)** Global cultivated area of the main species and their traits. **(B)** Global adoption percentages of the four principal crops (soybean, cotton, canola, and maize).

ares of the 25 million hectares of canola planted globally in 2002. Of the maize planted in 2002, 9% was transgenic (up significantly from 7% in 2001). If the global areas (conventional and transgenic) of these four crops are aggregated, the total area is 271 million hectares, of which almost 22% were genetically modified (up from 19% in 2001) *(128)*. Another commercialized transgenic product is the trypsin produced in seed corn by the Prodigene Company.

Notwithstanding the often-biased campaigns of ecologist groups against transgenic plants, the confidence in the benefits of this technology is augmenting among farmers around the world. This is reflected by the continuous increase in the global area in which transgenic plants are being cultivated and the increasing number of developing countries that are currently cultivating transgenic plants, including Argentina, Mexico, India, China, and Egypt. The transference of this technology to developing countries will depend on the consolidation of existing national institutions and the creation of new research institutes and companies capable of adapting this technology to the needs of local farmers.

An analysis of the risks and benefits of transgenic plants should contemplate field trials in which the diverse ecological variables would be analyzed carefully. Results of a long-term study of the performance of four different transgenic crops (oilseed rape, potato, maize, and sugar beet) grown in different habitats and monitored over a period of 10 yr demonstrated that in no case were the genetically modified plants found to be more invasive or more persistent than their conventional counterparts ***(129)***. An erroneous conclusion can be reached by extrapolating laboratory conditions to the field, as in the case of the monarch butterfly studies, which have generated a huge controversy ***(130,131)***. Insertion of transgenes into the chloroplast genome rather than into the nuclear genome has been proven to be an effective strategy to increase the level of herbicide tolerance ***(132)*** and insect resistance ***(133)*** (*see* Chapter 8) while avoiding the transfer of the transgene through the pollen.

An international consensus has been reached on the principles regarding evaluation of the food safety of genetically modified plants. The concept of *substantial equivalence* has been developed as part of a safety evaluation framework, based on the idea that existing foods can serve as a basis for comparing the properties of genetically modified foods with the appropriate counterpart. Application of the concept is not a safety assessment *per se*, but helps to identify similarities and differences between the existing food and the new product, which are then subject to further toxicological investigation. As an example, in the case of some Bt proteins, various studies have been performed on binding of the protein to tissues of the gastrointestinal tract of rodents and primates, including humans, and there was no evidence for the presence of specific receptors in mammalian tissues for these proteins, nor are there indications of an amino acid sequence homology to known protein food allergens, and no toxicity problems were identified ***(134)***.

6. Perspectives

Generation of transgenic plants from a new species has depended mainly on two factors: first, the development of transformation methods for cultivated plants and second, the knowledge of genes and their function. As mentioned previously, there are now transformation protocols available for the most important crops.

Knowledge of the function of individual genes in plants has advanced primarily as a result of the analysis of individual genes through the study of mutations. Most recently the study of genes has been accelerated with the sequencing of whole genomes. The genome of *Arabidopsis thaliana* was the first plant genome to be sequenced. The total genome consists of 125 megabases and 25,498 genes ***(135)***. The complete sequences of the Indica and japonica subspecies of rice have also been reported ***(136,137)*** and consist of 466 and 420 Mb, respectively. Nevertheless 80.6% of predicted *Arabidopsis thaliana* genes were found to have a homolog in rice, whereas only 49.4% of the predicted rice genes had a homolog in *A. thaliana*. The proportion of genes implied to be involved in the different cellular functions seems to be the same in rice and *Arabidopsis*, as suggested by analysis of transcription factors ***(137)***. The sequence of the rice genome as a model cereal will lead to a better understanding of the larger and more complex maize genome, currently undergoing sequencing.

Once whole-genome information is available for an organism, the next step is deciphering the function of all genes and the integration of these functions as a whole, by strategies referred to as functional genomics ***(138)***. Traditionally, gene expression has been analyzed through messenger RNA abundance, but only for one or a few genes at a time. With the knowledge of at least partial sequences of plant genomes it is possible to analyze simultaneously the expression of thousands of genes or whole genomes (expression profiles) under different conditions or from different tissues to determine which genes are related to each process or condition and understand the activity and biological roles of their encoded protein ***(139)***. This latter achievement is possible with the replacement of traditional hybridization methods for microarrays of specific DNA sequence information attached to a solid support via linkages that allow good access for hybridization. Finally, large-scale protein analysis has also begun and will contribute to the understanding of gene function, through microcharacterization, differential display, and protein–protein interactions ***(140)***.

With the advent of functional genomics, the discovery of new genes and their function in plant processes opens the opportunity to more effectively produced transgenic plants, based not only on monogenic traits but also on multigenic ones. The transduction pathways for response to pathogens are not completely elucidated and this knowledge will be crucial in the fight against pathogens through transgenic plants. The understanding of plant responses to drought and high salt tolerance and the development of transgenic strategies to combat these are important aspects, considering the extensive uncultivated areas resulting from high salinity and water shortage.

In spite of the positive results obtained from the commercial use of transgenic plants and the enormous potential of functional genomics to increase our capacity to genetically engineer new traits, including quantitative ones, in a more efficient and precise manner, the commercial future of transgenic plants remains uncertain. Functional genomics will allow the identification of components of quantitative traits as well as alleles that confer specific properties for a given plant species, however, it is now questioned whether the transfer by genetic engineering technology of genes from the very same species is acceptable. Moreover, the strict regulations, in terms of evaluating the potential impact of transgenic plants on human and animal health, biodiversity, and the environment have made the cost of gaining approval for their commercial use so high that only few large multinational companies will be able to comply with the regulatory process. Nevertheless, transgenic technology will continue to be essential for the study of regulatory and developmental processes that determine the final productivity of plants and therefore will become indispensable for generating new and more efficient plant breeding strategies.

References

1. Binns, A. and Campbell, A. (2001) Agrobacterium tumefaciens-mediated transformation of plant cells, in *Encyclopedia of Life Sciences*. Nature Publishing Group, London, UK, pp. 1–6.
2. Chilton, M. D. (2001) Agrobacterium. A memoir. *Plant Physiol.* **125,** 9–14.
3. Zaenen, I., Van Larebeke, N., Teuchy, H., Van Montagu, M., and Schell, J. (1974) Supercoiled circular DNA in crown-gall inducing *Agrobacterium* strains *J. Mol. Biol.* **86,** 109–127.
4. Chilton, M. D., Drummond, M. H., Merlo, D. J., et al. (1977) Stable incorporation of plasmid DNA into higher plant cells: the molecular basis of crown gall tumorigenesis. *Cell* **11,** 263–271.
5. Herrera-Estrella, L., Depicker, A., Van Montagu, M., and Schell, J. (1983) Expression of chimaeric genes transferred into plant cells using a Ti- plasmid-derived vector. *Nature* **303,** 209–213.
6. Herrera-Estrella, L., De Block, M., Messens, E., Hernalsteens, H. P., Van Montagu, M., and Schell, J. (1983) Chimeric genes as dominant selectable markers in plant cells. *EMBO J.* **2,** 987–995.
7. De Block, M., Herrera-Estrella, L., Van Montagu, M., Schell, J., and Zambryski, P. (1984) Expression of foreign genes in regenerated plants and their progeny. *EMBO J.* **3,** 1681–1689.
8. Bundock, P., den Dulk-Ras, A., Beijersbergen, A., and Hooykaas, P. (1995) Trans-kingdom T-DNA transfer from *Agrobacterium tumefaciens* to *Saccharomyces cerevisiae*. *EMBO J.* **14,** 3206–3214.
9. Gutiérrez-Mora, A., Santacruz-Ruvalcaba, F., Cabrera-Ponce, J. L., and Rodríguez-Garay, B. (2003) Mejoramiento gen⁄tico vegetal in vitro. E-Gnosis Vol1: Art 4. (online) Website: (www.e-gnosis.udg.mx/vol1/art4).
10. Crouzet, P. and Hohn, B. (2002) Transgenic plants. *Encyclopedia of Life Sciences*. Nature Publishing Group, London, UK, pp. 1–7.

11. Hamilton, C. M., Frary, A., Lewis C., and Tanksley, S. D. (1996) Stable transfer of intact high molecular weight DNA into plant chromosomes. *Proc. Natl. Acad. Sci. USA* **93,** 997–1001.
12. Trieu, A. T., Burleigh, S. H., Kardailsky, I. V., et al. (2000) Transformation of Medicago truncatula via infiltration of seedlings or flowering plants with *Agrobacterium. Plant J.* **22,** 531–541.
13. Sanford, J. C. (1988) The biolistic process. *Trends Biotechnol.* **6,** 299–302.
14. Hansen, G. and Wright M. S. (1999) Recent advances in the transformation of plants. *Trends Plant Sci.* **4,** 226–231.
15. Kohli, A., Leech, M., Vain, P., Laurie, D. A., and Christou, P. (1998) Transgene organization in rice engineered through direct DNA transfer supports a two phase integration mechanism mediated by the establishment of integration hot spots. *Proc. Natl. Acad. Sci. USA* **95,** 7203–7208.
16. Fromm, M. E., Taylor, L. P., and Walbot, V. (1986) Stable transformation of maize after gene transfer by electroporation. *Nature* **319,** 791–793.
17. Wilmink, A. and Dons, J. J. M. (1993) Selective agents and marker genes for use in transformation of monocotyledonous plants. *Plant Mol. Biol. Rep.* **11,** 165–185.
18. Haseloff, J., Siemering, K. R., Prasher, D. C., and Hodge, S. (1997) Removal of a cryptic intron and subcellular localization of green fluorescent protein are required to mark transgenic *Arabidopsis* plants brightly. *Proc. Natl. Acad. Sci. USA* **94,** 2122–2127.
19. Meinke, D. W., Cherry, M., Dean, C., Rounsley, S. D., and Koornneef, M. (1998) *Arabidopsis thaliana*: a model plant for genome analysis. *Science* **282,** 662–682.
20. Gidoni, D., Fuss, E., Burbidge, A., et al. (2003) Multi-functional T-DNA/Ds tomato lines designed for gene cloning and molecular and physical dissection of the tomato genome. *Plant Mol. Biol.* **51,** 8–98.
21. Jeon, J. S., Lee, S., Jung, K. H., et al. (2000) T-DNA insertional mutagenesis for functional genomics in rice. *Plant J.* **22,** 561–570.
22. Smith, C. J., Watson, C. F., Morris, P. C., et al. (1990) Inheritance and effect on ripening of antisense polygalacturonase genes in transgenic tomatoes. *Plant Mol. Biol.* **14,** 369–379.
23. Brummell, D. A. and Harpster, M. H. (2001) Cell wall metabolism in fruit softening and quality and its manipulation in transgenic plants. *Plant Mol. Biol.* **47,** 311–340.
24. Briggs, S. P. and Koziel, M. (1998) Engineering new plant strains for commercial markets. *Curr. Opin. Biotechnol.* **9,** 233–235.
25. Herrera-Estrella, L. (2000) Genetically modified crops and developing countries. *Plant Physiol.* **124,** 923–925.
26. Altenbach, S. B., Pearson, K. W., Meeker, G., Staraci L. C., and Sun, S. S. M. (1989). Enhancement of the methionine content of seed proteins by the expression of a chimeric gene encoding methionine-rich protein in transgenic plants. *Plant Mol. Biol.* **13,** 513–522.

27. Altenbach, S. B., Kuo, C. C., Staraci, L. C., et al. (1992) Accumulation of a Brazil nut albumin in seeds of transgenic canola results in enhanced levels of seed protein methionine. *Plant Mol. Biol.* **18,** 235–245.
28. Molving, L., Tabe, L. M., Eggum, B. O., et al. (1997) Enhanced methionine levels and increased nutritive value of seeds of transgenic lupinus (*Lupinus angustifolius* L.) expressing a sunflower seed albumin gene. *Proc. Natl. Acad. Sci. USA* **94,** 8393–8398.
29. Keeler, S. J., Maloney, C. L., Webber, P. Y., et al. (1997) Expression of the novo high-lysine alpha–helical coiled-coil proteins may significantly increase the accumulated levels of lysine in mature seeds of transgenic tobacco plants. *Plant Mol. Biol.* **34,** 15–29.
30. Falco, S. C., Guida, T., Locke, M., et al. (1995) Transgenic canola and soybean seeds with increased lysine. *BioTechnology* **13,** 577–582.
31. Chakraborty, S., Chakraborty, N., and Datta, A. (2000) Increased nutritive value of transgenic potato by expressing a nonallergenic seed albumin gene from *Amaranthus hypochondriacus*. *Proc. Nat. Acad. Sci. USA* **97,** 3724–3729.
32. Ye, X., Al-Babili, S., Klöti, A., et al. (2000) Engineering the provitamin A (β-carotene) biosynthetic pathway into (carotenoid-free) rice endosperm. *Science* **287,** 303–305.
33. Thelen, J. J. and Ohlrogge, J. B. (2002) Metabolic engineering of fatty acid biosynthesis in plants. *Metab. Eng.* **4,** 12–21.
34. Fraser, P. D., Romer, S., Shipton, C. A., et al. (2002) Evaluation of transgenic tomato plants expressing an additional phytoene synthase in a fruit-specific manner. *Proc. Natl. Acad. Sci. USA* **99,** 1092–1097.
35. Ogita, S., Uefuji, H., Yamaguchi, Y., Koizumi, N., and Sano, H. (2003) Producing decaffeinated coffee plants. *Nature* **423,** 823–824.
36. Estruch, J. J., Carozzi, N. B., Desai, N., Duck, N.B., Warren, G.W., and Koziel, M. (1997) Transgenic plants: an emerging approach to pest control. *Nat. Biotech.* **15,** 137–141.
37. Peferoen, M. (1997) Progress and prospects for field use of Bt genes in crops. *Trends Biotechnol.* **15,** 173–177.
38. Schuler, T. H., Poppy, G. M., Kerry, B. R., and Delhom, I. (1998) Insect-resistant transgenic plants. *Trends Biotechnol.* **16,** 168–175.
39. Vaeck, M., Reynaerts, A., Höfte, H., et al. (1987) Transgenic plants protected from insect attack. *Nature* **382,** 33–37.
40. Fischhoff, D. A., Bowdish, K. S., Perlack, F. J., et al. (1987) Insect tolerant transgenic tomato plants. *BioTechnology* **5,** 807–813.
41. Perlack, F. J., Deaton, R. W., Armstrong, T. A., et al. (1990) Insect resistant cotton plants. *BioTechnology* **8,** 939–943.
42. Perlack F. J., Stone, T. B., Muskopf, Y. M., et al. (1993) Genetically improved potatoes: protection from damage by Colorado potato beetles. *Plant Mol. Biol.* **22,** 313–321.
43. Koziel, M. G., Beland, G. L., Bowman, C., et al. (1993) Field performance of elite transgenic maize plants expressing an insecticidal protein derived from *Bacillus thuringiensis*. *BioTechnology* **11,** 194–200.

44. Stewart, C. N., Adang, M. J., All, J. N., Ramachandran, S., and Parrot, W. A. (1996) Insect control and dosage effect in transgenic canola containing a synthetic *Bacillus thuringiensis cryAc* gene. *Plant Physiol.* **112,** 115–120.
45. Stewart, C. N., Adang, M. J., All, J. N., Ramachandran, S., and Parrot, W. A. (1996) Genetic transformation, recovery and characterization of fertile soybean transgenic for a synthetic *Bacillus thuringiensis cryIAc* gene. *Plant Physiol.* **112,** 121–129.
46. Wün, J., Klöti, A., Burkhardt, P. K., et al. (1996) Transgenic Indica rice breeding line IR58 expressing a synthetic cryIA(b) gene from *Bacillus thuringiensis* provides effective insect pest control. *BioTechnology* **14,** 171–176.
47. Ostlie, K. (2001) Crafting crop resistance to corn rootworms. *Nat. Biotech.* **19,** 624–625.
48. Tu, J., Zhang, G., Datta, K., et al. (2000) Field performance of transgenic elite commercial hybrid rice expressing Bacillus thuringiensis δ-endotoxin. *Nat. Biotech.* **18,** 1101–1104.
49. Vázquez-Padrón, R. I. (2000) Insect-resistant tropical plants and new assessment about Cry proteins, in *Plant Genetic Engineering: Towards the Third Millennium* (Arencibia, A. D., ed.) Elsevier Science, Amsterdam, The Netherlands.
50. Lilley, C. J., Devlin, P., Urwin, P. E., and Atkinson, H. J. (1999) Parasitic nematodes, proteinases and transgenic plants. *Parasitol. Today* **15,** 414–417.
51. Kramer, K. J., Morgan, T. D., Throne, J. E., Dowell, F. E., Bailey, M., and Howard, J. A. (2000) Transgenic avidin maize is resistant to storage insect pests. *Nat. Biotech.* **18,** 670–674.
52. Burgess, E. P., Malone, L. A., Christeller, J. T., et al. (2002) Avidin expressed in transgenic tobacco leaves confers resistance to two noctuid pests, *Helicoverpa armigera* and *Spodoptera litura. Transgen. Res.* **2,** 199–214.
53. Powell, A. P., Nelson, R. S., Barun, D., et al. (1986) Delay of disease development in transgenic plants that express the tobacco mosaic virus coat protein gene. *Science* **232,** 738–743.
54. Shah, D. M., Rommens, C. M., and Beachy, R. N. (1995) Resistance to diseases and insects in transgenic plants: progress and applications to agriculture. *Trends Biotechnol.* **13,** 362–368.
55. Tennant, P. F., Gonsalves, C., Ling, K. S., et al. (1994) Differential protection against papaya ringspot virus isolates in coat protein gene transgenic papaya and classically cross-protected papaya. *Phytopathology* **84,** 1359–1366.
56. Gonsalves, D., Ferreira, S., Manshart, R., Fitch, M., and Slightom, J. (2000) Transgenic virus resistant papaya: new hope for controlling papaya ringspot virus in Hawaii. *Annu. Rev. Phytopathol.* **36,** 415–437.
57. Fitchen, J. H. and Beachy, R. N. (1993) Genetically engineered protection against viruses in transgenic plants. *Annu. Rev. Microbiol.* **47,** 739–763.
58. Wang, H. Z., Zhao, P. J., Xu, J. C., Zhao, H., and Zhang, H. S. (2003) Virus resistance in transgenic watermelon plants containing a WMV-2 coat protein gene. *Yi Chuan Xue Bao.* **30,** 70–75.

59. Lomonossoff, G.P. (1995) Pathogen-derived resistance to plant viruses. *Annu. Rev. Phytopathol.* **33,** 323–343.
60. Gutiérrez-Campos, R., Torres-Acosta, J. A., Saucedo-Arias, L. J., and Gómez-Lim, M. A. (1999) The use of cysteine proteinase inhibitors to engineer resistance against potyviruses in transgenic tobacco plants. *Nat. Biotech.* **17,** 1223–1226.
61. Zhang, L., Xu, J., and Birch, R. (1999) Engineered detoxification confers resistance against a pathogenic bacterium. *Nat. Biotech.* **17,** 1021–1024
62. De la Fuente, J. M., Mosqueda-Cano, G., Alvarez-Morales, A., and Herrera-Estrella, L. (1992) Expression of a bacterial phaseolotoxin-resistant ornithyl transcarbamylase in transgenic tobacco confers resistance to *Pseudomonas syringae* pv. *phaseolicola. BioTechnology* **10,** 905–909.
63. Song, W. Y., Wang, G., Chen, L., et al. (1995) A receptor kinase-like protein encoded by the rice disease resistant gene *Xa21. Science* **270,** 1804–1806.
64. Zhang, S., Song, W. Y., Chen, L., et al. (1998) Transgenic elite Indica rice varieties resistant to *Xanthomonas oryzae* pv. *Oryzae. Mol. Breeding* **4,** 551–558.
65. Tang, K., Sun, X., Hu, Q., et al. (2001) Transgenic rice plants expressing the ferredoxin-like protein (AP1) from sweet pepper show enhanced resistance to *Xanthomonas oryzae* pv. *oryzae. Plant Sci.* **160,** 1035–1042.
66. Hausler, R. E., Rademacher, T., Li, J., et al. (2001) Single and double overexpression of C(4)-cycle genes had differential effects on the pattern of endogenous enzymes, attenuation of photorespiration and on contents of UV protectants in transgenic potato and tobacco plants. *J. Exp. Bot.* **52,** 1785–1803.
67. Ku, M. S., Cho, D., Li, X., et al. (2001) Introduction of genes encoding C4 photosynthesis enzymes into rice plants: physiological consequences. *Rice Biotechnol.* **236,** 100–116.
68. Worrell, A. C., Bruneau, J. M., Summerfelt, K., Boersig, M., and Voelker, T. A. (1991) Expression of a maize sucrose phosphate synthase in tomato alters leaf carbohydrate partitioning. *Plant Cell* **10,** 1121–1130
69. Galtier, N., Foyer, C.H., Huber, J., Voelker, T.A., and Huber, S.C. (1993) Effects of elevated sucrose-phosphate synthase activity on photosynthesis, assimilate partitioning, and growth in tomato (*Lycopersicon esculentum* var UC82B). *Plant Physiol.* **101,** 535–543.
70. Holmberg, N. and Bülow, L. (1998) Improving stress tolerance in plants by gene transfer. *Trends Plant Sci.* **3,** 61–66.
71. Lillus, G. and Bülow, L. (1996) Enhanced NaCl stress tolerance in transgenic tobacco expressing bacterial choline dehydrogenase. *BioTechnology* **14,** 177–180.
72. Hayashi, H., Alia, Mustardy, L., Deshnium, P., Ida, M., and Murata, N. (1997) Transformation of *Arabidopsis thaliana* with the *codA* gene for choline oxidase; accumulation of glycinebetaine and enhanced tolerance to salt and cold stress. *Plant J.* **12,** 133–142.
73. Alia, Hayashi H. and Murata, N. (1998) Enhancement of the tolerance of *Arabidopsis* to high temperatures by genetic engineering of the synthesis of glycinebetaine. *Plant J.* **16,** 155–162.

74. Holström, K. O., Mäntylä, E., Wellin, B., et al. (1996) Drought tolerance in tobacco. *Nature* **379,** 683–684.
75. Garg, A. K., Kim, J. K., Owens, T. G., et al. (2002) Trehalose accumulation in rice plants confers high tolerance levels to different abiotic stresses. *Proc. Natl. Acad. Sci. USA* **99,** 15,898–15,903.
76. Thomas, J. C., Sepahi, M., Arendall, B., and Bonhert, H. J. (1995) Enhancement of seed germination in high salinity by engineering mannitol expression in *Arabidopsis thaliana. Plant Cell Environ.* **18,** 801–806.
77. Abebe T., Guenzi, A.C., Martin, B., and Cushman J.C. (2003) Tolerance of mannitol-accumulating transgenic wheat to water stress and salinity. *Plant Physiol.* **131,** 1748–1755.
78. Apse, M. P., Aarón, G. S., Snedden, W. S., and Blumwald, E. (1999) Salt tolerance conferred by overexpression of a vacuolar Na^+/H^+ antiport in *Arabidopsis. Science* **285,** 1256–1258.
79. Zhang, H. X. and Blumwald, E. (2001) Transgenic salt-tolerant tomato plants accumulate salt in foliage but not in fruit. *Nat. Biotech.* **19,** 765–768.
80. Ohta, M., Hayashi, Y., Nakashima, A., et al. (2002) Introduction of a Na^+/H^+ antiporter gene from *Atriplex gmelini* confers salt tolerance to rice. *FEBS Lett.* **532,** 279–282.
81. Kasuga, M., Liu, Q., Miura, S., Yamaguchi-Shinozaki, K., and Shinokazi, K. (1999) Improving plant drought, salt, and freezing tolerance by gene transfer of a single stress-inducible transcription factor. *Nat. Biotech.* **17,** 287–291.
82. Jaglo, K. R., Kleff, S., Amundsen, K. L., et al. (2001) Components of the *Arabidopsis* C-Repeat/Dehydration-responsive element binding factor cold-response pathway are conserved in *Brassica napus* and other plant species. *Plant Physiol.* **127,** 910–917.
83. Hsieh, T. H., Lee, J. T., Charng, Y. Y., and Chan, M. T. (2002) Tomato plants ectopically expressing *Arabidopsis* CBF1 show enhanced resistance to water deficit stress. *Plant Physiol.* **130,** 618–626.
84. Herrera-Estrella, L. (1999) Transgenic plants for tropical regions: Some considerations about their development and their transfer to the small farmer. *Proc. Natl. Acad. Sci. USA* **96,** 5978–5981.
85. Lee, J. A. (1998). The calcicole-calcifuge problem revisited. *Adv. Bot. Res.* **29,** 2–30.
86. Bar-Yosef, B. (1991) The hidden half, in *Plant Roots* (Waisel, Y., Eschel A., and Kafkati, V., eds.) Marcel Dekker, New York, NY, pp. 529–557.
87. López-Bucio, J., Guevara-García, A., Ramírez-Rodríguez, V., Nieto, M. F., De la Fuente, J. M., and Herrera-Estrella, L. (2000) Agriculture for marginal lands: plants toward the third millennium, in *Plant Genetic Engineering: Towards the Third Millenium,* (Arencibia, A. D., ed.), Elsevier Science, Amsterdam, The Netherlands, pp. 159–170.
88. Tyler, G. and Ström, L. (1995) Differing organic acid exudation patterns explain calcifuge and acidifuge behavior of plants. *Ann. Bot.* **75,** 75–78.

89. De la Fuente, J.M., Ramírez-Rodríguez, B., Cabrera-Ponce, J.L., and Herrera-Estrella L. (1997) Aluminum tolerance in transgenic plants by alteration of citrate synthesis. *Science* **276,** 1566–1568.
90. López-Bucio, J., Martínez-De la Vega, O., Guevara-García, A., and Herrera-Estrella, L. (2000) Enhanced phosphorous uptake in transgenic tobacco plants that overproduce citrate. *Nat. Biotech.* **18,** 450–453.
91. Koyama, H., Takita, E., Kawamura, A., Hara, T., and Shibata, D. (1999) Over expression of mitochondrial citrate synthase gene improves the growth of carrot cells in Al-phosphate medium. *Plant Cell Physiol.* **40,** 482–488.
92. Chaney, R. L., Malik, M., Li, Y. M., et al. (1997) Phytoremediation of soil metals. *Curr. Opin. Biotechnol.* **8,** 279–284.
93. Rugh, C. L., Wilde, H. D., Stack, N. M., Thompson, D. M., Summers, A. O., and Meagher, R. B. (1996) Mercuric ion reduction and resistance in transgenic *Arabidopsis thaliana* plants expressing a modified bacterial *merA* gene. *Proc. Natl. Acad. Sci. USA* **93,** 3182–3187.
94. Gisbert, C., Ros, R., De Haro, A., et al. (2003) A plant genetically modified that accumulates Pb is especially promising for phytoremediation. *Biochem. Biophys. Res. Commun.* **303,** 440–445.
95. Bennett, L. E., Burkhead, J. L., Hale, K. L., Terry, N., Pilon, M., and Pilon-Smits, E. A. (2003) Analysis of transgenic Indian mustard plants for phytoremediation of metal-contaminated mine tailings. *J. Environ. Qual.* **32,** 432–440.
96. Larrick, J. W. and Thomas, D. W. (2001) Producing proteins in transgenic plants and animals. *Curr. Opin. Biotechnol.* **12,** 411–418.
97. Goddjin, O. J. M. and Pen, J. (1995) Plants as bioreactors. *Trends Biotechnol.* **13,** 379–387.
98. Woodard, S. L., Mayor, J. M., Bailey, M. R., et al. (2003) Maize-derived bovine trypsin: characterization of the first large-scale, commercial product from transgenic plants. *Biotechnol. Appl. Biochem.* **38,** 123–130.
99. Giddings, G., Allison, G., Brooks, D., and Carter, A. (2000) Transgenic plants as factories for biopharmaceuticals. *Nat. Biotech.* **18,** 1151–1155.
100. Boothe, J. G., Parmenter, D. L., and Saponja, J. A. (1997) Molecular farming in plants: oilseeds as vehicles for the production of pharmaceutical proteins. *Drug Develop. Res.* **42,** 172–181.
101. Staub, J. M., García, B., Graves, J., et al. (2000) High-yield production of a human therapeutic protein in tobacco chloroplasts. *Nat. Biotech.* **18,** 333–338.
102. Ma, J., Hiatt, A., Hein, M., etal. (1995) Generation and assembly of secretory antibodies in plants. *Science* **268,** 716–719.
103. Daniell, H., Streatfield, S. J., and Wycoff, K. (2001) Medical molecular farming: production of antibodies, biopharmaceuticals and edible vaccines in plants. *Trends Plant Sci.* **6,** 219–226.
104. Lamphear, B. J., Streatfield, S. J., Jilka, J. M., et al. (2002) Delivery of subunit vaccines in maize seed. *J. Control Rel.* **85,** 169–180.
105. Mason, H. S., Lam, D. M., and Arntzen, C. J. (1992) Expression of hepatitis B surface antigen in transgenic plants. *Proc. Natl. Acad. Sci. USA* **89,** 11,745–11,749.

106. Ehsani, P., Khabiri, A., and Domansky, N. N. (1997) Polypeptides of hepatitis B surface antigen in transgenic plants. *Gene* **190,** 107–111.
107. Kong, Q., Richter, L., Yang, Y. F., Arntzen, C. J., Mason, H. S., and Thanavala, Y. (2001) Oral immunization with hepatitis B surface antigen expressed in transgenic plants. *Proc. Natl. Acad. Sci. USA* **98,** 11,539–11,544.
108. Hein, M. B., Yeo, T. C., Wang, F., and Sturtevant, A. (1995) Expression of colera toxin subunits in plants. *Ann. NY Acad. Sci.* **792,** 50–56.
109. Arakawa, T., Chong, D. K. X., Merrit, J. L., and Landgridge, W. H. R. (1997) Expression of colera toxin B subunit oligomers in transgenic potato plants. *Transgen. Res.* **6,** 403–413.
110. Arakawa, T., Chong, D. K. X., and Langridge, W. H. R. (1998) Efficacy of a food plant-based oral cholera toxin B subunit vaccine. *Nat. Biotech.* **16,** 292–297.
111. McGarvey, P. B., Hammond, J., Dienelt, M. M., et al. (1995) Expression of the rabies virus glycoprotein in transgenic tomatoes. *BioTechnology* **13,** 1484–1487.
112. Haq, T. A., Mason, H. S., Clements, J. D., and Arntzen, C. J. (1995) Oral immunization with a recombinant bacterial antigen produced in transgenic plants. *Science* **268,** 714–715.
113. Mason, H. S., Haq, T. A., and Clements, J. D. (1998) Edible vaccine protects mice against *Escherichia coli* heat-labile enterotoxin (LT): potatoes expressing a synthetic *LT-B* gene. *Vaccine* **16,** 1336–1343.
114. Mason, H. S., Warzecha, H., Mor, T., and Arntzen C. J. (2002) Edible plant vaccines: applications for prophylactic and therapeutic molecular medicine. *Trends Mol. Med.* **8,** 324–229.
115. Walmsley, A. M. and Arntzen C. J. (2003) Plant cell factories and mucosal vaccines. *Curr. Opin. Biotechnol.* **14,** 145–150.
116. Poirier, Y., Nawrath, C., and Somerville, C. (1995) Production of polyhydroxyalkanoates, a family of biodegradable plastics and elastomers, in bacteria and plants. *BioTechnology* **13,** 142–150.
117. Nawrath, C., Poirier, Y., and Somerville, C. (1994) Targeting of the polyhydroxybutyrate biosynthetic pathway to the plastids of *Arabidopsis thaliana* results in high levels of polymer accumulation. *Proc. Natl. Acad. Sci. USA* **91,** 12,760–12,764.
118. Maliyakal, E. J. and Keller, G. (1996) Metabolic pathway engineering in cotton: Biosynthesis of polyhydroxybutyrate in fiber cells. *Proc. Natl. Acad. Sci. USA* **93,** 12,768–12,773.
119. Comai, L., Facciotti, D., Hiatt, W. R., Thompson, G., Rose, R. E., and Stalker, D. M. (1985) Expression in plants of a mutant *aroA* gene from *Salmonella typhimurium* confers tolerance to glyphosate. *Nature* **317,** 741–745.
120. Fillatti, J. J., Kiser, J., Rose, R., and Comai, L. (1987) Efficient transfer of a glyphosate tolerance gene into tomato using a binary *Agrobacterium tumefaciens* vector. *BioTechnology* **5,** 726–730.
121. Shah, D. M., Horsch, R. B., Klee, H. J., et al. (1986) Engineering herbicide tolerance in transgenic plants. *Science* **233,** 478–481.

122. Padgette, S. R., Taylor, N. B., Nida, D. L., et al. (1996) The composition of glyphosate-tolerant soybean seeds is equivalent to that of conventional soybeans. *J. Nutr.* **126,** 702–716.
123. De Block, M., Botterman, J., Vandewiele, M., et al. (1987) Engineering herbicide resistance in plants by expression of a detoxifying enzyme. *EMBO J.* **6,** 2513–2518.
124. Melchiorre, M. N., Lascano, H. R., and Trippi, V. S. (2002) Transgenic wheat plants resistant to herbicide BASTA obtained by microprojectile bombardment. *Biocell* **26,** 217–223.
125. Senior, I. J., Moyes, C., and Dale, P. J. (2002) Herbicide sensitivity of transgenic multiple herbicide-tolerant oilseed rape. *Pest Manag. Sci.* **58,** 405–412.
126. Culpepper, A. S. and York, A. C. (1997) Weed management in no-tillage bromoxynil-tolerant cotton (*Gossypium hirsutum*). *Weed Technol.* **11,** 335–345.
127. Lee, H. J., Lee, S. B., Chung, J. S., et al. (2000) Transgenic rice plants expressing a *Bacillus subtilis* protoporphyrinogen oxidase gene are resistant to diphenyl ether herbicide oxyfluorfen. *Plant Cell Physiol.* **41,** 743–749.
128. James, C. (2002) Global status of commercialized transgenic crops: 2002. *ISAAA Briefs No. 23: Preview*, ISAAA, Ithaca, NY.
129. Crawley, M. J., Brown, S. L., Hails, R. S., Kohn, D. D., and Rees, M. (2001) Transgenic crops in natural habits. *Nature* **409,** 682–683.
130. Losey, J. E., Rayor, L. S., and Carter, M.E. (1999) Transgenic pollen harms monarch larvae. *Nature* **399,** 214.
131. Shelton, A. M. and Sears, M. K. (2001) The monarch butterfly controversy: scientific interpretations of a phenomenon. *Plant J.* **27,** 483–488.
132. Gray, A. J. and Raybould, A. F. (1998) Reducing transgene escape routes. *Nature* **392,** 653–654.
133. McBride, K. E., Svab, Z., Schaaf, D. J., Hogan, P. S., Stalker, D. M., and Maliga, P. (1995) Amplification of a chimeric *Bacillus* gene in chloroplasts leads to an extraordinary level of an insecticidal protein in tobacco. *BioTechnology* **13,** 362–365.
134. Kuiper, H. A., Kleter, G. A., Noteborn, H., and Kok, J. K. (2001) Assessment of the food safety issues related to genetically modified foods. *Plant J.* **27,** 505–528.
135. The *Arabidopsis* Genome Initiative. (2000) Analysis of the genome sequence of the flowering plant *Arabidopsis thaliana*. *Nature* **408,** 796–815.
136. Yu, J., Hu, S., Wang, J., et al. (2002) A draft sequence of the rice genome (*Oryza sativa* L. ssp. *indica*). *Science* **296,** 79–92.
137. Goff, S. A., Ricke, D., Lan, T. H, et al. (2002) A draft sequence of the rice genome (*Oryza sativa* L. ssp. *japonica*). *Science* **296,** 92–100.
138. Goga, O. and Tilghman, S. M. (2000) Exploring genome space. *Nature* **405,** 820–822.
139. Lockhart, D. J. and Winzeler, E. (2000) Genomics, gene expression and DNA arrays. *Nature* **405,** 827–836.
140. Pandey, A. and Mann, M. (2000) Proteomics to study genes and genomes. *Nature* **405,** 837–846.

II

Transformation

2

Plant Transformation

Agrobacterium-Mediated Gene Transfer

Abhaya M. Dandekar and Henry J. Fisk

Summary

Plant transformation is the process by which DNA is introduced into plant cells or tissues. The DNA can come from virtually any source. Gene transfer methodology has become part of an essential technology to manipulate plants for both scientific and commercial purposes. Transgenic plants, the products of this methodology, are useful for dissecting the mechanism(s) of plant gene regulation. This technology is also useful for identifying and evaluating agriculturally useful traits (genes) as well as for their introduction into commercially valuable crops. One of the most efficient methods for gene transfer employs *Agrobacterium tumefaciens* and takes advantage of the naturally evolved crown gall-inducing mechanisms of DNA transfer present in this common soil pathogen. Much has been learned about the mechanisms of this form of DNA movement and subsequent crown gall induction. This information has been applied to develop methods that result in the formation of gall-free, genetically transformed plants. This chapter describes a detailed protocol for *Agrobacterium*-mediated transformation of tobacco cells and their subsequent selection and regeneration into transgenic plants.

Key Words: *Agrobacterium tumefaciens*; genetic selection; plant transformation; regeneration; transgenes, transgenic plants.

1. Introduction

The production of transgenic plants involves the marriage of two critical yet distinct basic technologies. The first directs the introduction of new genetic material into plant cells (transformation); whereas the second uses methods based in tissue culture to regenerate the resulting transformed cells into transgenic plants. Of the various methods developed to introduce DNA into plant cells, most include a transformation step that is mediated by *Agro-*

From: *Methods in Molecular Biology, vol. 286: Transgenic Plants: Methods and Protocols*
Edited by: L. Peña © Humana Press Inc., Totowa, NJ

bacterium tumefaciens **(1–4)**. In nature, *Agrobacterium tumefaciens* is the causative agent of crown gall disease and was discovered at the turn of the last century. However, approx 75 yr passed before it was determined that this ubiquitous soil microorganism is capable of interkingdom DNA transfer **(5)**. The crown gall (tumor) represents a manifestation of the transfer and expression of bacterial DNA in plant cells. This highly evolved and elegant mechanism of transforming plant cells has been harnessed by plant biotechnologists for both knowledge and profit. For the purposes of this chapter, we highlight some of the salient features of *Agrobacterium*-mediated transformation of plant cells and their regeneration into transgenic plants as these features need to be understood from a basic perspective to carry out effectively the procedures presented. However, this is not an extensive review of the subject and the reader is urged to read recently published reviews on the use of *Agrobacterium* as a vector for gene transfer **(1–5)** and on *Agrobacterium* as an agent of disease **(6)**.

Agrobacterium is attracted to the amino acids, sugars and organic acids that are released from wounded plant tissues. It responds to these chemoattractants by seeking out the wounded cells that produced them and then by binding to them by a polar attachment mechanism **(4,7)**. During attachment, coordinated expression from a suite of genetic operons critical to the gene transfer process also begins **(8)**. These operons–*virB, virC, virD, virE,* and *virG*–are collectively termed the "*vir* regulon," and are coordinately regulated by a *virA*/*virG* two-component system. The wound phenolics and monosaccharides directly or indirectly cause the autophosphorylation of the *virA* transmembrane receptor kinase, which in turn activates the soluble cytoplasmic transcriptional factor virG through another phosphorylation event. Activated *virG* subsequently stimulates the transcription of the individual *vir* operons by binding to the upstream "*vir* box" *cis*/enhancer elements **(2,7)**.

Gene products that are generated from transcription of the *vir* operons perform functions that are critical to the transfer of a DNA fragment called T-DNA from the tumor-inducing (Ti) plasmid localized in the bacteria into plant cells. The gene products virD1 and virD2 are cooperatively responsible for cleavage of the T-strand delimited by the presence of border sequences (right and left border) **(8)**. The virD2 protein binds covalently to the 5'-end of the T-strand which is then coated to form a T-complex with the single strand binding protein virE2 either in the bacteria or *in planta* **(9)**. This T-complex is exported via a type 4 bacterial secretion system encoded by the *virB* operon and *virD4* **(9)**. Both virD2 and virE2 contain nuclear localization sequences that interact with the plant components that include an importin-α, a type 2C protein phosphatase and three cyclophilins (virD2-interacting factors), and vip1 and vip2 (virE2-interacting factors) which together help target the T-complex into the plant nucleus **(4,5)**. Once inside the nucleus the T-strand is integrated into the plant genome via nonhomologous recombination mediated by plant encoded proteins that are likely part of recombination and/or repair process in plants ***(10,11)***.

Instead of the naturally occurring single Ti plasmid, most laboratory strains of *Agrobacterium* used for transformation employ a binary system consisting of two plasmids ***(12)***. One plasmid contains the *vir* regulon sequences, the gene products of which work in *trans* to transfer the T-DNA from a separate plasmid. The oncogenes (gall-forming sequences) have been removed from the T-DNA and in their place engineered expression cassettes with genes from virtually any source may be substituted, usually by convenient insertion into multiple cloning sequences that have been incorporated into these plasmids. Different strains of *A. tumefaciens* display different levels of virulence (transformability), much of which stems from differences in the *vir* sequences ***(13)***.

Once a plant cell has incorporated the introduced DNA in a stable manner (i.e., covalently integrated within the host plant's genome), the next step is to regenerate a plant from the transformed cells. Position, frequency, and scope of regeneration events are critical to the isolation of transgenic plants ***(14)***. Most often, the major limiting step in the isolation of transgenic plants is a lack of regeneration occurring from within the transformed cell populations. There is a large amount of variability in the frequency and scope of regeneration among different angiosperm species as well as among different cultivars of any one species ***(15)***. The two pathways of regeneration that have been observed in most angiosperms are organogenesis and somatic embryogenesis (for review, *see* **ref.** ***16***). Organogenesis involves the regeneration of adventitious shoots or roots through the formation of organized, meristematic tissues. The second pathway involves the formation of embryos or embryo-like structures from somatic tissues. It has been suggested that somatic embryogenesis and organogenesis reflect different developmental events that are most likely mutually exclusive ***(16)***. This chapter presents a method for the organogenic regeneration of tobacco plants from leaf discs following *Agrobacterium*-mediated transformation that is loosely based on a landmark paper published nearly 20 yr ago ***(17)***.

2. Materials

Unless stated otherwise, all reagents and chemicals used in this protocol were of high purity and were analytical grade and/or tested for molecular biology or plant cell tissue culture applications. The water used was deionized and filtered through a Nanopure (Barnstead, Dubuque, IA) water purification system.

2.1. Supplies and Equipment

1. GA7 tissue culture boxes with lids (Magenta or equivalent).
2. Laminar flow hood.

3. Forceps.
4. Scalpels.
5. Sterile, disposable Petri dishes.
6. Sterile filter paper.
7. Cork borers (0.7 mm).
8. Cork borer sharpener.
9. Bunsen burners.
10. Inoculation loops.
11. Environmental shaker incubators (25°C).
12. Environmental growth chambers.
13. P20, P200, P1000, and P5000, Pipetman (or equivalent) micropipettors and appropriate tips.
14. 1.5-mL Microfuge tubes.
15. 15- and 50-mL capped centrifuge tubes (Falcon or equivalent).
16. 15% (v/v) Household bleach.
17. 70% Ethanol.
18. Laboratory sealing film (Parafilm or equivalent).
19. Heated water bath at 55°C.
20. Disposable 10-mL sterile syringes.
21. Acrodisc 0.2-μm syringe filter sterilization units (or equivalent).

2.2. Reagents, Solutions, and Media

1. 1/2X MSO, pH 5.8: half-strength Murashige and Skoog (MS) medium ***(18)*** solidified with 0.8% Phytagar (Invitrogen, Carlsbad, CA) (*see* **Note 1**).
2. *Agrobacterium* strains: any one of several common disarmed (non-gall-forming) laboratory strains (e.g., EHA 101, 105, C58, and LBA4404) containing an engineered binary transformation vector (*see* **Note 2**).
3. YEP medium, pH 7.2: 5.0 g/L of Bacto-yeast extract, 10.0 g/L of Bacto-peptone, 10 g/L of NaCl, 15 g/L of Bacto-agar.
4. Filter-sterilized MS20IM *Agrobacterium* induction medium, pH 5.25: MS salts and vitamins supplemented with 2 % (w/v) sucrose, 100 μ*M* acetosyringone, 1 m*M* betaine phosphate or proline, and 2.5 m*M* 2-(4-morpholino)ethanesulfonic acid (MES) (*see* **Note 3**).
5. Cocultivation medium, pH 5.8: MS medium supplemented with 4.5 μ*M* benzylaminopurine (BA), 0.5 μ*M* naphthalene acetic acid (NAA) and solidified with 0.8% (w/v) Phytagar (Gibco).
6. MSBN1.1 shoot regeneration medium, pH 5.8: identical to the cocultivation medium shown above with the exception that selective agents are used as appropriate (*see* **Note 4**).
7. MSHF rooting medium, pH 5.8: MS medium solidified with 0.8% Phytagar and supplemented with selective agents when appropriate.

3. Methods

3.1. Growth and Propagation of Tobacco (Nicotiana tabacum)

Most plants offer a number of tissues that will regenerate under the proper conditions. However, efficiencies may vary greatly. Plants or regenerable plant tissues grown under axenic conditions in culture offer the most consistent results with respect to regeneration, as some of the environmental conditioning that varies with season in plants grown outside of the laboratory has been eliminated. Material from cultures also leads to fewer downstream contamination problems. Shown below is a procedure for growing tobacco plants under axenic conditions. It should be noted that the methods have been optimized for the cultivar "Xanthi"; however, others, such as "SR1" also have been successfully transformed using this procedure.

1. Surface sterilize tobacco seeds by placing them in 15-mL conical centrifuge tubes and filling them with 10 mL of a 15% bleach solution plus one drop of Tween-20.
2. Shake the tubes continuously for 15 min on a gyratory shaker at 110 rpm.
3. Allow the seeds to settle, pipet off the Clorox solution, and rinse three times with sterile distilled water. Rinsing is accomplished by filling the centrifuge tube with 10 mL of sterile distilled water, then allowing the seeds to settle and pipetting off the rinse water. Remove all but 1 mL of water during the final rinse.
4. Dispense the last milliliter of water with seeds using a pipet onto 100 × 20 mm Petri dishes containing 25 mL of agar solidified 1/2X MSO.
5. Incubate plates at 26°C under soft fluorescent lights with a 16-h photoperiod.
6. After 10–14 d, transfer germinating green seedlings to Magenta boxes containing 50 mL of autoclaved MSHF (*see* **Note 5**).
7. Plants may be multiplied by removing expanded leaves from rooted plants, cutting the remaining stem between nodes, and inserting the resulting stem pieces into Magenta boxes containing MSHF. Individual plants may be maintained indefinitely without multiplication by simply propagating the shoot tip in a similar manner. Repeat subcultures to fresh medium once every 4 wk.

3.2. Growth of Agrobacterium and Preparation of Inoculum

Compared to other laboratory strains of bacteria such as *Escherichia coli*, *Agrobacterium* grows relatively slowly. To grow overnight cultures of sufficient densities consistently and conveniently, it is important to inoculate them with cells actively growing on solid medium.

1. Prepare a 50-mL culture tube containing 10 mL of YEP media containing the appropriate selective antibiotics.
2. Inoculate the tube with one loopful of active bacteria (*A. tumefaciens* containing a binary vector with the gene[s] of interest) taken from a selection plate kept at 4°C (*see* **Note 6**).

3. Grow 20–24 h at 25°C with agitation of 100–150 rpm. If an environmental shaker is unavailable, room temperature should be sufficient.
4. Determine the optical density of the cultures spectrophotometrically at 420 nm. Calculate the amount of culture needed to provide an optical density of 0.5 when diluted to 20 mL.
5. Centrifuge the appropriate amount of culture in a 50-mL Falcon tube for 15 min at 2500*g*.
6. Pour off the supernatant
7. Resuspend the pellet in 20 mL of MS20IM medium
8. Induce the *Agrobacterium* for transformation by shaking on a rotary shaker (100–150 rpm) for 5 h at 20–25°C (room temperature).

3.3. Preparation and Infection of Leaf Disks

The overall objective in preparing plant material is to maximize the number of wounded, cut surfaces for *Agrobacterium* attachment while maintaining enough healthy tissue that will later support efficient regeneration.

1. Remove expanded leaves from rooted plants growing axenically in culture and float them in 100-mm Petri dishes containing sterile MS20IM.
2. Cut disks from the leaves in dishes under MS20IM using a flame-sterilized 0.7-cm cork borer. Prepare leaf disks in batches of approx 50/plate (*see* **Note 7**).
3. Set aside approx 16 leaf disks to serve as controls for the transformation/regeneration procedure by transferring them directly to 100 × 15 mm Petri dishes containing cocultivation medium overlaid with sterile filter paper (8 disks/plate) after gently blotting away excess MS20IM using sterile filter paper.
4. Decant the MS20IM from the plates containing the remaining leaf disks using a sterile pipet and replace it with induced *A. tumefaciens* suspension. Incubate at room temperature (approx 25°C) for 10–20 min with occasional swirling.

3.4. Cocultivation

Agrobacterium attachment to plant tissue is completed during the earlier stages of cocultivation. The physical transfer of genetic material occurs later.

1. Remove each disk individually, gently blot off excess culture using sterile filter paper, and transfer to 100 × 15 mm Petri dishes containing cocultivation media overlaid with sterile filter paper. Place about 16 disks/plate.
2. For large scale experiments we routinely cut about 800 disks and inoculate Petri dishes with approx 24 disks/plate.
3. Seal all Petri dishes with laboratory sealing film (Parafilm or equivalent)
4. Incubate cultures at 20°C in the dark for 3 d (*see* **Note 8**).

3.5. Selection and Regeneration of Transgenic Tobacco Shoots

Several important events occur during selection and regeneration. Antibiotic(s) that do not affect plant cells are used to eliminate or arrest the growth of *A. tumefaciens*. Conditions are also optimized for the adventitious, organogenic regeneration of new plant tissues. To enrich the population of new growth with transgenic tissues, additional selective agents are incorporated into the regeneration medium for the purposes of genetic selection (*see* **Note 4**). Genetic selection is the process of selecting preferentially for those cells that have been transformed by the incoming transgenes. A selective advantage can be conferred on the transformed cells through the introduction of genes encoding antibiotic resistance or resistance to some metabolic inhibitor such as a herbicide. In the presence of the antibiotic or herbicide, the untransformed cells die whereas the transformed cells grow and multiply. If no form of genetic selection were used, then one would be faced with the option of screening every shoot that regenerated in a transformation experiment. In cases where the transformation frequency is high (i.e., the number of transformed cells or shoots arising from an explant), this would be feasible. However, for other species with lower transformation frequencies, this would become a laborious if not impossible task. Therefore, genetic selection is an essential component of any plant transformation protocol and has been accomplished by using various marker genes ***(14,16)***.

1. Subculture the disks to selective medium. All those infected with *A. tumefaciens* and half of the control disks (no infection with *A. tumefaciens*) should be transferred to MSBN1.1 regeneration medium containing the appropriate selective agents in 100 × 15 mm Petri dishes. The control disks under these conditions will provide an indication of nontransgenic regeneration ("escapes") under selection. Transfer the remaining control disks to regeneration medium (MSBN1.1) containing only the selective agent used to eliminate *Agrobacterium* (this is a control to evaluate overall regeneration frequency). In all cases, plate at a density of approx 8 disks/plate.
2. Maintain cultures at 20°C in low light (approx 45 µE/m^{2}s). Check regularly for contamination. If contamination is discovered, unaffected disks within the plate may be subcultured to fresh MSBN1.1.
3. All disks should be subcultured to fresh selection plates every 2–3 wk. The disks will expand and develop callus over time. Try to ensure that the expanded disks establish good contact with the media. Shoots will appear in 3–4 wk.

3.6. Rooting of Transgenic Shoots to Recover Complete Plantlets

The next step is to recover complete plants from any regenerated shoots through root organogenesis. In addition, the first meaningful screen to test for transformation is often the rooting procedure, as root organogenesis is usually more sensitive to the incorporated selective agents than shoot regeneration. Shoots recovered from selective regeneration procedures that do not root under selection are rarely transgenic and should be discarded.

1. Carefully remove regenerated shoots by cutting them at their base using a sterile scalpel and forceps and place them in GA7 boxes (about four shoots per vessel)

containing 50 mL of MSHF supplemented with selective agents. Roots should become visible within approx 10 d (*see* **Note 9**).

2. Subculture only the shoots that have rooted by cutting off the shoot with the top four internodes and introducing these individually into a GA7 box containing 50 mL of MSHF supplemented with the appropriate selective agents. These individual shoots may be considered as putative transformants.
3. Rooted shoots can be maintained and/or propagated to establish individual lines at monthly intervals as described in **Subheading 3.1.** Alternatively, the plants may be acclimatized and transferred to the greenhouse to produce seeds. It takes about 3 mo to set seed, depending on conditions.

3.7. Analysis of Transgenic Plants

Recovered plants are typically analyzed on a number of different levels to determine that they are in fact transgenic. Once plants grow large enough to provide enough tissue for analyses without compromising health, they may be assayed for transgene expression and molecularly for the presence of the appropriate sequences. The assay for gene expression is conducted using methods consistent with the transgene coding sequence and desired results. If such a procedure is impossible or inconvenient, polymerase chain reactions (PCRs) may also be performed. Plants that give a positive result must then be analyzed using a DNA blotting procedure (Southern) to confirm the presence of transgenes and their abundance (*see* **Note 10**).

4. Notes

1. Premixed tissue culture reagents are available commercially from a number of different sources. We routinely purchase MS salts and vitamins as a powder or concentrated stock solution from either Gibco or Sigma. Reagents from both sources provide consistent results.
2. When selecting a strain of *Agrobacterium* for the purpose of transformation, the genetic background is a factor that should be considered. Although it is well known that most dicot plants are susceptible to *A. tumefaciens* ***(13)***, resistance of the target plant tissues to this pathogen could be an important factor influencing its virulence and, ultimately, affect the efficiency of plant transformation. A growing body of evidence indicates, for most of the widely used strains of *A. tumefaciens*, wide variations in virulence that depends on the target plant tissue used ***(13)***. Many of these differences may stem from differences in interactions between the host plant and bacterial *vir* gene products.
3. This medium has been developed to provide *A. tumefaciens* optimal conditions for virulence induction. Environmental factors such as pH, temperature, and osmotic conditions strongly influence the expression and induction of virulence genes ***(14)***. The most direct effects on virulence induction are mediated by the presence of phenolic compounds such as acetosyringone (3',5'-dimethoxy-4'-

hydroxyacetophenone), sinapinic acid, coniferyl alcohol, caffeic acid, ethyl ferrulate, and methylsyringic acid, which are known inducers of virulence genes in *Agrobacterium* (reviewed by Kado *[19]*). The virulence induction is also influenced by the presence of other compounds such as monosaccharides ***(20)*** and opines ***(21)***. Betaine, proline, and other osmoprotective compounds have been shown to enhance synergistically the effect of phenolic compounds ***(15,22)***. Betaine has been shown to increase the expression of several virulence genes in *Agrobacterium* ***(15)***. Proline or betaine may help the bacteria to adapt to rapid changes in pH and osmotic pressure caused by the proximity of wounded plant cells, thus increasing the transformation efficiency ***(22)***.

4. Selective agents used for this purpose are usually prepared as stock solutions that are typically 500- to 1000-fold more concentrated than their working strength in cultures. They may be stored as filter-sterilized solutions in a freezer (–20°C) for up to 2 mo. Shown are the working concentrations (milligram/liter) of several antibiotics routinely used for selection during plant transformation procedures: kanamycin—100; tetracycline—5; gentamicin—20; cefotaxime—250–500; and carbenicillin—500. Kanamycin is commonly used to select for transgenic plant cells and tissues whereas the others are used to select for engineered strains of *A. tumefaciens* (tetracycline and gentamicin) or eliminate it (cefotaxime and carbenicillin) from cultures.
5. Growth of tobacco can vary widely depending on the cultivar and growth conditions. It may be advisable to use a larger culture container to allow for a reasonable amount of time to pass before it becomes necessary to subculture, or to maximize the leaf material available as source tissue for a transformation procedure. We routinely use glass household canning jars containing 100 mL of medium. The plants perform best if the vessels are capped with a sterile plastic cap. Avoid a glass cap and instead use, for example, the bottom of a disposable Petri dish and seal to the container with Parafilm (or equivalent).
6. To ensure that overnight cultures obtain an adequate cell density, it is important to use active inoculum. We routinely maintain the cultures as streaked bacteria on selective plates containing solidified YEP medium. The plates are incubated for approx 48 h at 28°C and then kept in a refrigerator (4°C). The bacteria should be subcultured to fresh plates every 4 wk.

   As an alternative to using a sterile loop to streak plates and inoculate liquid cultures, we routinely use sterile pipet tips. The barrel and ejector of the pipettor are sprayed with 70% ethanol and allowed to dry in a laminar flow hood. The pipettor is then used to place bacteria on a pipet tip that can then be used to streak a plate or ejected into a culture tube containing growth medium.
7. As an alternative to using a cork borer to prepare discs, the leaf tissue also may be cut into small squares with a scalpel and forceps. In either case, it is important to be as gentle as possible, because unnecessary wounding may lower regeneration frequencies. In addition, excessive drying may also result in adverse effects. Therefore, it is important to work quickly and minimize exposure of the leaf tissue to open air as much as possible.

8. We have observed that transformation frequencies trend upwards with increasing cocultivation time, up to 5 d. However, overgrowth of *A. tumefaciens* and subsequent losses of plant material owing to contamination result in cocultivation times exceeding 3 d. Overgrowth problems are the result of an interaction between inoculum concentration, cocultivation time, and plant species or cultivar. Therefore, concentration and time should be considered variables for optimization when establishing a transformation system.
9. If one of the transgenes contained within the binary vector is a scoreable marker, it may be possible to conduct a convenient preliminary screen for transformation prior to placing the shoots into rooting medium. After excising the regenerated shoots from the original explant, a very small piece of stem tissue may be taken from the basal region before it is placed in rooting medium. The cutaway stem tissue may then be used to assay for the expression of the scoreable marker. Decisions about moving forward with the corresponding shoots may then be conducted in a more informed manner.
10. It is important to confirm stable incorporation of the introduced gene(s) and its expression in the putatively transformed plants and their siblings. This is possible only if the incorporated DNA has been integrated into the genome of the transformed plant. In annual plants such as tobacco described here, this can be determined easily by backcrossing or selfing the plant to determine if the introduced gene is heritable. In the case of perennial species, often the long generation time makes this type of analysis impractical. Alternatively, transformation can be confirmed through a rigorous and comprehensive Southern analysis of the transformed tissue. Typically this analysis should be performed to reveal and identify different segments of the inserted T-DNA, that is, the presence of both internal and border fragments ***(23)***.

References

1. Rossi, L., Tinland, B., and Hohn, B. (1998) Role of virulence proteins of *Agrobacterium* in the plant, in *The Rhizobiaceae* (Spaink, H. P., Kondorosi, A., and Hooykaas, P. J. J., eds.), Kluwer Academic, Dordrecht, The Netherlands, pp. 303–320.
2. Zupan, J., Muth, T. R., Draper, O., and Zambryski, P. (2000) The transfer of DNA from *Agrobacterium tumefaciens* into plants: a feast of fundamental insights. *Plant J.* **23,** 11–28.
3. Gelvin, S. B. (2000) *Agrobacterium* and plant genes involved in T-DNA transfer and integration. *Annu. Rev. Plant Physiol. Plant Mol. Biol.* **51,** 223–256.
4. Tzfira, T. and Citovsky, V. (2002) Partners-in-infection: host proteins involved in the transformation of plant cells by *Agrobacterium. Trends Cell Biol.* **12,** 121–128.
5. Chilton, M.-D., Drummond, M. H., Merlo, D. J., et al. (1977) Stable incorporation of plasmid DNA into higher plant cells: the molecular basis of crown gall tumorigenesis. *Cell* **11,** 263—271.
6. Escobar, M. A. and Dandekar, A. M. (2003) *Agrobacterium tumefaciens* as an agent of disease. *Trends Plant Sci.* **8,** 380–386.

7. Winans, S. C. (1992) Two-way chemical signaling in *Agrobacterium*–plant interactions. *Microbiol. Rev.* **56,** 12–31.
8. Stachel, S. E. and Zambryski, P. C. (1985) VirA and VirG control the plant-induced activation of the T-DNA transfer process of *A. tumefaciens. Cell* **46,** 325–333.
9. Vergunst, A. C., Schrammeijer, B., der Dulk-Ras, A., de Vlaam, C. M. T., Regensburg-Tuink, T. J. G., and Hooykaas, P. J. J. (2000) VirB/D4-dependent protein translocation from *Agrobacterium* into plant cells. *Science* **290,** 979–982.
10. Ziemienowicz, A., Tinland, B., Bryant, J., Gloeckler, V., and Hohn, B. (2000) Plant enzymes but not *Agrobacterium* VirD2 mediate T-DNA ligation in vitro. *Mol. Cell Biol.* **20,** 6317–6322.
11. van Attikum, H., Bundock, P., and Hooykaas, P. J. J. (2001) Non-homologous end-joining proteins are required for *Agrobacterium* T-DNA integration. *EMBO J.* **20,** 6550–6558.
12. Hoekama, A., Hirsch, P. R., Hooykass, P. J. J., and Schilperoort, R. A. (1983) A binary plant vector strategy based on separation of *vir* and T region of the *Agrobacterium tumefaciens* Ti-plasmid. *Nature* **303,** 179–180.
13. De Cleene, M. and De Ley, J. (1976) The host range of crown gall. *Bot. Rev.* **42,** 389–466.
14. Fisk, H. J. and Dandekar, A. M. (1993) The introduction and expression of transgenes in crop plants. *Sci. Horticult.* **55,** 5–36.
15. Vernade, D., Herrera-Estrella, A., Wang, K., and Van Montagu, M. (1988) Glycine betaine allows enhanced induction of the *Agrobacterium tumefaciens vir* genes by acetosyringone at low pH. *J. Bacteriol.* **170,** 5822–5829.
16. Dandekar, A. M. (1994) Genetic transformation of angiosperms, in *Somatic Embryogensis in Woody Plants*, Vol. 1 (Jain, S. M., Gupta, P. K., and Newton, R. J., eds.), Kluwer Academic, Dordrecht, The Netherlands, pp. 193–225.
17. Horsch, R. B., Fry, J. E., Hoffman, N. L., Eichholtz, D., Rogers, S. G. and Fraley, R. T. (1985) A simple and rapid method for transferring genes into plants. *Science* **227,** 1229–1231.
18. Murashige, T. and Skoog, F. (1962) A revised medium for rapid growth and bio assays with tobacco tissue cultures. *Physiol. Plant.* **15,** 473–497.
19. Kado, C. I. (1991) Molecular mechanisms of crown gall tumorigenesis. *Crit. Rev. Plant Sci.* **10,** 1–32.
20. Ankenbauer, R. G. and Nester, E. W. (1990) Sugar-mediated induction of *Agrobacterium tumefaciens* virulence genes: structural specificity and activities of monosaccharides. *J. Bacteriol.* **172,** 6442–6446.
21. Veluthambi, K., Krishnan, M., Gould, J. H., Smith, R. H., and Gelvinm S. B. (1989) Opines stimulate induction of the *vir* genes of the *Agrobacterium tumefaciens* Ti plasmid. *J. Bacteriol.* **171,** 3969–3703.
22. James, D. J., Uratsu, S. L. Cheng, J., Negri, P., Viss P., and Dandekar. A. M. (1993) Conditions that induce Agrobacterium *Vir* genes also enhance apple cell transformation. *Plant Cell Rep.* **12,** 559–563.

23. Dandekar, A. M. (1992) Transformation, in *Biotechnologies of Perennial Fruit Crops* (Hammerschlag, F. and Litz, R., eds), C.A.B. International, Cambridge, MA, pp. 141–168.

3

Production of Hairy Root Cultures and Transgenic Plants by *Agrobacterium rhizogenes*-Mediated Transformation

Mary C. Christey and Robert H. Braun

Summary

Agrobacterium rhizogenes-mediated transformation results in the development of hairy roots at the site of infection. The production of hairy roots involves cocultivation of explants with *A. rhizogenes* and the subsequent selection of hairy roots on hormone-free medium. Hairy roots have many applications for research including secondary product production and for the study of biochemical pathways. In addition, transgenic plants regenerated from hairy roots often show an altered phenotype due to the presence of the *rol* genes. In this chapter we describe how to produce and grow hairy root cultures, how to regenerate shoots from these hairy roots, and how to conduct molecular analysis of these cultures.

Key Words: *Agrobacterium rhizogenes*; hairy roots; Ri phenotype; *rol* genes; transformation.

1. Introduction

Agrobacterium rhizogenes is a soil bacterium responsible for the development of hairy root disease on a range of dicotyledonous plants. This phenotype is caused by genetic transformation in a manner similar to the development of crown gall disease by *A. tumefaciens*. Infection of wound sites by *A. rhizogenes* is followed by the transfer, integration, and expression of T-DNA from the root-inducing (Ri) plasmid and subsequent development of the hairy root phenotype. Hairy roots can be induced on a wide range of plants and many can be regenerated into plants, often spontaneously. Transgenic plants have been obtained after *A. rhizogenes*-mediated transformation in 89 different taxa, representing 79 species from 55 genera and 27 families *(1)*.

From: *Methods in Molecular Biology, vol. 286: Transgenic Plants: Methods and Protocols*
Edited by: L. Peña © Humana Press Inc., Totowa, NJ

Cocultivation of explants with *A. rhizogenes* results in the production of hairy roots that are easily distinguished by their rapid, highly branching growth on hormone-free medium and plagiotropic root development. Plants regenerated from hairy roots often exhibit an altered phenotype characterized by several morphological changes including wrinkled leaves, shortened internodes, reduced apical dominance, reduced fertility, altered flowering, and plagiotropic roots ***(1)***. These characteristic phenotypic changes result from the transfer and expression of four loci (*rolA*, *B*, *C*, *D*) located on the T-DNA.

A. rhizogenes-derived hairy roots and plants have application for many areas of research. For example, hairy root cultures have been used extensively in root nodule research ***(2)***, for artificial seed production ***(3)***, for production of plant secondary metabolites ***(4)***, as an experimental system to study biochemical pathways ***(5)*** and responses to chemicals ***(6,7)***, and to study interactions with other organisms such as nematodes ***(8)***, mycorrhizal fungi, and root pathogens ***(7)***. Root cultures established by *A. rhizogenes*-mediated transformation are widely used as a source of useful compounds owing to their rapid growth in hormone-free medium and the relatively high production of secondary metabolites compared with the starting plant material. Hairy roots have been shown to produce a range of secondary metabolites including tropane alkaloids, indole alkaloids, terpenoids, aconites, and flavonoids ***(9)***. In addition, hairy roots have been used to express antibodies ***(10)***. Transgenic plants have been shown to express a wide variety of foreign genes including very complex proteins such as antibodies. Once established, use of hairy roots as a culture system to express these proteins offers many advantages for large-scale production. They are easy to grow, usually requiring no phytohormones for growth. These roots can be removed from the original explant/plant and established as long-term root clones capable of large increases in growth while maintaining their biosynthetic capacity. The hairy root phenotype is stable and characterized by profuse branching and high-density growth.

Over recent years there has been increased interest in the use of *A. rhizogenes* owing to the effect of *rol* genes on plant morphology and development. Some of these morphological changes such as increased flowering, altered architecture, and increased secondary product production are of horticultural use. In addition, *A. rhizogenes*-mediated transformation has been used to introduce a range of foreign genes of agronomic use ***(1)***.

This chapter describes how to produce and grow hairy root cultures successfully, how to regenerate shoots from these cultures, and how to conduct molecular analysis of these cultures and plants.

2. Materials

1. *A. rhizogenes* culture or glycerol stock containing 850 µL of bacterial culture and 150 µL of sterile glycerol.

2. Luria Bertani (LB) medium: 1% tryptone, 0.5% yeast extract, 0.5% NaCl, pH 7.0.
3. In vitro seedlings or shoots.
4. Hormone-free plant tissue culture medium, with and without antibiotics.
5. Shoot regeneration medium.
6. Tris-ethylenediaminetetraacetic acid (TE): 10 m*M* Tris-HCl, pH 7.5, 1 m*M* ethylenediaminetetraacetic acid (EDTA), pH 8.0.
7. Extraction buffer: 200 m*M* Tris-HCl, pH 7.5, 250 m*M* NaCl, 25 m*M* EDTA, 0.5% sodium dodecyl sulfate (SDS).
8. 2 m*M* dNTPs: Add 2 µL each of 100 m*M* dATP, 100 m*M* dCTP, 100 m*M* dTTP, and 100 m*M* dGTP (Roche) to 92 µL of sterile deionized water.
9. Oligonucleotide primers.
10. Electrophoresis equipment.
11. Thermal cycler machine.

3. Methods

The methods described in the following subheadings outline (a) *A. rhizogenes* culture, (b) cocultivation of explants to produce hairy root cultures, (c) regeneration of transgenic shoots, (d) molecular characterization of transgenic hairy roots and plants, and (e) transfer of plants to the greenhouse.

3.1. A. rhizogenes *Culture*

A culture of *A. rhizogenes* suitable for explant cocultivation is prepared as follows:

1. Use a single bacterial colony or glycerol stock to inoculate a flask of liquid LB medium containing the appropriate antibiotics for selection of the binary plasmid or 100 mg/L of streptomycin for wild-type *A. rhizogenes* strains (*see* **Note 1**).
2. Grow overnight at 28°C (*see* **Note 2**).
3. Dilute the culture 1:49 with antibiotic-free liquid LB.
4. Grow for a further 3–4 h before use. At this stage glycerol stocks can be established. Mix tubes thoroughly by vortexing and store immediately at –80°C.

In addition to wild-type *Agrobacterium* strains, foreign genes can be introduced into hairy roots by the use of binary vectors. The options available for binary vector components are covered in **ref. *11*** and the actual construction of a binary vector is outlined in **ref. *12***. Binary vectors are introduced into wild-type strains via the freeze–thaw method of transformation (*see* **Subheading 3.1.1.**).

3.1.1. Transformation by Agrobacterium

To prepare *A. rhizogenes* cells competent for transformation:

1. Inoculate 50 mL of LB broth with 5 mL of an overnight culture.
2. Grow for 4–5 h at 28°C.
3. Centrifuge the logarithmically growing cells at 1100*g* for 20 min at 4°C.
4. Gently wash the cells in 30 mL of sterile TE and pellet as before.

5. Resuspend gently in 5 mL of LB.
6. Aliquot into 500-µL lots and freeze in liquid nitrogen.
7. Store at –80°C. Note the efficiency for transformation will decline after 3 mo.

To transform competent cells:

1. Thaw cells slowly on ice.
2. Mix with 0.5–10 µg of plasmid.
3. Incubate on ice for 5 min.
4. Freeze in liquid nitrogen for 5 min.
5. Thaw in a water bath at 37°C for 5 min.
6. Immediately add 1 mL of LB and incubate with shaking for 3 h at 28°C.
7. Plate 200–500 µL of the mixtures onto LB plates with appropriate selection and incubate at 28°C for 48 h.
8. Select single colonies for PCR to confirm presence of the construct.

3.2. Production of Hairy Roots

3.2.1. In Vitro

For the production of hairy roots in vitro, most protocols follow the conventional in vitro explant cocultivation method as used for *A. tumefaciens*-mediated transformation. The major difference is that explants are placed on hormone-free medium to enable the selection of hairy root cultures. The choice of medium is dependent on plant species but is likely to be based on Murashige and Skoog medium ***(13)***. For studies where the aim is rapid production of hairy roots, highly susceptible materials such as Chinese cabbage or potato are excellent starting materials.

Various explant sources can be used. The important feature is that a cut surface is needed. For *Brassica* spp., in vitro seedling explants such as hypocotyls, leaves, and the cut end of the cotyledonary petiole are routinely used with *A. rhizogenes* strain A4T based on the protocol outlined below developed by Christey et al. ***(14)***.

1. Cut hypocotyls into approx 0.5-cm explants and leaves in halves or thirds.
2. Immerse explants briefly (10 s) in a diluted overnight *A. rhizogenes* culture. For cotyledonary petioles, only the cut end of the petiole of each cotyledon is dipped individually into the bacterial solution.
3. Blot explants on sterile filter paper.
4. Place explants horizontally onto hormone-free culture medium and coculture for approx 3 d.
5. Transfer explants to antibiotic-containing medium (e.g., 300 mg/L of Timentin or 200 mg/L of cefotaxime) to suppress *A. rhizogenes* growth. If binary vectors are being used, explants can be transferred at this stage to selection medium or selection can be delayed for 7–10 d.
6. Transfer explants to fresh antibiotic-containing medium every 3–4 wk.

7. Once good root growth is noted from the explants, excise the roots from the explants and transfer to individual containers. It is important to ensure that explants are kept well apart to enable distinguishing independent root cultures. Hairy root cultures can be very fast growing and can quickly grow over each other.
8. Once established, subculture roots by cutting a 1.5 cm square of culture and transfer to fresh medium every 6–8 wk or as required. Tall Petri dishes (9 cm diameter, 2 cm tall) or pots are preferred as the extra height provides more space for the plagiotropic roots to growth into. All culture manipulations are conducted at 25°C with a 16-h/d photoperiod, provided by Cool White fluorescent lights, 20 $\mu E/m^2/s$. However, hairy roots also grow well in the dark.

Hairy roots can also be initiated from the stem of in vitro cultures by using a needle or other sharp tool to wound the stem of in vitro shoots. The site is then infected with the bacterial culture. Another method used successfully involves upturned seedlings ***(15)***. Explants are excised from in vitro seedlings by cutting the hypocotyl approx 0.5 cm below the cotyledons. Explants are placed inverted on culture medium so that the upper-leaf surfaces of the cotyledons are in contact with the medium. A 1- to 3-μL drop of diluted *A. rhizogenes* culture is placed onto the cut surface of the hypocotyl. In both cases, hairy roots are excised as they appear and cultured as described above.

3.2.2. In Vivo

In vivo methods involve wounding of the stem or petiole of greenhouse plants with a needle or toothpick dipped in bacterial solution or injection with a needle. High humidity is essential for the production of hairy roots so wound sites are often covered with gauze to maintain the high humidity needed to enable hairy root development. In the absence of appropriate humidity, tumor structures may develop instead, as noted in wasabi (Christey, unpublished observations). In this case, hairy roots were visible lower down the stem at the soil level where humidity was higher.

3.2.3. A. rhizogenes *Strain Selection*

Opines are carbon compounds produced by the crown galls and hairy roots induced by *A. tumefaciens* and *A. rhizogenes*, respectively. These novel condensation products of plant metabolic intermediates are used as nutritional sources by the *Agrobacterium* strains that induced the growths ***(16)***. *A. rhizogenes* strains are characterized by the type of opine they engineer plant cells to produce and that they degrade to use as a growth substance. Agropine (e.g. strains A4T, 15834, TR105), mannopine (e.g., strains TR7, 8196), mikimopine (e.g., A5, A6), and cucumopine (e.g., strains 2588, 2657) type *A. rhizogenes* strains are available. It is advisable to test a range of *A. rhizogenes* strain types with different genotypes and explant sources to determine the most infective combination for the plant species or cultivar of interest.

3.2.4. Identification of Hairy Roots by Morphology

In the first few weeks after cocultivation, it may be difficult to distinguish hairy roots from normal roots owing to the regeneration of normal roots also. However, after several weeks the difference in morphology between normal and hairy roots should be readily apparent. If binary vectors are being used, the inclusion of selection agents or reporter genes such as green fluorescent protein (GFP) aids the identification and selection of hairy root cultures from normal roots compared with wild-type strains. Hairy root cultures are characterized by rapid branching growth on hormone-free medium (**Fig. 1A,B**). With brassicas, these roots are thicker than normal roots, have more root hairs, and the extra branching is easily noted (**Fig. 1A**). Hairy root morphology can vary between species. Inclusion of seedling roots or noninoculated controls should enable the clear differences to be seen. Molecular techniques such as PCR (*see* **Subheading 3.4.**) can also be used to confirm the root cultures are of *A. rhizogenes* origin.

3.2.5. Improvement of Hairy Root Production

As noted with *A. tumefaciens*-mediated transformation, there is a wide range of transformation frequencies obtained when using *A. rhizogenes*, which varies between species and between cultivars. The methods available to increase production of hairy roots are similar to those used for increasing rates of *A. tumefaciens*-mediated transformation as reviewed in Christey and Braun *(17)* for vegetable brassicas. To improve transformation rates many factors can be studied. These cover two main areas concentrating on manipulation of cultural conditions, that is, both bacterium and explant, and on genetic factors.

1. Bacterial factors include, for example, testing a range of *Agrobacterium* strains or using acetosyringone (*see* **Note 3**) to increase *vir* gene expression. Plant factors include selection of a suitable genotype. In vegetable brassicas, there are clear effects of species and cultivar on *A. rhizogenes*-mediated transformation rates. Three quantitative trait loci (QTL) for transgenic root production have been identified in broccoli *(18)*.
2. Susceptibility of the chosen genotype to *A. rhizogenes* is an important prerequisite for the production of hairy roots. In addition, the actual explant used must be susceptible to *A. rhizogenes* to enable efficient transformation. The use of the *gfp* reporter gene aids optimizing transformation conditions as gene expression can be monitored rapidly and nondestructively as shown with broccoli *(19)*. This approach enables the rapid evaluation of a range of variables.
3. In *A. tumefaciens*-mediated transformation of brassicas, several factors are important in the successful production of transgenic plants including explant source and age, cocultivation time, delayed introduction of selection, explant preculture, and so on *(17)*. Manipulation of cocultivation conditions can also increase transforma-

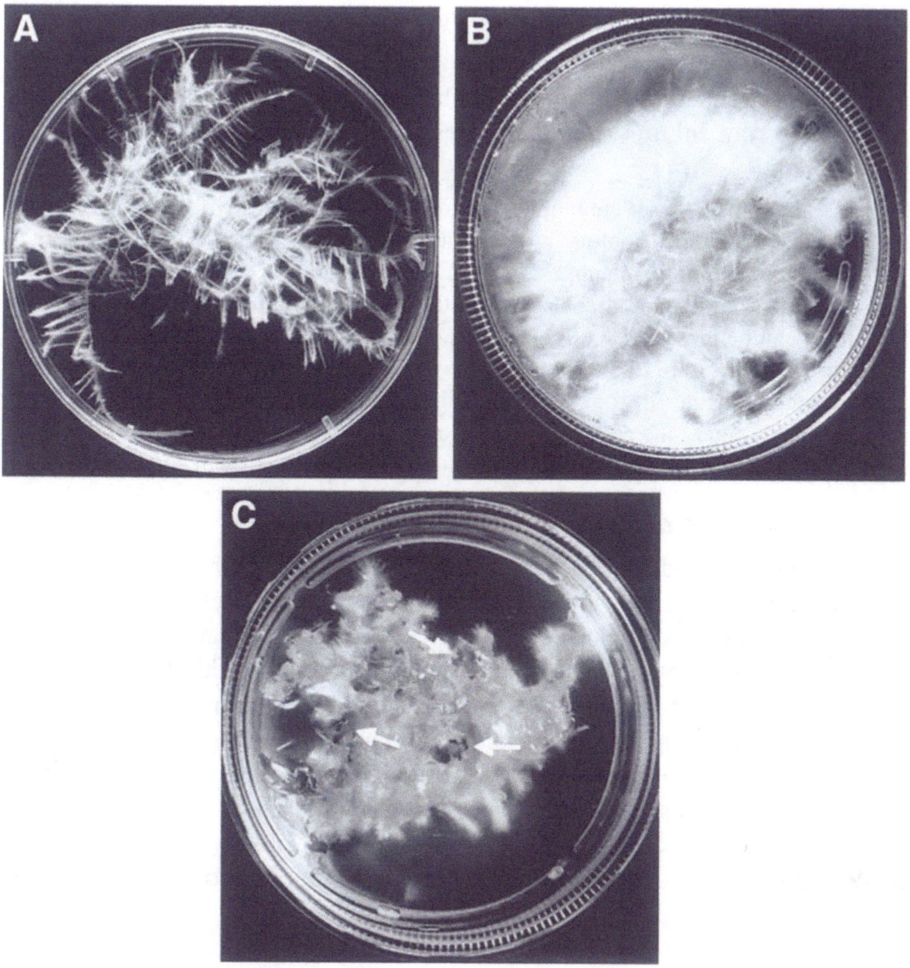

Fig. 1. **(A)** A Chinese cabbage hairy root culture showing the characteristic hairy root phenotype. **(B)** A hairy root culture of forage rape 3 wk after transfer to hormone-free medium. **(C)** Callus induction and shoot regeneration from forage kale hairy roots 3 wk after transfer to regeneration medium.

tion rates. The rate of hairy root production in broccoli was improved by the use of acetosyringone in the bacterial culture medium, addition of a *B. campestris* feeder layer (*see* **Note 4**), and use of acetosyringone and mannopine in the cocultivation medium ***(20)***. Increased rates of hairy root production have also been obtained by inclusion of 2,4-dichlorophenoxyacetic acid (2,4-D) in the medium used to resuspend the *A. rhizogenes* prior to inoculation ***(15)***.

3.2.6. Establishment of Liquid Cultures

Although growth on solid medium is sufficient for most studies, for secondary product production and studies on biochemical pathways or the effect of additives to root cultures, liquid cultures or scaleup to larger bioreactors may be required.

1. Inoculation of flasks containing liquid culture medium involves transferring a small amount of root culture into the flask, which is then shaken. It is important to ensure an antibiotic to suppress *A. rhizogenes* growth is still present, as residual *A. rhizogenes* remains associated with the hairy root cultures for a long time. Cultures can be grown in the light or dark. The issues and problems associated with the scaleup culture of hairy root cultures are covered in detail by Doran ***(21)***.
2. To study changes in root growth attributable to the presence of different additives or to study the effect of cultural changes on root growth, fresh weight, dry weight, or image analysis ***(22)*** can be used to quantify growth changes.

3.3. Regeneration of Shoots From Hairy Root Cultures

A. rhizogenes-mediated transformation is widely used for the production of transgenic shoots after regeneration from hairy root cultures. This method of transformation is used for species where *A. rhizogenes*-mediated transformation is more efficient than *A. tumefaciens* transformation. In addition, the altered phenotypes obtained from *A. rhizogenes*-mediated transformation (**Fig. 2**) such as increased flowering, shortened stature, and increased secondary products are of interest for horticultural purposes. Although the presence of the *rol* genes usually results in an altered phenotype, several studies have shown segregation of Ri and tumor-inducing (Ti)-T-DNA and thus the recovery of phenotypically normal transgenic plants ***(15,23)***. In addition, regeneration of plants with normal phenotype does occur from hairy roots and is probably caused by *rol* gene silencing ***(23)***.

1. Regeneration of shoots from hairy root cultures can occur spontaneously on hormone-free medium but often requires the transfer of roots to callus induction and shoot regeneration medium ***(24)***. The actual hormone combination required is dependent on the plant of interest and literature searches will indicate the combination that is most suitable.
2. Healthy root sections of 1–2 cm are excised from an actively growing root culture and placed on the appropriate callus induction medium. In contrast to growth on hormone-free medium, on callus induction medium roots will stop growing and start to thicken and produce callus (**Fig. 1C**).
3. This callus can be subbed every 2–3 wk onto the same medium or transferred to shoot regeneration medium. As shoot regeneration may be inhibited by the presence of kanamycin in some plants, it may be advisable to eliminate this from the culture medium. However, an antibiotic such as Timentin should be included to suppress *A. rhizogenes* growth.

Fig. 2. Transgenic **(left)** and control **(right)** forage kale plants 8 wk after transplanting to the field, showing the effect of the Ri phenotype. (From **ref. *1*** with permission. Copyright © 2001 by the Society for In Vitro Biology [formerly the Tissue Culture Association].)

4. Once shoot buds are initiated they can be transferred to hormone-free medium for further development.

3.4. Identification of Hairy Roots by PCR

PCR analysis provides confirmation that the selected root cultures are actually of *A. rhizogenes* origin. The production of the hairy root phenotype involves the integration of *rol* genes from the Ri plasmid. A PCR analysis for *rolB* or *rolC* is routinely used to confirm the presence of these genes. In addition, PCR analysis should be conducted for the gene(s) of interest if binary vectors are being used as DNA is inserted independently from the T-DNA regions of the Ri and Ti plasmids. PCR for *A. rhizogenes virG* gene can be used to confirm lack of bacterial infection.

3.4.1. PCR Analysis

DNA is isolated from in vitro hairy root cultures or shoots using a method modified from **ref. *25***.

1. Macerate plant material using disposable pestles for 15 s at room temperature (*see* **Notes 5** and **6**).
2. Add 400 µL of extraction buffer and grind for a further 15 s.
3. Centrifuge for 2 min at 16,000*g*.
4. Add 300 µL of the supernatant to 300 µL of room temperature isopropanol.

5. Invert samples gently, incubate for 20 min at –20°C, and centrifuge for 5 min at 16,000*g*.
6. Air-dry DNA pellets for 30 min and resuspend in 100 µL of sterile water.
7. Use 1 µL of this template in a 25 µL PCR reaction containing 2.5 µL of 2 m*M* dNTPs, 1 µM of each primer (*see* **Note 7**), 1X PCR buffer (Roche), 1 U of *Taq* polymerase (Roche), and 16.4 µL of water. Cover samples with paraffin oil if required.
8. PCR conditions for *rolB* and *rolC* are: 94°C for 30 s, 68°C for 30 s, and 72°C for 30 s for 40 cycles in an Eppendorf Mastercycler personal thermal cycler.
9. Run 5 µL of the PCR reaction on a 1% agarose (Invitrogen) gel and visualize by ethidium bromide staining.

This crude method works well for many tissue types; however, some plant material may not amplify well. A titration of template (e.g., 0.5, 1, or 2 µL per 25-µL PCR reaction) or further purification (*see* **Note 8**) may be necessary.

3.5. Transfer of Plants to Greenhouse Conditions

Once in vitro shoots have adequate root and shoot growth they can be successfully transferred to greenhouse conditions to enable more accurate observation of phenotype and also for determination of fertility and seed collection. As plants in tissue culture are used to high humidity it is important to ensure a gradual introduction to the lower humidity and higher light intensity of a greenhouse. This can be obtained by use of a mist bed. Alternatively, a plastic bag can be placed over the pot after transfer of the plant to soil as outlined below.

1. Water the soil mix in the pot (7 cm^3) well.
2. Mist the plant with water immediately prior to transfer to soil.
3. Wash excess agar off the roots carefully before transfer to soil.
4. Place plant in the pot and immediately cover the plant and pot with a plastic bag and secure with a rubber band (**Fig. 3A**).
5. Place plants under a greenhouse bench. If the plants are not in direct sun they will not dry out for a least one wk.
6. After one wk, cut a small hole in the top corner of each bag.
7. Make the hole gradually larger over the next week until the bag top is completely open. During this hardening off time check the mix daily for dryness.
8. Remove bags totally 2 wk after exflasking (**Fig. 3B**). Plants may now be moved onto a greenhouse bench but may be prone to wilting in hot weather.
9. Repot plants into larger containers as required.

4. Notes

1. Antibiotics are added to media after autoclaving from filter sterilized stock solutions.
2. *A. rhizogenes* must be grown at 28°C not 37°C.
3. Acetosyringone (200 µ*M* for brassicas, but level varies widely) is added to the cocultivation medium after autoclaving from a 20 m*M* filter sterilized stock dis-

Fig. 3. Transfer of hairy root derived plants to soil. **(A)** Plants immediately after transfer to soil. **(B)** Plants 2 wk after transfer to soil.

solved in hot water. In experiments where acetosyringone is added to the bacterial medium, LB medium containing 5 m*M* 2-(*N*-morpholino)ethanesulfonic acid (MES), pH 5.6, is used.

4. Feeder layers are established from rapidly growing cell suspension cultures. Approximately 1.5 mL of cells is plated onto the cocultivation medium either 1 d or immediately prior to use. A piece of sterile filter paper is placed over the feeder layer immediately prior to use and the cocultivated explants are placed on top of the filter paper to prevent accidental transformation of the feeder layer and not the explant source of interest. Various plant sources are suitable for use as a feeder layer including the species being transformed or other species with high transformation ability such as tobacco.
5. It is important to ensure agar is removed from hairy root cultures as it can inhibit the PCR reaction.
6. Disposable plastic pestles can be reused a number of times as long as abrasive agents such as sand are not used. Wash pestles in detergent, rinse thoroughly and soak overnight in 1 M HCl to ensure degradation of nucleic acids. Rinse three times in double distilled water and autoclave before reuse.
7. Primers used for *rolB* are: 5'AAAGTATGCTACCATTCCCCA3' and 5'CCCA TAAGCCACGACATCATA3' which produce a 393-bp fragment with strain A4T. The primers for *rolC* are 5'CGACCTGTGTTCTCTCTTTTTC AAGC3' and 5'GCACTCGCCATGCCTCACCCAACTCACC3', which produce a 514-bp internal fragment with strain A4T *(26)*. Controls should include DNA from the binary vector, the *Agrobacterium* strain used and DNA from a nontransgenic plant.
8. A simple reprecipitation step can often overcome template problems. Add an additional 100 µL of water to the existing sample and vortex-mix for 30 s. Centri-

fuge down debris at top speed in a microcentrifuge and transfer the supernatant to a fresh tube. Add 2X starting volume of ice-cold 100% ethanol and 10% 0.5 *M* NaCl. Mix thoroughly and centrifuge at top speed for approx 10 min. Carefully pour off the ethanol and briefly rinse the pellet with 70% ethanol. Air-dry and resuspend the pellet as before.

Acknowledgments

The authors thank Jill Reader for assistance with development of the exflasking method, Hannah Wensink for supplying Fig. 1A and Robert Lamberts for photography.

References

1. Christey, M. C. (2001) Use of Ri-mediated transformation for production of transgenic plants. *In Vitro Cell Dev. Biol. Plant* **37,** 687–700.
2. Quandt, H.-J., Phler, A., and Broer, I. (1993) Transgenic root nodules of *Vicia hirsuta*: a fast and efficient system for the study of gene expression in indeterminate-type nodules. *Mol. Plant Microbe Interact.* **6,** 699–706.
3. Uozumi, N. and Kobayashi, T. (1997) Artificial seed production through hairy root regeneration, in Hairy Roots: Culture and Applications (Doran, P. M., ed.), Harwood Academic, Amsterdam, The Netherlands, pp. 113–122.
4. Hamill, J. D. and Lidgett A. J. (1997) Hairy root cultures opportunities and key protocols for studies in metabolic engineering, in *Hairy Roots: Culture and Applications* (Doran, P. M., ed.), Harwood Academic, Amsterdam, The Netherlands, pp. 1–30.
5. Braun, R. H., Eady, C., Christey, M. C., Shaw, M., Pither-Joyce, M., and McCallum, J. (2002) The use of hairy root cultures for the study of sulfur metabolism in plants, in Microbes and Molecules 2002, A combined meeting of the NZSBMB, NZSPP and NZMS, November 26–29, 2002, University of Canterbury, Christchurch, New Zealand, p. 208.
6. Downs, C. G., Christey, M. C., Davies, K. M., King, G. A., Sinclair B. K., and Stevenson, D. G. (1994) Hairy roots of *Brassica napus*: II. Glutamine synthetase overexpression alters ammonia assimilation and the response to phosphinothricin. *Plant Cell Rep.* **14,** 41–46.
7. Mugnier, J. (1997) Mycorrhizal interactions and the effects of fungicides, nematicides and herbicides on hairy root cultures, in *Hairy Roots Culture and Applications* (Doran, P. M., ed.), Harwood Academic, Amsterdam, The Netherlands, pp. 123–132.
8. Kifle, S., Shao, M., Jung, C., and Cai, D. (1999) An improved transformation protocol for studying gene expression in hairy roots of sugar beet (*Beta vulgaris* L.). *Plant Cell Rep.* **18,** 514–519.
9. Bais, H. P., Loyola-Vargas, V. M., Flores, H. E., and Vivanco, J. M. (2001) Root-specific metabolism: the biology and biochemistry of underground organs. *In Vitro Cell. Dev. Biol. Plant* **37,** 730–741.

10. Wongsamuth, R. and Doran, P. M. (1997) Production of monoclonal antibodies by tobacco hairy roots. *Biotech. Bioeng.* **54,** 401–415.
11. Hellens, R. and Mullineaux, P. (2000) A guide to *Agrobacterium* binary Ti vectors. *Trends Plant Sci.* **5,** 446–451.
12. Gleave, A. P. (1992) A versatile binary vector system with a T-DNA organisational structure conducive to efficient integration of cloned DNA into the plant genome. *Plant Mol. Biol.* **20,** 1203–1207.
13. Murashige, T. and Skoog, F. (1962) A revised medium for rapid growth and bioassay with tobacco tissue cultures. *Physiol. Plant.* **15,** 473–497.
14. Christey, M. C., Sinclair, B. K., Braun, R. H., and Wyke, L. (1997) Regeneration of transgenic vegetable Brassicas (*Brassica oleracea* and *B. campestris*) via Ri-mediated transformation. *Plant Cell Rep.* **16,** 587–593.
15. Puddephat, I. J., Robinson, H. T., Fenning, T. M., Barbara, D. J., Morton, A., and Pink, D. A. C. (2001) Recovery of phenotypically normal transgenic plants of *Brassica oleracea* upon *Agrobacterium rhizogenes*-mediated co-transformation and selection of transformed hairy roots by GUS assay. *Mol. Breed.* **7,** 229–242.
16. Dessaux, Y., Petit, A., and Tempe, J. (1993) Chemistry and biochemistry of opines, chemical mediators of parasitism. *Phytochemistry* **34,** 31–38.
17. Christey, M. C. and Braun, R. H. (2004) Production of transgenic vegetable Brassicas, in *Biotechnology in Agriculture and Forestry*, Brassica *Biotechnology* (Pua, E. C. and Douglas C. J., eds), Springer-Verlag, Berlin, Germany.
18. Cogan, N. O. I., Lynn, J. R., King, G. J., Kearsey, M. J., Newbury, H. J., and Puddephat, I. J. (2002) Identification of genetic factors controlling the efficiency of *Agrobacterium rhizogenes*-mediated transformation in *Brassica oleracea* by QTL analysis. *Theor. Appl. Genet.* **105,** 568–576.
19. Cogan, N., Harvey, E., Robinson, H., et al. (2001) The effects of anther culture and plant genetic background on *Agrobacterium rhizogenes*-mediated transformation of commercial cultivars and derived doubled-haploid *Brassica oleracea. Plant Cell Rep.* **20,** 755–762.
20. Henzi, M. X., Christey, M. C., and McNeil, D. L. (2000) Factors that influence *Agrobacterium rhizogenes*-mediated transformation of broccoli (*Brassica oleracea* L. var. *italica*). *Plant Cell Rep.* **19,** 994–999.
21. Doran, P. M., ed. (1997) *Hairy Roots: Culture and Applications*, Harwood Academic, Amsterdam, The Netherlands.
22. Coles, G. D., Abernethy, D. J., Christey, M. C., Conner, A. J., and Sinclair, B. K. (1991) Monitoring hairy root growth by image analysis. *Plant Mol. Biol. Rep.* **9,** 11–18.
23. Christey, M. C., Braun, R. H., and Reader, J. K. (1999) Field performance of transgenic vegetable brassicas (*Brassica oleracea* and *B. rapa*) transformed with *Agrobacterium rhizogenes. SABRAO J. Breed. Genet.* **31,** 93–108.
24. Christey, M. C. (1997) Transgenic crop plants using *Agrobacterium rhizogenes*-mediated transformation, in *Hairy Roots: Culture and Applications* (Doran, P. M., ed.), Harwood Academic, Amsterdam, The Netherlands, pp. 99–111.

25. Edwards, K., Johnstone, C., and Thompson, C. (1991) A simple and rapid method for the preparation of plant genomic DNA for PCR analysis. *Nucleic Acids Res.* **19,** 1349.
26. Scorza, R., Zimmerman, T. W., Cordts, J. M., Footen, K. J., and Ravelonandro, M. (1994) Horticultural characteristics of transgenic tobacco expressing the *rolC* gene from *Agrobacterium rhizogenes*. *J. Am. Soc. Horticult. Sci.* **119,** 1091–1098.

4

Stable Transformation of Plant Cells by Particle Bombardment/Biolistics

Julie R. Kikkert, José R. Vidal, and Bruce I. Reisch

Summary

Particle bombardment, or biolistics, is a commonly used method for genetic transformation of plants and other organisms. Millions of DNA-coated metal particles are shot at target cells or tissues using a biolistic device or gene gun. The DNA elutes off the particles that lodge inside the cells, and a portion may be stably incorporated in the host chromosomes. A protocol for the generation of transgenic grapevines via biolistic transformation of embryogenic cell suspension cultures is detailed in this chapter. In a typical experiment, transient gene expression averaged nearly 8000 "hits" per bombarded plate. Five months after bombardment, there were nearly five putative transgenic embryos per bombarded plate. About half of the embryos were regenerated into confirmed transgenic plants. The basic bombardment procedures described are applicable to a wide range of plant genotypes, especially those for which embryogenic cell cultures are available. All users of particle bombardment technology will find numerous useful tips to maximize the success of transformation.

Key Words: Ballistics; biolistic; biotechnology; embryogenic cells; gene gun; genetic engineering; grapevine; microcarrier; microparticle bombardment; microprojectile bombardment; particle acceleration; particle bombardment; particle gun; plant transformation; *Vitis*.

1. Introduction

Particle bombardment employs high-velocity microprojectiles to deliver substances into cells and tissues. For genetic transformation, DNA is coated onto the surface of micron-sized tungsten or gold particles by precipitation with calcium chloride and spermidine. Once inside the cells, the DNA elutes off the particles. If the foreign DNA reaches the nucleus, then transient expres-

From: *Methods in Molecular Biology, vol. 286: Transgenic Plants: Methods and Protocols*
Edited by: L. Peña © Humana Press Inc., Totowa, NJ

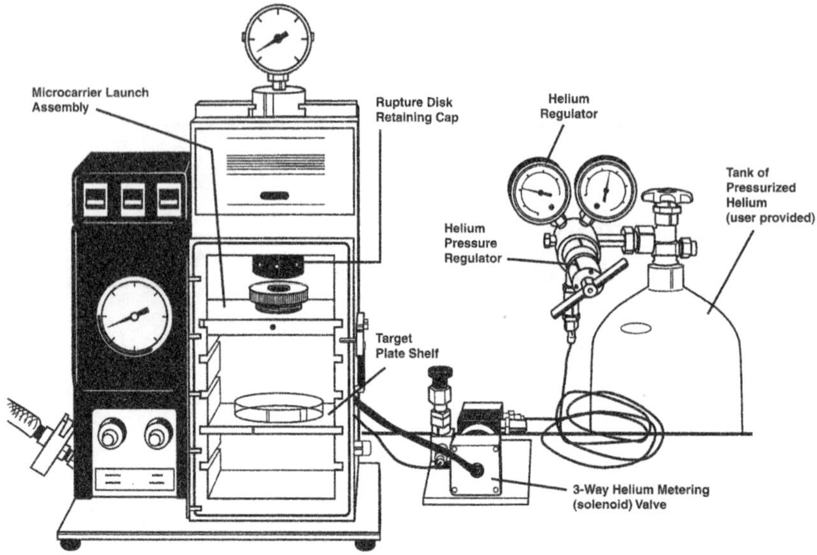

Fig. 1. Components of the Biolistic® PDS-1000/He particle delivery system. (Drawing courtesy of Bio-Rad Laboratories, Hercules, CA.)

sion will likely result and the transgene may be stably incorporated into host chromosomes. Sanford and colleagues at Cornell University developed the original bombardment concept *(1,2)* and coined the term "biolistics" (short for "biological ballistics") for both the process and device. "Biolistics" is a registered trademark of E. I. du Pont de Nemours and Co. and has been used to market the devices now sold by Bio-Rad Laboratories, Hercules, CA. However, as there are several homemade "gene guns" or "particle guns," the process often is called by other names such as microprojectile bombardment, particle bombardment, particle acceleration, or ballistics.

The most widely used device for plant transformation is the Biolistic® PDS-1000/He Particle Delivery System *(3)* marketed by Bio-Rad Laboratories (**Fig. 1**). The system employs high-pressure helium released by a rupture disk to propel a macrocarrier sheet loaded with millions of DNA-coated metal particles (microcarriers) toward target cells (**Fig. 2**). A stopping screen halts the macrocarrier, and the microcarriers continue toward the target and penetrate the cells.

Because of its physical nature and simple methodology, the biolistic process can be used to deliver substances into a wide range of intact cells and tissues from a diversity of organisms. In plant research, the major applications have been transient gene expression studies, production of genetically transformed

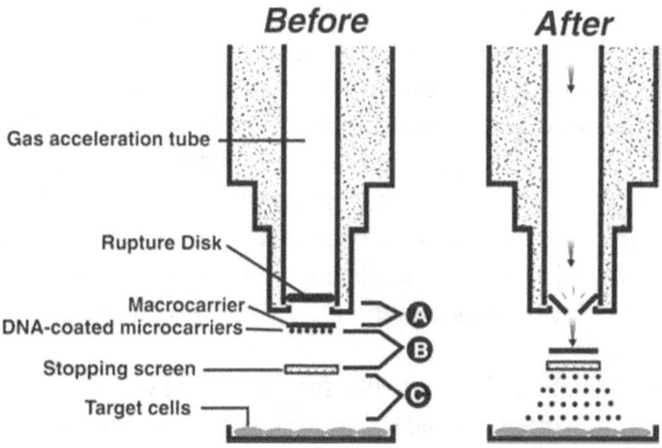

Fig. 2. The Biolistic® bombardment process. The gas acceleration tube is filled with helium gas until the maximum pressure of the rupture disk is reached. When the disk ruptures, the ensuing helium shock wave launches a plastic macrocarrier onto which the DNA-coated microcarriers have been dried. The macrocarrier flies downward until it impacts a stopping screen. On impact, the macrocarrier is retained by the stopping screen, while the microcarriers are launched and continue downward at high velocity until they impact and penetrate the target cells.

The velocity of the macrocarriers is dependent on the helium pressure in the gas acceleration tube, the distance from the rupture disk to the macrocarrier (gap distance) (A), the macrocarrier travel distance to the stopping screen (B), the distance between the stopping screen and target cells (C), and the amount of vacuum in the bombardment chamber. (Drawing courtesy of Bio-Rad Laboratories, Hercules, CA.)

plants, and inoculation of plants with viral pathogens *(2,4,5)*. Many "firsts" were achieved through the application of biolistic technology including chloroplast and mitochondria transformation, as well as nuclear transformation of important monocot species such as wheat, corn, and rice *(2)*. Although other technologies have since been proven in these arenas, Sanford in the year 2000 *(2)*, stated the following: "I believe it is accurate to say that most of the presently grown transgenic crop acreage in the entire world was created through the use of the biolistic process—having been originally transformed with the gene gun."

As with any plant transformation method, several parameters need to be optimized for the process to be maximally effective. With biolistics, the parameters can be grouped as physical, biological, and environmental *(4–7)*. Physical parameters include the composition and size of the microcarriers, the attachment of DNA to the microcarriers, and several instrument parameters.

During development of the PDS-1000/He, instrument settings were varied over a wide range and tested with numerous organisms *(7,8)*. A vacuum of 28.0 in Hg (94.8 kPa), a helium pressure of about 1100 psi (7584.2 kPa), a gap distance of 6.5–10.0 mm, and a macrocarrier travel distance of 6.0–10.0 mm are near optimal for most plant transformation applications. Gold particles in the range of 0.7–1.0 μm mean diameter generally result in the highest rates of stable transformation, but the less expensive, more heterogeneous tungsten particles are also widely used. Consistent coating of DNA to the particles and spread of the particles onto the macrocarrier are critical, and proficiency develops with practice.

The first biological parameter to consider is a gene construct in the form of a circular or linear plasmid or a linear expression cassette (promoter–gene–terminator). It is important to match the promoter and/or other regulatory sequences with the plant tissue, so that the gene will be expressed at desired levels. Other biological parameters include tissue type, cell size, cell culture age, mitotic stage, general cellular health, target tolerance of vacuum, cell density, and cell turgor pressure. The physiological status of the target influences receptivity to foreign DNA delivery and susceptibility to injury that may adversely affect the outcome of the transformation process. For recovery of transgenic plants, it is very important to target cells that are competent for both transformation and regeneration. Furthermore, the ability of bombarded cells to regenerate plants depends on the type and concentration of the selection agent. In some cases, it is best to start with a low concentration of the selective agent and increase it after 2 or 4 wk of cell culture.

Environmental factors such as temperature; humidity; and light intensity, quality, and duration have a direct effect on tissue physiology and thus transformation success *(6)*. In addition, some explants may require a "healing" period after bombardment under special regimens of light, temperature, and humidity *(6)*. Humidity also is important in microcarrier preparation and bombardment. High humidity can cause the microcarriers to clump and/or to bind irreversibly to the macrocarrier, thus reducing transformation rates. High humidity may also affect stocks of alcohol used during the DNA/microcarrier coating steps. Some researchers use cold temperatures while coating macrocarriers with DNA, whereas our laboratory uses room temperature. We are not aware of a published study on the effect of temperature on microcarrier coating or bombardment.

There has been much discussion over the advantages and disadvantages of the biolistic process as compared to *Agrobacterium* (*see* Chapter 2) for the production of transgenic plants. The physical nature of the biolistic process eliminates concerns about using another biological organism in the transformation process. In grapevines, there is often a hypersensitive response to *Agrobacterium* that causes plant cell death *(9)*. Biolistics obviates both the need to kill *Agrobacterium* after transformation and the occurrence of false positives arising from growth of *Agrobacterium* in the host tissues. Operation of the biolistic device is easy and there are only a few instrument parameters to adjust. Because the Biolistic® PDS-1000/He unit is commercially available, the user benefits from convenience, ease of use, technical support, and standardization with other labs. Furthermore, plasmid construction is often simplified and cotransformation with multiple transgenes ***(10)*** is routine, because plasmid DNA is simply mixed together before coating onto the microcarriers. The use of linear expression cassettes (also called clean gene technology) eliminates the chance that extraneous plasmid backbone DNA will be inserted into the target as can happen with whole plasmids or *Agrobacterium* ***(11)***. Biolistics is the method of choice for the study of transient gene expression and for plastid transformation ***(5)***. Furthermore, biolistics is the only successful method of transformation currently available for certain genotypes ***(5)***.

Some disadvantages of biolistics are that the transformation efficiency may be lower than with *Agrobacterium* and the device and consumables are costly. Many researchers have strayed from biolistics because of the tendency for complex integration patterns and multiple copy insertions that could cause gene silencing. Some laboratories have overcome this problem by reducing the quantity of DNA loaded onto the microcarriers and/or by use of linear cassettes ***(11)***. Random integration is also a concern and is being addressed by several groups ***(5)***, the most promising being the use of the Cre–Lox system for targeted integration ***(12)***.

As many parameters need to be optimized for any transformation method, often the experience of the investigator and nearby colleagues determines which method is chosen. The user must weigh the advantages and disadvantages of the various methods available. Patents and licensing availability should also be considered. Particle bombardment technology is covered by several patents held by E. I. du Pont de Nemours and Co. and PowderJect Vaccines, Inc. Use of particle bombardment for commercial purposes may require a commercial license from the appropriate patent holder. There are also patents held by different companies for the use of particle bombardment for certain plant species such as *Zea mays*. Thus, patents rights must be investigated thoroughly. In comparison, patent rights for *Agrobacterium*-mediated transformation are less clear and are tied up in the legal system. Thus, obtaining a license for *Agrobacterium* is more difficult for those outside of the patent-holding companies.

Our laboratory has successfully employed biolistics to obtain transgenic grapevine plants. This chapter details a protocol for transformation of *Vitis vinifera* L. 'Chardonnay' embryogenic suspension cultures in which numerous transformation parameters have been optimized ***(13)***. Bombardment with gold particles coated with plasmid pBI426 (double CaMV 35S promoter,

Alfalfa mosaic virus (AMV) leader sequence, *uidA* gene, nos terminator) resulted in an average of 7883 ± 1928 β-glucuronidase (GUS) positive blue spots per Petri plate at 2 d and 46 ± 32 at 95 d. A total of 447 embryos were harvested from 84 bombarded plates on selection medium within 5 mo after cobombardment with two separate plasmids. This represents more than 5 putative transgenic embryos per bombarded plate. From those, 242 plants were regenerated, which corresponds to a 54% rate of conversion of embryos to regenerated plants. The cotransformation frequency of genes on different plasmids was in the range of 50% in the group of regenerated plants ***(13)***. The basic cell handling and bombardment procedures have been used for numerous other genotypes ***(7)***; however, the media and environmental conditions for cell growth, transformant selection, and plant regeneration must be optimized for each.

2. Materials

All reagents should be tissue culture or molecular biology grade.

2.1. Culture and Preparation of Plant Cells

1. Plant material: embryogenic *Vitis vinifera* L. 'Chardonnay' cell suspension cultures ***(13)*** (*see* **Note 1**).
2. Medium for cell suspension cultures: (GM + NOA medium ***[14]***): Murashige and Skoog (MS) ***(15)*** basal medium (macro- and microelements, vitamins, and inositol) with 18 g/L of maltose hydrate, 4.6 g/L of glycerol, and 5 μ*M* β-napthoxyacetic acid (NOA). Adjust pH to 5.8 with KOH before autoclaving. To prepare 100 mL of a 1 m*M* NOA stock solution, dissolve 20.2 mg of NOA in 2 mL of 1 *M* KOH. Stir briefly and add 90 mL of Type I water. Continue stirring for 1 h. Bring to final volume and filter sterilize. Store at 4°C; stock is good for 1 yr. Use 5 mL of stock per liter of media.
3. 250-, 500-, and 1000-mL Erlenmeyer flasks, capped with aluminum foil and autoclaved.
4. Double-screen mesh (1.1 mm^2 pore size) in a polypropylene funnel to filter cell suspensions, autoclaved.
5. Disposable 10- and 25-mL plastic pipets, cotton-plugged, sterile.
6. Compound microscope, glass slides, and cover slips.
7. Magnetic stir plate and autoclaved stir bar.
8. Graduated 12- or 15-mL conical centrifuge tube.
9. 100-mL media bottle with screw cap lid, autoclaved.
10. 1-mL sterile polyethylene transfer pipet.
11. Büchner funnel (8 cm in diameter, autoclaved), size arm flask (1 L, autoclaved), and vacuum source.
12. 7-cm diameter Whatman no. 2 filter papers, autoclaved.
13. Bombardment medium (1/2 MS-HF [hormone-free] medium with osmotica [*see* **Note 2**]): MS medium with half-strength macro- and microelements, full-strength vitamins and inositol, 30 g/L of sucrose, 0.125 *M* mannitol, 0.125 *M* sorbitol, and

2.5 g/L of Phytagel (Sigma, St. Louis, MD). Adjust pH to 5.8 with KOH before autoclaving. Dispense in 10-mL aliquots on top of a sterile, circular filter paper (S&S Sharkskin, 9 cm in diameter, VWR International, South Plainfield, NJ, cat. no. 28314-028) that is contained in a 100 × 15 mm Petri plate. The filter paper should have a small tab of tape attached (homemade) so that once the medium is solidified; forceps can be used to pick up the whole unit by the tab. The sterile medium can be stored in sterile plastic bags at room temperature for 1 mo.

14. Sterile forceps.

2.2. Preparation of DNA-Coated Microcarriers

2.2.1. Sterilization of Macrocarriers and Holders

1. Macrocarriers for biolistic device (Bio-Rad).
2. Macrocarrier holders (Bio-Rad).
3. 70 and 95% ethanol.
4. Glass beaker and glass Petri plate (autoclaved).
5. Sterile Kimwipes or paper towels.
6. Sterile forceps with fine point tips (curved tips work well).
7. Desiccant in glass Petri dishes. A sterile filter paper or inverted plastic Petri plate with holes (homemade) should be placed over the desiccant to provide a stable, dust-free platform for loading DNA-coated particles onto the macrocarriers. We use Drierite brand desiccant, which changes from blue to pink as it absorbs water. Bake at 180°C for approx 4 h to restore blue color and desiccating ability.

2.2.2 Sterilization of Gold Particles

1. Microcarriers: gold particles, 0.75 µm in diameter (Analytical Scientific Instruments, El Sobrado, CA) (*see* **Note 3**).
2. Small glass vial or tube (1–3 mL).
3. Oven that will reach 180°C.
4. 500-µL micropipettor and tips.
5. Isopropanol, HPLC grade.
6. 1.5-mL microcentrifuge tubes, autoclaved, Treff Lab, Degersheim, Switzerland (cat. no. 96.7246.9.02) (*see* **Note 4**).
7. Sterile type I water.
8. Glycerol (50% v/v): Mix glycerol 1:1 with type I water and autoclave.

2.2.3 Coating Gold Particles With DNA

1. Micropipettors and tips (5- to 500-µL range).
2. 1.5-mL microcentrifuge tubes, autoclaved, Treff brand (*see* **item 6** in **Subheading 2.2.2.**).
3. Plasmid DNA at 1 µg/µL in sterile TE buffer (1 m*M* Tris-HCl, pH 7.8, 0.1 m*M* disodium ethylenediaminetetraacetic acid [EDTA]) (*see* **Note 5**).
4. 2.5 *M* $CaCl_2$, filter-sterilized: To make 50 mL, dissolve 18.38 g of calcium chloride dihydrate in type I water. Filter sterilize and store at 4°C in small aliquots.

5. 0.1 *M* spermidine free base, filter-sterilized. Solid spermidine is very hygroscopic. Therefore, take a 1-g unopened bottle of spermidine free base (Sigma cat. no. S-0266), add 1 mL of type I water, adjust the volume to 68.9 mL, vortex to mix thoroughly, filter sterilize and store at –20°C in 1.2-mL aliquots in 1.5-mL microcentrifuge tubes with screw-cap lids. The stock is good for 1 mo (*see* **Note 6**). Discard individual tubes after first use.
6. Continuous vortex mixer such as the Vortex Genie-2 Mixer with 15.2-cm platform head (cat. no. 58815-178 and 58815-214; VWR, International, South Plainfield, NJ).
7. HPLC grade isopropanol.
8. Ultrasonic water bath cleaner (Model B1200R-1; Branson Ultrasonics Corporation, Danbury, CT, or similar unit).

2.3. Bombardment

1. Biolistic® PDS-1000/He Instrument (Bio-Rad).
2. Helium gas cylinder; high pressure (2400–2600 psi [16,547.4–17,926.4 kPa]); grade 4.5 or 5.0 (99.995% or higher purity).
3. Vacuum pump; oil-filled rotary vane, with a pumping speed of 90–150 L/min (3–5 ft^3/min).
4. Rupture disks (1100 psi [7,584.2 kPa], Bio-Rad), sterilize with isopropanol (*see* **Note 7**).
5. Stopping screens (Bio-Rad), sterilize by autoclaving.
6. Safety glasses.
7. Hair net and latex gloves.
8. Opaque plastic box sterilized with 70% ethanol to store bombarded plates.

2.4. Postbombardment Reduction of Medium Osmoticum

1. Medium (1/2 MS-HF) without osmotica: MS medium with half-strength macro- and micro- elements, full-strength vitamins and inositol, 30 g/L of sucrose, and 2.5 g/L of Phytagel. Adjust pH to 5.8 with KOH and autoclave. Dispense in 10- and 20-mL aliquots into 100 × 15 mm Petri plates (*see* **Note 8**).
2. Sterile forceps.

2.5. Analysis of Transient and Long-Term GUS Expression

1. GUS histochemical staining solution: To prepare 200 mL, combine the following components: 150 mL of type I water, 0.744 g of EDTA, disodium salt, dihydrate, 1.76 g of sodium phosphate monobasic, 0.042 g of potassium ferrocyanide, and 0.2 mL of Triton X-100. Adjust the volume to 198 mL, and the pH to 7.0. Add 100 mg of 5-bromo 3-chloro 3-indolyl β-D-glucuronic acid (X-Gluc) that has been dissolved in 2 mL of dimethyl sulfoxide (DMSO). Filter sterilize and store at –20°C; stock is good indefinitely.
2. Sterile forceps.
3. Petri plates, 100 × 15 mm diameter, sterile.
4. Incubator, 37°C.

5. Stereomicroscope.
6. Plastic sheet with an imprinted grid (homemade).
7. Cell counter.

2.6. Embryo Selection, Germination, and Regeneration

1. Kanamycin monosulfate (Km) stock (25 mg/mL, pH 5.8, filter-sterilized). Prepare in type I water. Store at –20°C in small aliquots. Frozen stock is good indefinitely. Warm to add to autoclaved media (*see* **Subheading 2.6., item 2**) that has been cooled to 50–55°C.
2. Selective medium: 1/2 MS-HF medium with 30 g/L of sucrose, 3 g/L of activated charcoal, 7 g/L of Bacto-agar (Difco, Detroit, MI), and 10 or 15 mg/L of Km (added after autoclaving). Adjust pH to 5.8 with KOH and autoclave. Dispense in 20-mL aliquots into 100 × 15 mm Petri plates.
3. Embryo germination medium: 1/2 MS-HF (Km-free) with 30 g/L of sucrose, 3 g/L of activated charcoal, and 2.5 g/L of Phytagel. Adjust pH to 5.8 with KOH and autoclave. Dispense 20 mL per 100 × 15 mm Petri plate or 30 mL per baby food jar.
4. Plant growth medium: woody plant medium (WPM) ***(16)***, pH 5.8, with 20 g/L of sucrose and 2.5 g/L of Phytagel. Dispense 50 mL per Magenta GA7 vessel (Magenta Corp., Chicago, IL).
5. Parafilm (American National Can, Menasha, WI).
6. Venting Tape (Scotch brand no. 394; 3M Corporation, Minneapolis, MN).

3. Methods

Preparation for bombardment (*see* **Note 9**) should begin 6 d in advance (**Table 1**). All steps should be carried out in a laminar flow hood to avoid microbial contamination.

3.1. Culture and Preparation of Plant Cells

1. Maintain embryogenic suspension cells in GM+NOA medium in 250- or 500-mL Erlenmeyer flasks at 120 rpm, in the dark at 23 ± 1°C. Each week, the medium should be refreshed by removing and replacing one half of the spent medium with fresh medium using a sterile plastic 10- or 25-mL pipet (*see* **Note 10**). Cells should be poured through a funnel with sterile screen mesh (*see* **Subheading 2.1., item 4.**) to remove large clumps as needed.
2. Use cells for bombardment 4 d after subculture. The cell suspension culture should be checked immediately before use for microbial contamination by placing a sample on a glass slide with cover slip and observing it under a compound microscope. Fungal strands or bacteria can be easily recognized (*see* **Note 11**).
3. Pour all cells needed for bombardment through a sterile screen mesh in a funnel positioned over the mouth of a 1-L sterile Erlenmeyer flask. Add a sterile stir bar and place the flask on a magnetic stir plate (in a laminar flow hood). Turn the stir plate on a low setting to mix the cells.

Table 1
Flow Chart of Steps for Particle Bombardment Transformation

Time	Activity (in sequential steps)
	Week prior to bombardment
(–) 6 d	Sterilize supplies (Whatman and Sharkskin filter papers, funnels, flasks, water, etc.).
(–) 5 d	Prepare media needed for transformation procedure. GM+NOA suspension culture medium. 1/2 MS-HF bombardment medium with osmotica. 1/2 MS-HF medium without osmotica. 1/2 MS-HF selective medium.
(–) 4 d	Subculture or refresh medium of embryogenic cell suspensions.
	Week of bombardment
(–) 1 d	Set gene gun parameters (distances as described in **Fig. 2**). Weigh gold particles (microcarriers) and place in an oven overnight. Sterilize macrocarriers, holders and stopping screens. Assemble macrocarriers into holders.
Key d	Bombardment day (suggested day, Tuesday). Examine embryogenic cell suspension for contamination using a microscope. Prepare cells on filter paper for bombardment. Sterilize microcarriers. Coat microcarriers with DNA. Bombard cells. Incubate cells in the dark at 23± 1°C.
(+) 1 d	Transfer cells to medium without osmotica. First transfer approx 16 h after bombardment. Second transfer approx 24 h after bombardment.
(+) 2 d	Transfer cells to selective medium. Analysis of reporter gene (i.e., GUS assay) for transient expression.
(+) 3 d	Examine GUS-positive blue spots per filter paper.
	Postbombardment weeks
(+) 30 d	Transfer cells to fresh selective medium. Reporter gene assay for transient expression.
(+) 60 d	Check plates for development of embryos. Transfer embryos to germination medium. Transfer remaining cells to fresh selective medium. Reporter gene assay for long-term expression.
(+) 90 d	Items and procedure as in (+) 60 d. Transfer germinated embryos to plant growth medium.

4. To standardize cell density for bombardment, place a 10-mL sample of the cell suspension in a graduated 12- or 15-mL conical centrifuge tube and allow cells to settle for 15 min (30 min if cell suspension is very fine). Record the settled cell volume and discard the sample in the centrifuge tube. Adjust the density of the cell suspension in the flask to be used for bombardment to 0.2 mL of settled cell volume per 10-mL sample by adding or removing GM+NOA medium.
5. For each plate to be bombarded, place a sterile Whatman no. 2 filter paper in a Büchner funnel positioned on a 1-L side-arm flask. Using a sterile transfer pipet, remove 1 mL of GM+NOA medium from the small media bottle and place on the Whatman no. 2 filter paper to moisten it. While continuing to stir the cell culture, use a sterile 10-mL pipet to collect 5-mL of cells from the culture flask and then spread as a single layer onto the filter. Apply a slight vacuum to draw off excess liquid and to help spread the cells (*see* **Note 12**).
6. Transfer the filter paper with attached cells to bombardment medium using sterile forceps.

3.2. Preparation of DNA Coated Microcarriers

3.2.1 Sterilization of Macrocarriers and Holders

1. Place macrocarrier holders in a glass beaker and macrocarriers in a glass Petri dish. Fill containers with 70% ethanol and let stand for 15 min (*see* **Note 13**).
2. Remove the macrocarrier holders from the 70% ethanol with sterile forceps and place on sterile Kimwipes or paper towels in a laminar flow hood to dry.
3. Using sterile forceps, remove the macrocarriers from the 70% ethanol and dip them briefly in 95% ethanol. Place on sterile Kimwipes or paper towels in a laminar flow hood to dry.
4. Assemble macrocarriers into the holders using sterile forceps and place the units in glass Petri plates with desiccant.

3.2.2. Sterilization of Gold Particles

This protocol prepares enough particles for 60 shots.

1. Weigh 30 mg of gold particles and place into a glass vial.
2. Heat particles in an oven at 180°C for 12 h (*see* **Note 14**).
3. After cooling, add 0.5 mL of isopropanol and vortex-mix vigorously for 2 min.
4. Soak for 15 min, vortex-mix (1 min), and transfer into a 1.5-mL microcentrifuge tube.
5. Pellet by centrifugation at 13,000*g* for 1 min.
6. Carefully remove the supernatant with a pipet and discard.
7. Add 0.5 mL of sterile type I water and resuspend particles by vortex-mixing vigorously for 30 s.
8. Centrifuge for 1 min and discard the supernatant as before.
9. Repeat the water wash for a total of three times.
10. Resuspend particles in 0.5 mL of 50% (v/v) glycerol/type I water. Vortex-mix vigorously for 1 min. Particles are ready for use, or may be stored in 50-μL aliquots at 4°C for 1 mo.

3.2.3 Coating Gold Particles With DNA

This protocol is for six shots.

1. Vortex-mix gold particles vigorously for 2 min and dispense 50 µL of particles into a 1.5-mL microcentrifuge tube. Vortex-mix for 5 s before each subsequent particle dispensement and just prior to adding the DNA (*see* **Note 15**).
2. Add the following components sequentially and quickly to the tube:
   a. 5 µL of 1 µg/µL plasmid DNA (for cotransformation with two plasmids, use 2.5 µL of each); gently finger vortex.
   b. 50 µL of 2.5 *M* $CaCl_2$; gently finger vortex.
   c. 20 µL of 0.1 *M* spermidine; gently finger vortex.
3. Incubate on a continuous vortex mixer for 10 min.
4. Pellet by centrifugation at 13,000*g* for 5 s. Remove and discard the supernatant.
5. Add 140 µL of isopropanol, finger vortex, and centrifuge as previously; then remove and discard the supernatant.
6. Resuspend in 48 µL of isopropanol by gentle pipetting up and down or finger vortexing.
7. Dip the microcentrifuge tubes into an ultrasonic cleaner three times for 1 s each.
8. Finger vortex to homogenate the DNA-coated microcarriers in the suspension and spread 6 µL in a circle approx 1 cm in diameter onto the center of a macrocarrier/holder assembly, which is contained in a Petri plate with desiccant (*see* **Note 16**).

3.3. Bombardment

1. Read the instrument manual and follow the manufacturer's directions and safety precautions. All users should wear safety glasses. A hair net and latex gloves are recommended to reduce the risk of microbial contamination to the plant samples.
2. Set the PDS-1000/He to the following parameters (*see* **Note 17**): 1300 psi (8963.2 kPa) helium (200 psi [1378.9 kPa] above the desired rupture disk value), 1 cm distance between the rupture disk and macrocarrier, 1 cm macrocarrier flight distance, 12 cm of target cell distance, 28-in. Hg (94.8 kPa) vacuum. Sterilize the chamber and all components with 70% ethanol (some components may be autoclaved per the manufacturer's instructions).
3. Place a rupture disk that has been dipped in isopropanol into the retaining cap. Place cap on the end of the gas acceleration tube and tighten.
4. Insert a sterile stopping screen into the support. Load a macrocarrier/holder unit with the microcarriers facing down, on top of the fixed nest. Tighten the macrocarrier cover lid and reposition the microcarrier launch assembly in the bombardment chamber.
5. Place uncovered Petri plate containing target cells into the chamber and close the door.
6. Activate the PDS-1000/He unit by first pressing the vacuum switch. When the pressure reaches 28 in. Hg (94.8 kPa) move the vacuum switch to "hold" (*see*

Note 18). Press the "fire" button until the rupture disk bursts. After bombardment, release the vacuum by moving the switch to "vent." Remove the Petri dish with bombarded cells from the chamber; replace the lid and place in an opaque plastic box. Discard the used rupture disk, macrocarrier, and stopping screen.

3.4. Postbombardment Reduction of Medium Osmotic Potential

1. Incubate all Petri plates (bombarded cells and nonbombarded controls) in the dark at 23 ± 1°C for 2 d to allow cell repair and DNA integration.
2. Approximately 16 h after bombardment, begin to reduce the osmotic potential of the culture medium by transferring the cells and bombardment medium below as a unit (using Sharkskin filter paper with attached tabs) to Petri plates containing 10 mL of 1/2 MS-HF medium without osmotica.
3. At approx 24 h postbombardment, transfer the cells and bombardment medium (using Sharkskin filter paper with attached tabs) to Petri plates containing 20 mL of 1/2 MS-HF medium without osmotica, leaving the 10 mL of medium from the previous transfer behind (discard).

3.5. Analysis of Transient and Long-Term GUS Expression

Transient GUS expression is assayed in a portion of the plates 48 h after bombardment. A plate of negative control cells (nonbombarded or bombarded without the *uidA* gene) should be assayed as well. This assay is destructive (*see* Chapter 14) (*see* **Note 19**). The analysis should be repeated in other plates on a monthly basis for 3–6 mo to evaluate rates of long-term GUS expression as an indication of stable transformation.

1. Using sterile forceps, transfer filter papers with cells to empty Petri plates and place 600 µL of X-gluc solution on top of the cells.
2. Incubate at 37°C overnight. Transformed cells will turn blue.
3. Count the number of blue spots per plate using a stereomicroscope. A black grid on transparent plastic (homemade) placed either above or below the cells aids counting. When transformation rates are high, only a portion of the cells on the plates needs to be counted.

3.6. Embryo Selection, Germination, and Regeneration (see Note 20)

1. Two days after bombardment, cells should be transferred to selective medium with 10 mg/L of Km. Using sterile forceps, lift the original Whatman no. 2 filter paper supports with cells from the bombardment medium and place on top of selection medium. Wrap the Petri plates with Parafilm and incubate at 27 ± 0.1°C in the dark for embryo induction. After 4 wk (and every 4 wk thereafter) transfer the cells with supporting filter paper to fresh selective medium with 15 mg/L of Km. Putative Km-resistant embryos should be visible beginning approx 6–8 wk after bombardment.

2. Harvest individual embryos with a 1–2 cm long radicle from Km-selective medium and place directly on embryo germination medium in Petri plates. Wrap the plates with Parafilm and incubate embryos for 4 wk at 23 ± 1°C with low light intensity (10 μE/m^2/s), 14:10-h light/dark (L/D) photoperiod (*see* **Note 21**).
3. Transfer embryos every 4 wk to fresh embryo germination medium in baby food jars. Wrap jars with Venting tape and incubate at 23 ± 1°C with increased light intensity (50 μE/m^2/s), 14:10–h L/D photoperiod.
4. Transfer germinated embryos with elongated roots and open green cotyledons to Magenta boxes containing plant growth medium. Incubate embryos at 23 ± 1°C for root elongation and shoot formation. Transfer to fresh medium every 4 wk.
5. Maintain regenerated plants on plant growth medium in Magenta boxes at 23 ± 1°C for multiplication. Transfer shoots to fresh medium every 6 to 8 wk.

4. Notes

1. Embryogenic cell cultures are often the best tissue to use for biolistic transformation because they can be spread to provide a uniform target of cells, and because they have a high capacity to regenerate into plants. We use proembryogenic cells that are finely divided because they spread easily on the filter papers. Small cell clusters also are effective for selection of transformants as fewer nontransformed escapes result.
2. Supplementing the bombardment medium with osmotica (mannitol/sorbitol) resulted in higher rates of stable transformants for all suspension cultured cells we have tested. However, the benefits of osmotica are less clear when intact tissues such as leaves or whole embryos are used. It is believed that plasmolysis of the cells reduces damage by preventing leakage of protoplasm from bombarded cells ***(17,18)***. Partial drying of cells has also been used ***(19)***.
3. Bio-Rad also sells gold particles in different sizes, with 0.6 μm and 1 μm being most applicable for plant cell transformation. Tungsten particles work well for many plant species and are much less expensive. However, the size is heterogeneous and tungsten may degrade DNA or be toxic to plant cells ***(20)***. See Bio-Rad bulletin US/EG Bulletin 2015 for a discussion of particle types/sizes (available at Website: http://www.bio-rad.com).
4. DNA and tungsten particles may stick to the sides of certain brands of microcentrifuge tubes, resulting in loss of particles. We have not tested all brands, but know that Treff tubes work well.
5. DNA should be very pure (free of RNA or protein) or microprojectiles may clump. We purify DNA by CsCl gradient centrifugation or a plasmid purification kit (Qiagen, Valencia, CA).
6. Spermidine stocks can degrade even when frozen, causing dramatic reductions in transformation efficiency. Fresh stocks should be made monthly.
7. Rupture disks come in a range of bursting pressures from 450 to 2200 psi. The most commonly used for plant tissues are 1100 psi. Rupture disks of higher psi impart higher velocity to the macro- and microcarriers, but also cause more tissue damage. These may be appropriate for more sturdy tissue such as leaves.

8. To dispense 10 mL onto the plates, the medium must be spread by swirling the plates, or by pipetting extra medium and then removing medium until only 10 mL remains.
9. It is important to design bombardment experiments to be performed comfortably by the operator so that the experiment is not rushed or critical details overlooked. In our laboratory, with two people working together it is possible to bombard a maximum of 50–60 plates of suspension cultured cells in 1 d. One person prepares the target cells and adds them to Petri plates with bombardment medium, and the second person prepares the DNA-coated microcarriers and the biolistic device. They then work together to perform the bombardment.
10. The cells in the flasks should be divided into multiple flasks as the population increases. There is no specific formula for dividing the cell culture; rather, the transfer technician should develop an eye as to how dense the population should be to maintain a creamy white or light yellow color and a small cell cluster size.
11. Contamination of the original cell culture can be a source of frustration because whole experiments can be lost after the work of bombardment. At each weekly subculture of the cell suspension, samples of media and cells should be streaked onto Petri plates with bacterial growth medium and/or plant growth medium and incubated both at 25°C and 37°C. Just prior to preparing the cells for bombardment, a sample of the cells and growth medium should be placed on a glass slide with a cover slip and examined with a compound microscope. Use phase-contrast optics if available or move the condenser out of focus to observe cells and possible microbes better. To gain experience in observing microorganisms in culture, researchers should practice looking at plant cell cultures contaminated with various organims as well as those known to be clean.
12. The bore of a 5-mL pipet is too small and cell clumps cause blockage. Attempt to minimize cells lost off of the edge of the filter paper while also achieving a uniform spread across the whole filter paper. It takes some practice to achieve a uniform layer of cells on the filter paper.
13. Macrocarriers and holders may be assembled and autoclaved as a unit. However, we have occasionally experienced shrinkage of macrocarriers after autoclaving, resulting in premature slipping of the macrocarriers from the holders. Thus, we prefer alcohol sterilization. Macrocarriers should be kept free of dirt and oil (from fingers).
14. We follow the protocol suggested by Sawant et al. ***(21)***, in which heating gold particles was shown to reduce particle agglomeration and significantly enhance transformation.
15. Particles settle out of suspension quickly. When removing aliquots, work quickly and vortex-mix often. As stated by Birch and Franks ***(22)***: "The importance of consistent technique in precipitating the DNA onto the microprojectiles and loading the accelerating apparatus should not be underestimated. Two operators of a single apparatus may obtain a 100-fold difference in transformation frequencies because of slight variations in technique at this stage."

16. Finger vortex-mix each time before aliquoting microcarriers. It is important to place macrocarriers in a desiccator to dry immediately after they are loaded. Exposure to high humidity during and after drying may result in clumping of the particles and tight (sometimes irreversible) binding to the macrocarrier ***(23)***. Use DNA-coated macrocarriers within 2 h after preparation.
17. The gene gun settings are critical for success and should be checked before each bombardment. We use a prototype of the Bio-Rad instrument in which the settings are adjustable over a larger range. However, the settings we describe here can be achieved with the Bio-Rad unit. We use a small plastic ruler to measure the distances. Higher particle velocities are obtained with higher helium pressures, and shorter rupture membrane to macrocarrier and macrocarrier to target cell distance. One must be cautious in interpreting transient expression assays because the factors that increase particle velocity also increase the shockwave to the tissue and may actually decrease stable transformation. The settings we use are standard in our laboratory for cell suspension cultures. With intact tissues it may be desirable to increase helium pressure, decrease target cell distance, or bombard each sample multiple times to improve penetration of the particles into the tissues. The reader is referred to several reviews for further discussion on the optimization of biolistic parameters ***(4,5,7,22)***.
18. Leaving the Petri plate at or near 28 in. Hg (94.8 kPa) can allow medium to boil and flip out of the plate. This problem can be avoided by using slightly lower vacuum, by increasing the concentration of gelling agent in the medium, or by letting medium set for 2 wk before use.
19. The green fluorescent protein (*gfp*) gene is another commonly employed reporter gene whose assay by UV light is nondestructive to the cells (refer to Chapter 15).
20. The procedures and growth media we describe here have been used for *V. vinifera* cultivars 'Chardonnay,' 'Merlot,' and 'Pinot Noir' in our laboratory. Other grapevine species and cultivars have not been tested with this protocol. Researchers should use the optimal embryo and plant growth medium for the genotypes they are working with. Similarly, the type and concentration of selective agent needs to be optimized for each genotype and tissue (even for each cell culture line).
21. Embryos could be incubated either at 4°C in the dark for 2 wk for chilling treatment ***(24)*** and then incubated at 23 ± 1°C with low light intensity (10 $\mu E/m^2/s$), 14:10-h light/dark (L/D) photoperiod, for an additional 2 wk, or incubated at 23 ± 1°C with low light intensity for 4 wk. In our laboratory, we did not find statistical differences between the two treatments.

Acknowledgments

Our research was supported by Research Grant US-2759-96 from BARD, The United States–Israel Binational Agricultural Research and Development Fund, as well as grants from the USDA-Viticulture Consortium-East, the New York Wine & Grape Foundation, and the Kaplan Fund. J. R. Vidal was supported by a postdoctoral grant from the Spanish Ministry of Education and Science.

References

1. Sanford, J. C., Klein, T. M, Wolf, E. D., and Allen, N. (1987). Delivery of substances into cells and tissues using a particle bombardment process. *Particulate Sci. Technol.* **5,** 27–37.
2. Sanford, J. C. (2000) The development of the biolistic process. *In Vitro Cell. Dev. Biol. Plant* **36,** 303–308.
3. Kikkert, J. R. (1993) The Biolistic® PDS-1000/He device. *Plant Cell Tiss. Org. Cult.* **33,** 221–226.
4. Southgate, E. M., Davey, M. R., Power, J. B., and Marchant, R. (1995). Factors affecting the genetic engineering of plants by microprojectile bombardment. *Biotechnol. Adv.* **13,** 631–651.
5. Taylor, N. J. and Fauquet, C. M. (2002) Microparticle bombardment as a tool in plant science and agricultural biotechnology. *DNA Cell Biol.* **21,** 963–977.
6. McCabe, D. and Christou, P. (1993) Direct DNA transfer using electric discharge particle acceleration (ACCELL™ technology). *Plant Cell Tiss. Org. Cult.* **33,** 227–236.
7. Sanford, J. C., Smith, F. D., and Russell, J. A. (1993) Optimizing the biolistic process for different biological applications. *Methods Enzymol.* **217,** 483–509.
8. Sanford, J. C., DeVit, M. J., Russell, J. A., et al. (1991) An improved, helium-driven biolistic device. *Technique* **3,** 3–16.
9. Perl, A., Lotan, O., Abu-Abied, M., and Holland, D. (1996) Establishment of an *Agrobacterium*-mediated transformation system for grape (*Vitis vinifera* L.): The role of antioxidants during grape-*Agrobacterium* interactions. *Nat. Biotechnol.* **14,** 624–628.
10. Francois, I. E. J. A., Broekaert, W. F., and Cammue, B. P. A. (2002) Different approaches for multi-transgene-stacking in plants. *Plant Sci.* **163,** 281–295.
11. Fu, X., Duc, L. T., Fontana, S., et al. (2000) Linear transgene constructs lacking vector backbone sequences generate low-copy-number transgenic plants with simple integration patterns. *Transgen. Res.* **9,** 11–19.
12. Srivastava, V. and Ow, D. (2001) Biolistic mediated site-specific integration in rice. *Mol. Breed.* **8,** 345–350.
13. Vidal, J. R., Kikkert, J. R., Wallace, P. G., and Reisch, B. I. (2003) High-efficiency biolistic co-transformation and regeneration of 'Chardonnay' (*Vitis vinifera* L.) containing *npt-II* and antimicrobial peptide genes. *Plant Cell Rep.* **22,** 252–260.
14. Mauro, M. C., Toutain, S., Walter, B., et al. (1995) High efficiency regeneration of grapevine plants transformed with the GFLV coat protein gene. *Plant Sci.* **112,** 97–106.
15. Murashige, T. and Skoog, F. (1962) A revised medium for rapid growth and bioassays with tobacco tissue cultures. *Physiol. Plantarum* **15,** 473–497.
16. Lloyd, G. and McCown, B. (1980) Commercially-feasible micropropagation of mountain laurel, *Kalmia latifolia*, by use of shoot-tip culture. *Int. Plant Prop. Soc. Proc.* **30,** 421–427.

17. Russell, J. A., Roy, M. K., and Sanford, J. C. (1992) Major improvements in biolistic transformation of suspension-cultured tobacco cells. *In Vitro Cell. Dev. Biol.* **28P,** 97–105.
18. Vain, P., McMullen, M. D., and Finer, J. J. (1993) Osmotic treatment enhances particle bombardment-mediated transient and stable transformation of maize. *Plant Cell Rep.* **12,** 84–88.
19. Finer, J. J. and McMullen, M. D. (1991) Transformation of soybean via particle bombardment of embryogenic suspension culture tissue. *In Vitro Cell. Dev. Biol.* **27P,** 17–182.
20. Russell, J. A., Roy, M. K., and Sanford, J. C. (1992) Physical trauma and tungsten toxicity reduce the efficiency of biolistic transformation. *Plant Physiol.* **98,** 1050–1056.
21. Sawant, S. S., Singh, P. K., and Tuli, R. (2000) Pretreatment of microprojectiles to improve the delivery of DNA in plant transformation. *BioTechniques* **29,** 246–248
22. Birch, R. G. and Franks, T. (1991) Development and optimisation of microprojectile systems for plant genetic transformation. *Aust. J. Plant Physiol.* **18,** 453–469.
23. Smith, F. D., Harpending, P. R., and Sanford, J. C. (1992) Biolistic transformation of prokaryotes: factors that affect biolistic transformation of very small cells. *J. Gen. Microbiol.* **138,** 239–248.
24. Martinelli, L. and Mandolino, G. (1994) Genetic transformation and regeneration of transgenic plants in grapevine (*Vitis rupestris* S.). *Theor. Appl. Genet.* **88,** 621–628.

5

Electroporation

Introduction and Expression of Transgenes in Plant Protoplasts

Henry J. Fisk and Abhaya M. Dandekar

Summary

An optimized protocol for the electroporation-based transfection of tobacco protoplasts is described that routinely results in transgene expression frequencies approaching 90%. The overall efficiency of the procedure depends collectively on numerous key parameters, including protoplast viability; DNA concentration, purity, and topology; carrier DNA; and electrical conditions such as ionic strength of the electroporation buffer, electric field strength, pulse duration, and capacitance. Individual methodologies that address each one of these parameters are presented in sufficient detail to enable successful reproduction of this method along with notes that describe helpful tips.

Key Words: Electroporation; plant transformation; protoplast; transfection; transient expression.

1. Introduction

Electroporation refers to a technique that utilizes short, high-intensity electric fields to permeabilize reversibly the lipid bilayers of cell membranes (for detailed articles that address both theoretical and practical aspects of electroporation, the reader is referred to **refs. *1–5***). It is widely believed that the electric pulse causes extensive compression and thinning of the plasmalemma. The resulting transient formation of pores permits free diffusion of various classes of macromolecules including dyes *(6)*, antibodies *(7)*, RNA and viral particles *(8)*, and DNA (representative citations are presented throughout this chapter). The majority of reports in which electroporation was used involve the transfection of DNA, frequently for stable transformation of various plant species, but also to capitalize on the advantages offered by transient expression. Transient

From: *Methods in Molecular Biology, vol. 286: Transgenic Plants: Methods and Protocols*
Edited by: L. Peña © Humana Press Inc., Totowa, NJ

expression from electroporated plant cells has been used to define functional elements within a promoter ***(9,10)***, to examine the effects of antisense RNA on gene expression ***(11)***, to study the translocation of proteins into both plastids ***(12)*** and nuclei ***(13)*** of intact protoplasts, to examine cell cycle specific gene expression ***(14)*** and to study responses to plant hormones ***(10,15)***. These few examples demonstrate the general utility of the technique to a broad group of plant research topics.

As a method of DNA transfer, electroporation is convenient and the results are consistently duplicated as a daily routine. In most cases it is more efficient than other methods designed for the same purpose, such as particle bombardment. In addition, it does not suffer from host-range limitations imposed by biology-based systems such as those employing *Agrobacterium tumefaciens* or toxicity problems sometimes encountered using a polyethylene glycol based procedure ***(16,17)***. Finally, electroporation coupled with a transient expression assay is rapid, allowing for the reproducible detection of gene products within hours of the introduction of DNA. This is in contrast to a stable transformation strategy that involves months to regenerate transformants and suffers from uncontrollable large variations in gene expression because of "positional effects" (reviewed in **ref.** ***18***). In the context of a transformation program where stable integration of genetic material is required, transient expression may be used to rapidly demonstrate functionality of new transgene sequences before they are used to generate transformants by some other method of DNA introduction.

An electroporation-based transfection system consists of a number of potentially important variables, including method of protoplast preparation, electric pulse strength and duration, ionic concentration, and composition of the electroporation buffer and DNA purity, concentration and topology. We have analyzed the importance of these variables in addition to a few others with the goal of identifying and optimizing the parameters necessary to increase expression frequency among a population of tobacco protoplasts. This chapter describes optimized conditions for an electroporation-based transient expression assay that routinely results in a nearly 90% expression frequency. This procedure should prove useful for studies that require maximal expression while maintaining good protoplast health.

2. Materials

Unless stated otherwise, all reagents and chemicals used in this protocol were of high purity and were analytical grade and/or tested for molecular biology or plant cell tissue culture applications. The water used was deionized and filtered through a Nanopure (Barnstead, Dubuque, IA) water purification system.

2.1. Expression Vector Preparation

1. Cesium chloride.
2. 10 mg/mL of ethidium bromide (store at room temperature in a dark bottle).
3. 30°C water bath.
4. Refrigerated super-speed centrifuge.
5. Sorvall SS34 rotor (or equivalent).
6. Pasteur pipets.
7. Beckman Quick-Seal ultracentrifuge tubes (or equivalent).
8. Ultracentrifuge.
9. VTi65.2 ultracentrifuge rotor (Beckman, Fullerton, CA) or equivalent.
10. 21-gage hypodermic needles.
11. Siliconized 30-mL Corex centrifuge tubes (or equivalent).
12. Water-saturated butanol or isoamyl alcohol.
13. Ethanol (100%).
14. Tris-ethylenediaminetetraacetic acid (TE) buffer: 10 m*M* Tris-HCl, 1 m*M* EDTA, pH 8.0.
15. Spectrophotometer with UV capabilities.
16. Buffered phenol–chloroform (1:1; *see* **Note 1**).
17. 3 *M* Sodium acetate.

2.2. Carrier DNA Preparation

1. Type I calf thymus DNA (Sigma Chemical, St. Louis, MO).
2. TE buffer (*see* **Subheading 2.1.**).
3. Branson Cell Disruptor 200 Sonifier (or equivalent).
4. IBI horizontal agarose gel electrophoresis apparatus (or equivalent).
5. SeaKem LE agarose (or equivalent).

2.3. Protoplast Preparation

1. Filter-sterilized protoplasting media, pH 5.6: 0.4 *M* mannitol, 3 m*M* 2-(*N*-morpholino) ethanesulfonic acid (MES), 1 m*M* calcium chloride, 1 μ*M* potassium iodide, 1 m*M* potassium nitrate, 0.2 m*M* KH_2PO_4, 1% (w/v) Cellulase Onozuka RS (Karlan Research Products, Santa Rosa, CA), 0.1% Pectolyase Y23 (Karlan Research Products), and 0.5% bovine serum albumin (BSA; Sigma Chemical) (*see* **Note 2**).
2. Sterile screw-cap conical centrifuge tubes (15 mL, Falcon or equivalent).
3. Centrifuge with 80–100*g* capabilities.
4. Nalgene (or equivalent) vacuum filter sterilization units (0.2 μm).
5. Sterile disposable plastic Petri plates (100 × 15 mm).
6. Orbital shaker.
7. Inverted microscope.
8. HNaM wash buffer, pH 7.1: 10 m*M* HEPES, 80 m*M* sodium chloride, 0.04 m*M* calcium chloride, and 0.4 *M* mannitol.
9. Nylon mesh filter (80 μm).

10. Autoclavable nylon mesh filter frame (approx 6 cm diameter).
11. Narrow-stemmed funnel.
12. Sterile, wide-tipped pipets.
13. Fuchs–Rosenthal hemocytometer.
14. Upright compound microscope.
15. HNaS electroporation buffer: HNaM buffer described previously with 0.4 *M* sucrose in place of mannitol.

2.4. Electroporation

1. Capacitance discharge-type electroporation apparatus (e.g., Gene Pulser available from Bio-Rad Laboratories, Hercules, CA; *see* **Note 3**).
2. Rainin Pipetman (or equivalent).
3. Sterile pipet tips.
4. Electroporation cuvets (0.4-cm electrode gap).
5. Sterile disposable plastic Petri plates (60 × 15 mm).
6. Incubation medium (liquid Murashige and Skoog medium [MS] *[19]* supplemented with 1000 mg/L of casein hydrolysate, 2 mg/L of 2,4-dichlorophenoxyacetic acid and 0.1 mg/L of kinetin).

3. Methods

3.1. Expression Vector Preparation

The first critical procedure for this method is purification of the target DNA, usually in the form of a plasmid expression vector. However, the step prior to this is isolation of relatively large quantities (0.5–1 mg) of plasmid DNA, which may be obtained by any one of several methods *(20)*. One that we recommend is amplifying the plasmid DNA in 200 mL overnight cultures of *Escherichia coli* and using a modified alkaline lysis procedure for extraction. Described below is the procedure for the isolation of purified, supercoiled plasmid DNA fractions after CsCl/ethidium bromide (EtBr) density gradient equilibrium centrifugation. The plasmid DNA contained in these fractions is then extracted in buffered phenol–$CHCl_3$ (1:1) just prior to a final concentration adjustment with TE buffer (*see* **Notes 4** and **5**).

1. For every milliliter of plasmid DNA solution, add 1 g of solid CsCl. Warm the solution in a 30°C water bath and mix gently until the salt is dissolved.
2. Add ethidium bromide (EtBr; *see* **Note 6**) to the DNA–CsCl and adjust to a final concentration of 740 µg/mL. To achieve the proper density, this must be accomplished by adding 0.8 mL of EtBr solution (10 mg/mL in water) for every 10 mL of DNA–CsCl solution.
3. Centrifuge the solution at approx 7500*g* for 5 min at room temperature. The scum formed at the top are complexes formed between bacterial proteins and EtBr. A deep red pellet may also be observed.

4. Remove the clear, red solution from below the scum using a Pasteur pipet and place in Beckman Quick-Seal ultracentrifuge tubes (or equivalent). Fill any remaining space in the top of the tubes with light paraffin oil and seal the tubes following the manufacturer's instructions.
5. Centrifuge the tubes in a VTi65.2 (or equivalent) at approx 325,000*g* for 6 h at 20°C (*see* **Note 7**; other rotor/speed/time combinations are possible).
6. Three red bands should be visible in the tubes following centrifugation (*see* **Note 8**). The bottom band consists of supercoiled plasmid. Carefully collect this fraction by first gently introducing an air hole into the top of the tube by insertion of a 21-gage needle. With a gloved finger covering the hole, gently "drill" a second hole into the bottom of the tube with the same needle. Be careful not to disrupt any pellet (RNA) that may have formed. Roll the tube between your fingers to allow air to enter the top hole and to regulate the flow of solution out the bottom of the tube. Just before the band of supercoiled DNA begins to exit through the hole, place a Dnase-free container (e.g., siliconized 30-mL Corex tubes) under the hole and collect the entire band.
7. Remove the EtBr from the DNA solution by repeated extractions with either butanol saturated with water or isoamyl alcohol. Add an equal volume of the solvent to the DNA solution, mix thoroughly by inverting the tubes, and then centrifuge in a benchtop centrifuge for 3 min. Remove the upper phase and repeat the extraction until the pink coloration is no longer visible in the aqueous phase (usually four to six times). Use a white background to help make this determination.
8. Precipitate the DNA to remove the CsCl. Dilute the DNA with three volumes of water, gently mix and then add two volumes of ethanol. Gently mix again by inverting the tube several times and let stand for at least 15 min on ice. Pellet the DNA by centrifuging at 10,000*g* for 15 min at 4°C. Wash the pellet twice with 70% ethanol.
9. Dissolve the DNA pellet in approx 1 mL of TE.
10. Measure the absorbance of the solution at 260 nm and calculate the concentration (*see* **Note 9**).
11. Extract the DNA solution with buffered phenol/$CHCl_3$. Add an equal volume of the solvent, vortex-mix, and then centrifuge for 5 min in a benchtop microfuge. Recover the lower aqueous phase and place in a fresh tube.
12. Precipitate the DNA by adding one tenth volume of 3 *M* sodium acetate and two volumes of ice-cold ethanol. Mix by gently inverting and store on ice or at –20°C for at least 15 min. Centrifuge at 10,000*g* at 5°C for 15 min. Wash the pellet twice with 70% ethanol. Decant the supernatant and store the tubes inverted until there are no signs of liquid remaining.
13. Rehydrate the DNA pellets with enough TE to achieve a final concentration of approx 4 µg/mL (based on the spectrophotometric analysis in **step 10**).
14. Measure the absorbance at 260 and 280 nm to determine the final concentration and purity (*see* **Note 10**). If necessary, change the concentration through TE addition or precipitation/rehydration.

3.2. Carrier DNA Preparation

Carrier DNA may serve as an alternative substrate for endogenous DNase activities present in recipient protoplasts, thereby protecting the transfected plasmid from inactivation through digestion *(2)*.

1. Mix 0.1 g of calf thymus DNA with 5 mL of TE in a 15-mL Corex tube.
2. Place on ice and let stand for approx 4 h with occasional mixing.
3. With a standard benchtop, probe-type sonicator set to approximately one-half power, subject the DNA solution to approx 30 20-s pulses while on ice, with 40-s intervals in between each pulse to allow for cooling.
4. Analyze 0.5–1.0 μL of the DNA by agarose gel electrophoresis. By comparison to a size standard, the sheared DNA should be less than or equal to 2 kb. If not, adjust the intensity/duration of the sonication accordingly and repeat.
5. Desalt the DNA with an ethanol precipitation (*see* **Subheading 3.1., step 12**). Rehydrate the pellet in approx 8 mL of TE.
6. Analyze the DNA solution spectrophotometrically for concentration and purity (**Subheading 3.1., step 14**). The final concentration should be approx 10 μg/μL.

3.3. Protoplast Preparation

Protoplast yields in practice can be highly variable and depend largely on the species and the nature of the starting material (e.g., suspension cells vs mesophyll cells). However, two important determinants that may be controlled experimentally are enzyme composition and time of digestion. In many cases, a compromise must be made between overall yield and yield of viable protoplasts that offer good transient expression results following an electroporation procedure. The following is a method that begins with tobacco suspension cells (*Nicotiana tabacum* L. cv Xanthi) and should be used only as a general guide for other plant materials. If the electroporation procedure will be used routinely, then establishment of cell suspension cultures is highly recommended for convenience and for increased yields.

1. Harvest early- to mid-log phase cells from fine suspensions by transferring the suspension to sterile, conical centrifuge tubes and centrifuging at 100*g* for 5 min.
2. Resuspend the decanted pellets in filter-sterilized protoplasting media.
3. Transfer the digesting cells into a 100 × 15 mm Petri dish and incubate in the dark overnight (approx 14 h) at 25°C without agitation (*see* **Note 11**).
4. Release the protoplasts by gently swirling the plates (40 rpm) on a rotary shaker for approx 2 h.
5. Gently filter the solution through sterile 80-μm nylon mesh into centrifuge tubes (*see* **Note 12**) and collect the protoplasts by centrifugation at 80–100*g* for 7 min. Use sterile, wide tipped pipets to help prevent protoplast lysis.
6. Resuspend the decanted pellets in 10 mL of HNaM buffer.
7. Quantify the protoplast concentration using a hemocytometer.

8. Recentrifuge the protoplasts as in **step 5** and resuspend them in an amount of HNaS electroporation buffer necessary to achieve a final concentration of 1×10^6 protoplasts/mL (*see* **Note 13**).

3.4. Electroporation

The most critical variables with respect to efficient transfection and subsequent gene expression are directly related to the electroporation step. Most of them are electrical in nature (e.g., field strength, capacitance, and pulse number) but may also be related to the type and amount of DNA transfected. All of these variables in addition to several others were thoroughly investigated to derive the optimized protocol shown below. However, as with protoplast preparation, it should be used as a general guide as other variables (most notably plant material) may result in the need for some slight modifications (*see* **Notes** listed throughout this section).

1. Place both the plasmid DNA and the carrier DNA into the bottom of a prechilled electroporation cuvet. The final DNA concentrations in each 0.8-mL electroporation reaction should be 50 µg/mL of plasmid and 150 µg/mL of carrier (*see* **Note 14**). The combined volume of the DNA solutions should not exceed 20 µL.
2. Add 0.78 mL of protoplasts in HNaS to the cuvets with DNA and place on ice for 10 min (*see* **Note 15**).
3. While the protoplast/DNA suspensions are chilling, prepare the electroporation apparatus. Set the machine to deliver a 550-V/cm pulse from a 500-µF capacitor (*see* **Note 16**).
4. To ensure even distribution between the electrodes following the 10 min on ice, gently mix the protoplast/DNA mixture by passing the solution in and out of a trimmed, sterile pipet tip (*see* **Note 15**).
5. Deliver three electric pulses to the cuvets as rapidly as the electroporator can attain full charge and then immediately replace them on ice for an additional 10 min (*see* **Note 17**).
6. Allow the cuvets to warm slowly by placing them at room temperature for an additional 10 min.
7. Gently transfer the protoplasts to 60 × 15 mm Petri dishes containing 6 mL of incubation medium (*see* **Note 18**). Seal with Parafilm and place in the dark at 25°C for approx 20 h (*see* **Note 19**).
8. Assay for gene expression using methods consistent with the transfected coding sequence and desired results (e.g., extraction or *in situ* analyses of isolated protoplasts).

4. Notes

1. Equal amounts of phenol and chloroform are equilibrated by extracting several times with 0.1 *M* Tris-HCl, pH 7.6, and then stored under Tris buffer in a dark bottle at 4°C.

2. The use of BSA is optional, but recommended. It is included to serve as an alternative substrate for any proteases that may be present in the hydrolytic enzyme preparations used to digest plant cell walls. Enzyme solutions containing BSA should be filter sterilized only in units that employ vacuum because the viscosity makes it too difficult to force them through syringe–filter combinations.
3. There are numerous different kinds of electroporation devices available from commercial suppliers of scientific equipment. When choosing an electroporator, look for equipment that is designed with capacitors and is capable of delivering electric field strengths (up to at least 1 kV/cm) and pulse lengths (1–40 ms) that are variable by the operator. Alternatively, a device may be constructed by competent individuals using widely available parts. However, we do not recommend this approach if the apparatus will be operated by users with limited knowledge of the hazards of a high-voltage device, as "homemade" versions usually do not include safety shields or other equipment intended to protect the operator.
4. Purifying plasmid DNA by density gradient equilibrium centrifugation is probably the most time consuming of the methods currently available. We have also tested the much easier Plasmid Maxi Kit purification system (Qiagen, Valencia, CA) based on anion exchange chromatography. In four side-by-side comparisons, gene expression from the anion-exchange-purified plasmid was, on average, only 40% of that obtained with DNA from density gradients.
5. Although counterintuitive, a phenol–chloroform extraction following purification through a density gradient did substantially enhance detectable gene expression, but for unknown reasons. Spectrophotometric analysis of the DNA before and after the extraction revealed no discernable differences in purity.
6. Ethidium bromide is a mutagen and toxic. Wear proper protective clothing, gloves, and goggles where appropriate. Dispose with hazardous waste.
7. It is most efficient to carry out the centrifugation overnight. However, a delayed start time must be used. It is important to be ready to begin the next step as soon as possible after centrifugation is complete. This minimizes diffusion of the separated components as well as loosening of solid impurities that adhere to the walls of the ultracentrifuge tubes.
8. Three bands and a pellet should be visible within the tube following centrifugation. The pellet consists of bacterial RNA and the top band is protein. The two middle bands consist of DNA with the usually thicker bottom band consisting of supercoiled plasmid and the top of nicked or linear DNA. Only the closed plasmid band should be recovered. Not only is the upper DNA band possibly contaminated with chromosomal sequences, but we have also found that linearized plasmid resulted in up to a 50% loss in detectable transient gene expression in side-by-side comparisons with its supercoiled counterpart. For applications where the desired outcome is stable transformation, linearized plasmid may be more effective *(3)*.
9. An absorbance reading of 1 at 260 nm corresponds to a DNA concentration of approx 50 μg/mL. Note that spectrophotometry can not easily detect RNA con-

tamination. A fluorimetric analysis is more accurate, but probably not necessary for most applications.

10. The 260/280 reading ratio for pure DNA preparations is 1.8. Higher ratios may indicate contamination with RNA while lower numbers will occur from the presence of phenol or proteins.
11. The time of cell wall digestion is highly variable and depends on starting material and enzyme activity. Progress of the digestion should be monitored the first few times with an inverted microscope to view the cells for optimal results.
12. Protoplasts are fragile. Care must be taken when filtering the protoplast/enzyme solution to maintain sterility and preserve cell integrity. The nylon membrane should be placed in a cylindrical filter frame (two pieces that screw together with the filter in between) and autoclaved wrapped in foil. Set up a clamp to hold a sterile screw-cap tube at a slight angle and place a sterile funnel in the tube. The funnel should have a stem long enough to reach more than halfway to the bottom of the tube. Prewet the bottom side of the filter with HNaM and place the filter/frame in the funnel. It is best to do all manipulations with sterile forceps or gloves. In addition to filtering away undigested tissues and other cellular debris, the overall goal of the approach is to ensure that the protoplasts do not drop long distances in air (i.e., from the filter to the bottom of the tube) and instead more gently travel along the walls of the funnel and tube.
13. Sucrose is a more effective osmoticum than mannitol with respect to electroporation and subsequent gene expression analyses. However, living protoplasts are more buoyant in sucrose, making sedimentation by gentle centrifugation problematic. Therefore, it is not introduced until final adjustment of protoplast concentration, just prior to electroporation.
14. Detectable gene expression increases proportionately with increasing plasmid DNA concentrations. We have tested concentrations as high as 200 µg/mL with no discernible toxicity effects. We have also obtained detectable expression results with concentrations as low as 12.5 µg/mL. The recommended concentration of 50 µg/mL economizes the use of purified plasmid while maintaining good transgene expression levels.
15. If standard laboratory pipet tips are used to measure and transfer the protoplasts, use a razor blade to cut the ends off of the tips prior to autoclaving. This effectively widens the opening through which the protoplasts must pass and helps minimize shearing forces that may contribute to protoplast lysis.
16. Proper selection of electric field strength (EFS) and capacitor size is critical. In general, we have found that EFS values ranging from 500 to 800 V/cm delivered from either a 500- or 960-µF capacitor were optimal. The expression maxima was broader for this EFS range when delivered from the smaller capacitor, presumably because the conditions are less punishing with respect to cell viability (i.e., when compared to a 500-µF capacitor, pulse durations are roughly twice as long for any given EFS when delivered from one rated at 960 µF). The highest expression levels were observed with a 960-µF capacitor, but transfection efficiency was substantially lower because of cell death (**Fig. 1.**).

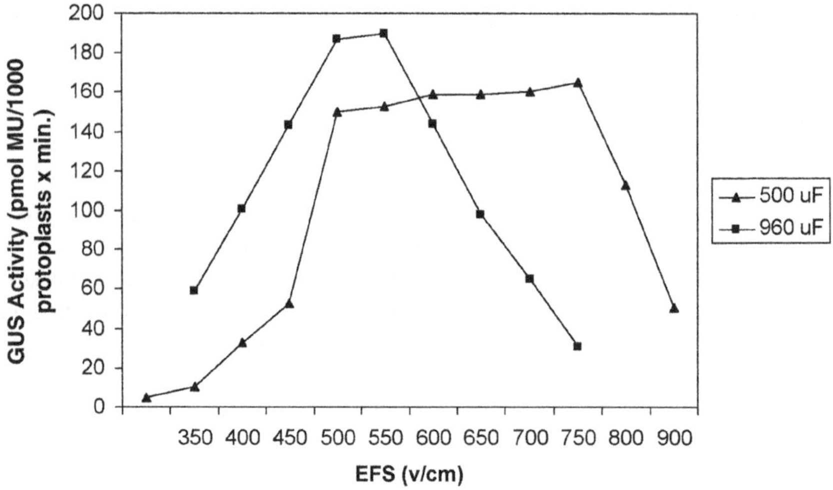

Fig. 1. Effects of EFS delivered from a 500-μF or a 960-μF capacitor on the transient expression of GUS in electroporated tobacco protoplasts. The concentration of vector DNA was 50 μg/mL and that of the carrier, 150 μg/mL. Both species were transfected into 1×10^6 protoplasts suspended in 80 m*M* NaCl (HNaSM) using three pulses. The values shown represent the means of four replicates per treatment from two separate experiments (two replicates per experiment).

Ionic strength of the electroporation buffer is also an important consideration and will influence the selection of EFS and capacitor settings. However, we have found that within the 80–120 m*M* range, salt composition is a more important variable. For example, in side-by-side tests, NaCl always outperformed KCl.

17. Expression levels may increase by a factor of approximately fivefold with each successive pulse up to a maximum of three.
18. Extra care must be taken during protoplast transfer as they may be even more fragile following electroporation.
19. A 20- to 24-h incubation period is the minimum time required for maximal activity. However, there is no benefit in waiting longer. We have analyzed protoplasts up to 68 h following transfection and observed an insignificant change in expression levels when compared to the shorter time intervals.

References

1. Fromm, M., Callis, J., Taylor, L. P., and Walbot, V. (1987) Electroporation of DNA and RNA into plant protoplasts, in *Methods in Enzymology*, Vol. 153 (Wu, R. and Grossman, L., eds.), Academic Press, London, UK, pp. 351–366.
2. Joersbo, M. and Brunstedt, J. (1991) Electroporation: mechanism and transient expression, stable transformation and biological effects in plant protoplasts. *Physiol. Plant.* **81,** 256–264.

3. Bates, G. W. (1994) Genetic transformation of plants by protoplast electroporation. *Mol. Biotech.* **2,** 135–145.
4. Dillen, W., Van Montagu, M., and Angenon, G. (1998) Electroporation-mediated DNA transfer to plant protoplasts and intact plant tissues for transient gene expression assays, in *Cell Biology*, Vol. 4 (Celis, J. E., ed.), Academic Press, London, UK, pp. 92–99.
5. Chen, G. Y., Conner, A. J., Wang, J., Fautrier, A. G., and Field, R. J. (1998) Energy dissipation as a key factor for electroporation of protoplasts. *Mol. Biotech.* **10,** 209–216.
6. Joersbo, M., Brunstedt, J., and Floto, F. (1990) Quantitative relationship between parameters of electroporation. *J. Plant Physiol.* **137,** 169–174.
7. Maccarrone, M., Veldink, G. A., Agro, A. F., and Vliegenthart, J. F. G. (1995) Lentil root protoplasts—a transient expression system suitable for coelectroporation of monoclonal antibodies and plasmid molecules. *Biochim. Biophys. Acta* **1243,** 136–142.
8. Valat, L., Toutain, S., Courtois, N., et al. (2000) GFLV replication in electroporated grapevine protoplasts. *Plant Sci.* **155,** 203–212.
9. Higo, K. and Higo, H. (1996) Cloning and characterization of the rice CatA catalase gene, a homologue of the maize Cat3 gene. *Plant Mol. Biol.* **30,** 505–521.
10. Kao, C.-Y., Cocciolone, S. M., Vasil, I. K., and McCarty, D. R. (1996) Localization and interaction of the *cis*-acting elements for abscisic acid, VIVIPAROUS1, and light activation of the *C1* gene of maize. *Plant Cell* **8,** 1171–1179.
11. Ecker, J. R. and Davis, R. W. (1986) Inhibition of gene expression in plant cells by expression of antisense RNA. *Proc. Natl. Acad. Sci. USA* **83,** 5372–5376.
12. Teeri, T. H., Patel, G. K., Aspegren, K., and Kauppinen, V. (1989) Chloroplast targeting of neomycin phosphotransferase II with a pea transit peptide in electroporated barley protoplasts. *Plant Cell Rep.* **8,** 187–190.
13. Fisk, H. J. and Dandekar, A. M. (1998) Nuclear localization of a foreign gene product in tobacco results in increased accumulation due to enhanced stability. *Plant Sci.* **133,** 177–189.
14. Nagata, T., Okada, K., Kawazu, T., and Takebe, I. (1987) Cauliflower mosaic virus 35 S promoter directs S phase specific expression in plant cells. *Mol. Gen. Genet.* **207,** 242–244.
15. Salmenkallio, M., Hannus, R., Teeri, T. H., and Kauppinen, V. (1990) Regulation of α-amylase promoter by gibberellic acid and abscisic acid in barley protoplasts transformed by electroporation. *Plant Cell Rep.* **9,** 352–355.
16. Kao, K. N. and Michayluk, M. R. (1974) A method for high-frequency intergeneric fusion of plant protoplasts. *Planta* **115,** 355–367.
17. Tyagi, S., Spörlein, B., Tyagi, A. K., Herrmann, R. G., and Koop, H. U. (1989) PEG- and electroporation-induced transformation in *Nicotiana tabacum*: influence of genotype on transformation frequencies. *Theor. Appl. Genet.* **78,** 287–292.
18. Fisk, H. J. and Dandekar, A. M. (1993) The introduction and expression of transgenes in plants. *Sci. Hortic.* **55,** 5–36.

19. Murashige, T. and Skoog, F. (1962) A revised medium for rapid growth and bio assays with tobacco tissue cultures. *Physiol. Plant.* **15,** 473–497.
20. Sambrook, J., Fritsch, E. F., and Maniatis, T. (1989) *Molecular Cloning: A Laboratory Manual*, 2nd Edit., Cold Spring Harbor Laboratory Press, Cold Spring Harbor, NY, pp. 1.21–1.39.

6

Floral Dip

Agrobacterium-Mediated Germ Line Transformation

Steven J. Clough

Summary

Many researchers use the flowering plant *Arabidopsis thaliana* to study gene function and basic plant biology. This easy-to-grow, small plant is ideal for genetic studies as it has a relatively simple genome compared to crop plants and its genetic material has been recently sequenced. Another very useful feature of *Arabidopsis* is that it is extremely simple to transform genetically. The ability to insert genes of interest stably into a given plant is essential to understand and verify gene function. Transformation is also a means of introducing specific traits that are difficult or impossible to introduce by conventional breeding techniques. This chapter provides detailed explanations on the floral dip protocol, a simple method to transform *Arabidopsis* by inoculating immature flowers with *Agrobacterium tumefaciens*.

Key Words: *Agrobacterium tumefaciens*; *Arabidopsis thaliana*; floral dip; T-DNA; transformation; vacuum infiltration.

1. Introduction

Most methods used to genetically transform plants involve tissue culturing steps. However, tissue culture methods can lead to unwanted genetic changes such as alterations in cytosine methylation, induction of point mutations *(1)*, and various chromosomal aberrations *(2)*. Therefore, a transformation method involving intact plant tissue would be the method of choice.

In the late 1980s, Feldmann and Marks reported that they succeeded in genetically transforming *Arabidopsis* by inoculating seed with *A. tumefaciens* *(3)*. Approximately 1 out of 10,000 seeds harvested was stably transformed. Transformation rates greatly improved when Bechtold et al. *(4)* inoculated

From: *Methods in Molecular Biology, vol. 286: Transgenic Plants: Methods and Protocols*
Edited by: L. Peña © Humana Press Inc., Totowa, NJ

whole plants that were at the flowering stage. Southern blot analysis revealed a large degree of variation of T-DNA insertion patterns among the transformed progeny, suggesting that random germline cells (ovules or pollen) were the targets of transformation in Bechtold et al.'s protocol ***(4,5)***. The studies of Bechtold et al. also suggested that the transformation observed by Feldmann and Marks was apparently because of the rare event in which *Agrobacterium* managed to expand its colonization from inoculated seed to the developing flower bud. Three manuscripts, from three separate laboratories, provided evidence strongly supporting that the ovule (or ovule chromosomal material) is the primary target of germline transformation via floral inoculation ***(6–8)***.

2. Materials

1. Potting soil such as Sunshine Mix LC1 (Sungro Horticulture, Quincy, MI).
2. Plastic window screening (optional).
3. *Agrobacterium* strain such as GV3101 ***(9)*** carrying a disarmed helper plasmid and the appropriate binary vector carrying the gene(s) of interest.
4. Appropriate antibiotics to maintain the presence of the binary vector.
5. Luria Bertani (LB) medium (10 g of tryptone, 5 g of yeast extract, 5 g of sodium chloride per liter).
6. Sucrose (may be food-grade sugarcane).
7. Surfactant such as Silwet L77 (Lehle Seeds, Round Rock, TX; Website: www.arabidosis.com) (skin and eye protection required).
8. Plastic plant tray with clear cover.
9. Microfuge tubes (1.5 or 2 mL).
10. Bleach.
11. Tween-20.
12. 15-mL plastic conical centrifuge tubes.
13. 100 × 15 and 150 × 15 mm Petri dishes.
14. MS salts: Murashige and Skoog Basal Medium (Sigma, St. Louis, MO, cat. no. M-5519).
15. Tissue-culture tested agar (such as Sigma cat. no. A-1296).

3. Methods

Transformation of *Arabidopsis* via floral dip inoculation is accomplished in four simple steps:

1. Growth of plants to flowering.
2. Plant inoculation with *A. tumefaciens* carrying the appropriate vector constructs.
3. Plant growth to maturity.
4. Selection of transformed seed.

This protocol assumes that researchers already have the appropriate genetic constructs made and transformed into *A. tumefaciens*. Researchers can expect between 0.5 and 3% of harvested seed to be transformed.

3.1. Growth of Plants to Flowering

Plants may be grown as individuals or as a group of multiple plants per pot. The goal is to produce healthy plants with many clusters of immature flower buds. Growth under 9-h daylight with occasional fertilization will generate large plants with many flower buds and high numbers of transformants. Transformants may be recovered more rapidly, but in lower numbers, from plants grown under 22 to 24 h light (*see* **Note 1**).

1. Plant *Arabidopsis* seed in a well-aerated soil, such as Sunshine Mix LC1. To prevent excess soil from falling into the inoculum solution during the inoculation step, the soil may be mounded beyond the top of the pot and covered snuggly with plastic window screening material (or netted cloth) fastened with a rubber band ***(10)***.
2. Sow a single seed in a 2-in. square pot; alternatively, up to about 16 seeds may be evenly dispersed on a 3.5-in. square pot of moist soil (*see* **Note 2**).
3. Cover the seeded pots with a plastic cover and place at 4°C for 3 d, as many ecotypes require a short vernalization period for maximal seed germination.
4. After this cold treatment, move pots to a growth chamber or fluorescent light shelf with light intensity between 100 and 180 μmol/m^2 s and temperature about 21–23°C. Keep soil moist and covered with a clear plastic dome until the four-leaf stage of growth. Then remove the cover and water every week or two as needed, allowing the soil to dry slightly (does not feel wet to the touch) between waterings.
5. Begin adding low levels of a commercial houseplant fertilizer to each watering after about 1 mo of growth.
6. Continue growing plants to flowering with the goal of having as many unopened flower buds as possible at the time of inoculation. There are two options in terms of when to inoculate. One is to inoculate the primary florescence after it has developed for about 1–2 wk and has developed numerous immature bud clusters. The other option is to cut off the primary florescence to encourage a flush of secondary florescences that are inoculated once they have high numbers of immature flower bud clusters (*see* **Note 3**).

3.1.1. Inoculation—Culturing Agrobacterium

1. At 4 to 10 d before inoculating the plants, streak a bacterial culture on solid medium (LB plus 16 g of agar) from a fresh culture or from a –80°C stock (14% glycerol in a liquid growth medium such as LB). Single colonies will appear in 2–3 d if grown at 25°C.
2. After the appearance of colonies, plates may be stored at 4°C for at least 1 wk.
3. Two days before plant inoculation, use a single colony from the recently streaked plate to start a 5-mL culture in liquid LB. Grow at cells 25°C, 250 rpm.
4. One day before plant inoculation, use this 5-mL culture to seed a new liquid LB culture. The amount of *Agrobacterium* that one grows for the inoculation depends on the size of plants, the number of pots to inoculate, and whether or

not one wishes to reuse the inoculum for multiple pots. For example, if inoculating plants in 3.5-in.^{2} pots, grow a 50-mL culture seeded with 2 mL of the overnight 5-mL culture. For plants in smaller pots, grow about 25 mL, seeded with 1 mL of the overnight 5-mL culture. Grow bacteria at 25°C with shaking at about 250 rpm (*see* **Note 4**).

3.1.2. Inoculation—Preparing the Inoculum

1. The next day (culture has been growing at least 18 h and is very turbid) pellet the cells by centrifuging at 6000*g* for 5 min.
2. Discard the supernatant and resuspend the cells in 5% sucrose solution (500 mL for a 50-mL overnight culture, 250 mL for a 25-mL culture). Resuspension is facilitated by using a wooden applicator stick (or disposable plastic inoculation loop) to disrupt the pellet, followed by vigorous vortex-mixing or shaking.
3. Once cells are completely dispersed, add the surfactant Silwet L77 at 0.03%, or 300 µL/L (see **Note 5**).
4. Thoroughly mix the surfactant into solution (*see* **Note 6**). There is no need to check the cell density of the inoculum (*see* **Note 7**).

3.1.3. Inoculation—Inoculating the Plants

Avoid watering the plants the day before inoculation to prevent soil or mud from contaminating the inoculum.

1. For a large plant or multiple plants in a 3.5-in.^{2} pot, pour 500 mL of inoculum into a 500-mL beaker.
2. Invert the pot and fully submerge the florescences into the bacterial suspension.
3. Leave florescences soak in solution for about 5 s with gentle agitation to allow the solution to penetrate deep into the flower buds and into the developing ovaries. Some researchers prefer to submerge the rosettes completely into the inoculum to ensure that flower buds within the leaf rosette are also inoculated. However, care should be taken to avoid prolonged contact with the soil. If the inoculum remains free of soil, it may be reused for additional inoculations.
4. One may repeat the inoculation or use other inoculum application methods, such as vacuum infiltration or spraying (*see* **Note 8**).
5. After inoculation, lay plants on their sides in a tray, cover with a clear plastic cover to retain humidity, and move to a low-light area until the next day.

3.2. Growth to Maturity and Seed Harvest

1. The day after inoculation, place the plants back to normal growing conditions, and remove the plastic cover.
2. Water and fertilize as needed to nurture plants to maturity.
3. As plants approach seed set, wrap a sheet of wax paper around the plant(s) and tape it to the pot to prevent entanglement with neighboring plants.
4. When plants stop producing new siliques and start browning, reduce watering to hasten maturation.

5. Harvest seed when siliques are completely dried (no longer green). A simple method to harvest is to spill seed onto a piece of paper. To harvest, prefold the paper in half, rotate it 90° and fold in half again such that the folds make an X that crosses the middle of the paper. Gently pull the dried florescences through one's fingers over the folded piece of paper. The seed and other dry plant material will fall onto the paper. With the paper slightly creased at the folds, gently blow off the plant debris leaving the heavier seed within the folds of the paper. Pour the seed into a labeled collection tube (microfuge tube).
6. Store the seed under desiccation.

3.2.1. Selection of Transformants—Seed Sterilization

Plate 2000–3000 seeds per treatment to recover about 10–90 transformants (0.5–3.0% transformation rates are typical).

1. Fifty seeds weigh approx 1 mg; therefore, weigh 40–60 mg of seeds for each plating.
2. To sterilize seeds:
   a. Pour seeds into a 15-mL conical centrifuge tube.
   b. Treat for 30 s in 2 mL of 95% ethanol.
   c. Decant alcohol.
   d. Soak 5 min in 2 mL 50% of bleach–0.05% Tween-20.
   e. Discard the bleach solution.
   f. Rinse three times with 5 mL of sterile water.

Alternatively, seeds may be sterilized under chlorine gas (*see* **Note 9**).

3.2.2. Selection of Transformants—Plating Seeds on Selective Medium

1. Add 4 mL of 0.1% sterile agarose to the sterilized seed.
2. Pour the agarose–seed suspension onto 150 × 15 mm selection plates (1/2X MS salts, 0.8% tissue-culture tested agar, and antibiotic such as kanamycin at 50 μg/mL).
3. Disperse the seeds evenly across the plates with the aid of a sterile plastic inoculating loop, glass rod, or spreader.
4. Leave the lids partially open to allow plates to dry.
5. Seal the dried plates with Parafilm or a porous medical tape.
6. Place at 4°C for 3 d to ensure maximum germination.
7. After this vernalization treatment, put plates at approx 22°C, under 16 to 22 h light periods for 10–14 d.
8. If plates accumulate large amounts of condensation, shake the water off the lids within a sterile hood.

An alternative selection to kanamycin is the use of a binary construct containing the *bar* gene conferring resistance to phosphinothricin (*see* **Note 10**).

3.2.3. Selection of Transformants—Transplanting

Transformed plants are easy to identify after about 10–14 d growth on the selective medium. Transformed seedlings will be green and healthy, whereas the nontransformants will be chlorotic, stunted, and dying. The root system of a transformant will also be much longer than that of the nontransformants. Transplant the transformants to soil to allow plants to reach maturation. To transplant:

1. Break the agar around the roots with a pair of tweezers and pull the plantlet from the plate by lifting it with the tweezers without squeezing the plant.
2. Once removed from the plate, remove any excess large chunks of agar that may be adhering to the roots by gently squeezing the chunks with one's fingers.
3. Place the plantlet into the soil and water heavily with a squirt bottle such that the wet soil gradually fills in around the root (*see* **Note 11**).
4. Cover the flat of transplants with a clear plastic cover and place under low-light conditions overnight.
5. The next day, place plants under normal growing conditions.
6. Wait one more day to remove the cover.
7. Continue watering as needed.

Various methods can be used to verify transformation, but Southern blot analysis is the best *(11)*. The seedlings recovered on selection plates are called T1s and should be hemizygous for the transgene(s). The seed produced from the transplanted T1 plants will be T2s and may be homozygous, hemizygous, or null for the transgene(s). When performing Southern blots, one must keep in mind that *Agrobacterium* is a prolific colonizer of plants and may persist for several generations after inoculation (*see also* Chapter 24). Therefore, bands on the Southern blot can be mistaken for transformed plant DNA when they are actually the detection of the tumor-inducing (Ti) plasmid from colonizing *Agrobacterium*. Proper controls (lanes for digested and undigested total DNA from the *Agrobacterium* used to transform) will help distinguish true transformants from *Agrobacterium* contamination and should always be included.

Transformation via the germline works very well for *Arabidopsis*, but few researchers have had success with other plants. *See* **Note 12** for a short discussion on this topic (*see also* Chapter 7).

4. Notes

1. Some researchers have had good transformation success using plants grown under 22 to 24 h light and inoculating the small primary florescence. However, plants grown under this lighting regimen are weaker and, therefore, more sensitive to the high amounts of sugar, bacteria, and surfactant in the inoculum. Therefore, if a laboratory wishes to pursue the use of these weaker plants, it is recommended that the laboratory optimize the inoculum on test plants to determine ideal concentrations of these ingredients before conducting large-scale experiments. One should keep in mind that the total seed yield (and therefore the number of transformed plants recovered) will most likely be lower than what would be recovered from larger plants grown under short day lengths.

2. One may use a slightly moistened fine paint brush (size 000) to pick up and deposit individual seed. Alternatively, when planting many seed of the same lot, one may suspend the seed in 0.1% sterile agar and use a Pasteur pipet to disperse the seed evenly over the soil surface. Another method to aid seed sowing is to add dry sand to the seed and sprinkle this sand–seed mix over the pots.
3. The growth stage at which one inoculates the plants is one of the most critical factors favoring the recovery of large numbers of transformants ***(8,12)***. One needs to balance the need for rapid production vs high yield. Rapid production was discussed earlier (*see* **Note 1**). For maximum seed yield, one may produce large plants yielding more seed by planting single seed in large pots (i.e., one plant per 3.5-in.^{2} pot) and growing under short day lengths. If one is faced with using an older plant, opened flowers and already developed siliques may be removed just after (or before) inoculation, as these will not produce transformed seed ***(8)***.
4. The precise manner in which *A. tumefaciens* is cultured is not critical, as long as one takes care to grow it in the presence of the selectable marker associated with the binary vector. However, it is recommended to seed liquid cultures with an abundance of cells (i.e., growing a small culture first to use as seed for the larger culture). It is also recommended to subject *Agrobacterium* to as little antibiotic pressure as possible. Usually, the only antibiotic that is required is the one for selection of the binary plasmid (the plasmid carrying the T-DNA). For example, it is not necessary to grow in the presence of antibiotics such as rifampicin that are used to restrict growth of possible contaminants, as long as normal sterile techniques are used when culturing. Likewise, there is no need to add antibiotics for maintenance of a disarmed Ti-plasmid, as these large plasmids are usually stable without antibiotic pressure. It is recommended that researchers avoid the use of tetracycline as the selectable marker of their binary vector if using a C58-derived *A. tumefaciens* strain such as GV3101. These strains contain a native, chromosomally located Tet-resistance gene that is normally repressed, but growth on tetracycline relieves this repression ***(13)***. Because the *Agrobacterium* Tet resistance gene product is highly efficient, these spontaneous Tet-resistant cells will rapidly take over a culture, and the binary will be lost leading to little or no transformation. Another note is that *Agrobacterium* will transcribe off the 35S promoter (however, *Escherichia coli* will not), allowing kanamycin to be used to maintain the binary vector if the vector contains the *nptII* gene under control of the 35S promoter. Many binary vectors used for *Arabidopsis* transformation have this construct as the 35S*:nptII* allows for kanamycin selection of the transgenic plants. In addition, there in no need to add acetosyringone or any other inducer of the *vir* regulon ***(12)***. Cell density also appears to not be too critical for successful transformation. However, it is recommended to grow cells to stationary phase, as stationary-phase cells perform somewhat better than log-phase cells ***(12)***, and it is simpler to grow a culture to stationary phase than to try to achieve an exact cell density.
5. For some plants, especially thin weak plants, 0.03% Silwet L77 may be too toxic. If surfactant toxicity is a concern, one should lower the concentration to 0.02%.

As an alternative to Silwet L77, some laboratories (not tried by the author) have used other surfactants such as 0.02% Tween-20 or Triton X-100 with acceptable results.

6. The surfactant must be thoroughly mixed into solution after it is added to the bacterial suspension. A long stirring rod works well for this task. If not mixed well, the viscous surfactant will rest at the bottom of the beaker instead of going into solution. If a plant florescence touches a high concentration of unmixed surfactant, it will rapidly wilt and become necrotic within 1 or 2 d after inoculation.
7. The exact cell density of the inoculum is not critical and can vary from an OD_{600} of 0.1–2.0 ***(12)***. As long as the *Agrobacterium* cultures were healthy, grown from a fresh overnight liquid culture, and not under multiple antibiotic pressures, the cultures should be able to reach stationary phase (OD_{600} of approx 1.0 or above) within 24 h. A 1:10 dilution of these cells should transform well.
8. Repeat dip inoculations have been shown to increase the percentage of transformants recovered; however, repeat inoculations are somewhat harmful to the plants ***(12)***. Therefore, repeat inoculations should not be more frequently than once a week. Some laboratories prefer to use vacuum infiltration to aid infiltration of bacteria into the buds. However, the rates of transformation are not necessarily any higher and vacuum infiltration is messier, more time consuming, and harsher on the plants. In addition, the total number of seed recovered from vacuum infiltrated plants will most likely be lower than that recovered from plant inoculation by dipping, as dipped plants are less damaged by inoculation. If one wishes to vacuum infiltrate, it is highly recommended to use screening material over the soil at time of sowing (*see* **Subheading 3.1.**) and that one use a beaker–pot combination that allows the pot to rest on the top of the beaker during inoculation such that the plant is fully submerged but the soil is not. Another alternative inoculation method is to apply the *Agrobacterium* solution by fine aerosol sprays. Transformation rates can occur at about the same efficiency by spraying the inoculum onto flowering plants as what can be recovered by dipping ***(12)***. Spray application is the most practical means to inoculate large numbers of plants, such as for the production of large collections of T-DNA tagged lines. Wear skin and eye protection.
9. To sterilize by chlorine gas, pour the seed into labeled microfuge tubes. Common laboratory markers tend to bleach out making labels difficult to read after sterilization. Therefore, one should either pretest the labeling marker for fading, or label the tubes with stickers and pencil. Working in a chemical exhaust hood, place tubes (lids open) into small racks inside a large bell jar. Wear protective gear for eyes and skin and pour 100 mL of bleach into a 250-mL beaker. Place the beaker in the center of the bell jar. Carefully trickle 3.5 mL of concentrated HCl along the inner walls of the beaker. As soon as the HCl is added, seal the bell jar and allow sterilization to proceed for 4–16 h in the chemical exhaust hood. After sterilization, quickly open the bell jar within the chemical exhaust hood, remove the beaker of bleach–HCl and place it out of the way in the exhaust hood. Quickly reseal the bell jar to minimize risk of contamination and bring to a sterile hood.

Quickly open the bell jar, place the racks of open tubes of seed on the sterile bench, reseal the bell jar, and return the bell jar immediately to the chemical exhaust hood. Open the bell jar and allow it to vent in the chemical exhaust hood for several days to remove residual fumes and to minimize the exposure of the lab to chlorine gas. The spent bleach–HCl should be poured slowly down the drain along with plenty of water. Allow the chlorine-treated seed to degas for 30 min or more in the sterile hood. To plate, shake the seed onto a 150 × 15 mm selection plate containing a 3-mL puddle of 0.1% agar. Rinse the microfuge tube with 1 mL of 0.1% agar, and evenly disperse the seed–0.1% agar across plate with the aid of a sterile plastic inoculating loop or glass bacterial spreader.

10. The *bar* gene provides resistance to the herbicide phosphinothricin. To use phosphinothricin (also known as glufosinate ammonium, Basta, Liberty, or Ignite) as the selectable marker, spread the seeds (nonsterile) over wet soil, cover, cold treat 3 d, and grow under normal conditions. Spray the plantlets shortly after they emerge with 300 μ*M* phosphinothricin and then again every 2 or 3 d until transformants are easily recognizable among nontransformed dead/dying plants.
11. Heavy watering promotes transplanting success. Start by making an approx 1-in. deep slot in the center of each pot of soil (use one 2-in.^{2} pot per transplant). Place the plant root into this slot such that the leaves are at the soil line. Using the force of water from a squirt bottle, encircle the plantlet with water until the soil collapses around the plantlet and the soil is saturated.
12. Comparison of *Arabidopsis* to plants that have shown to be resistant to transformation via floral inoculation provides some insight regarding which factors determine successful transformation. One such plant that is difficult to transform via flower inoculation is soybean. At the University of Illinois, we screened more than 100,000 seeds from *Agrobacterium*-inoculated soybean plants and did not recover a single transformed seed (Clough, Desfeux, Bent, Maughn, and Vodkin, unpublished). The soybean flower has several differences compared to the *Arabidopsis* flower. One observation is that the *Arabidopsis* flower produces a vase-like gynoecium that develops over a 2-wk period and remains open until the stigma seals it off a few days before anthesis. This developing ovary is loosely wrapped by the petals and sepals. It appears to be easy for the *Agrobacterium* to slip pass the protective folds provided by the petals and sepals to enter the opened, developing ovary. Desfeux et al. showed that there is a 3- to 4-d window of opportunity that begins about 6 d prior to anthesis ***(8)***. Once inside the ovary, the *Agrobacterium* transforms the ovules, or the ovule chromosomes, prior to or during fertilization ***(6–8)*** and there are many potential targets as each locule can contain about 20–30 ovules. In soybean, the flower develops quite differently. These differences include: the gynoecium does not develop as a vase (but somewhat like a hotdog bun); the petals and sepals form a very tight, waxy seal over the ovary; there are only four possible ovules per fruit; the ovary is sealed about 10 d before anthesis; and the flowers abort with little irritation ***(8,14,15)***. In plants such as soybean, methods need to be developed to

have the *Agrobacterium* pass the tightly folded petals and sepals and into the ovary without provoking abortion. One possible strategy to overcome these barriers would be to hand-inoculate flowers with a fine needle. Although this method might ensure the inoculum penetrates the sepal–petal barrier, it is labor intensive and may lead to a high abortion rate. The use of vacuum and the addition of a higher concentration of surfactants might aid the entry of inoculum into an immature flower, but these treatments might also lead to more aborted flowers owing to the extra irritation. Another possible strategy would be to add an antioxidizing agent to reduce the plant defense response against *Agrobacterium*. Olhoft et al. showed that the antioxidants cysteine and dithiothreitol were very effective at increasing the transformation events in soybean using tissue culture methods ***(16)***. Although researchers are experiencing difficulties adapting the germline protocol for the transformation of crop plants, it remains a worthy challenge to continue the efforts (*see* Chapter 7).

Disclaimer

Mention of trade names or commercial products in this publication is solely for the purpose of providing specific information and does not imply recommendation or endorsement by the US Department of Agriculture.

Acknowledgments

The author would like to thank Joseph Nicholas for proofreading this chapter and for offering helpful suggestions.

References

1. Phillips, R. L., Kaeppler, S. M., and Olhoft, P. (1994) Genetic instability of plant tissue cultures: breakdown of normal controls. *Proc. Natl. Acad. Sci. USA* **91,** 5222–5226.
2. Singh, R. J. (2003) Chromosomal aberrations in cell and tissue culture, in *Plant Cytogenetics*, (Singh, R. J., ed.), CRC Press, Boca Raton, FL, pp. 307–326.
3. Feldmann, K. A. and Marks, M. D. (1987) *Agrobacterium* mediated transformation of germinating seeds of *Arabidopsis thaliana*: a non-tissue culture approach. *Mol. Gen. Genet.* **208,** 1–9.
4. Bechtold, N., Ellis, J., and Pelletier, G. (1993) *In planta Agrobacterium* mediated gene transfer by infiltration of adult *Arabidopsis thaliana* plants. *C. R. Acad. Sci. (Paris) Life Sci.* **316,** 1194–1199.
5. Mollier, P., Montoro, P., Delarue, M., Bechtold, N., Bellini, C., and Pelletier, G. (1995) Promoterless *gusA* expression in a large number of *Arabidopsis thaliana* transformants obtained by the *in planta* infiltration method. *C. R. Acad. Sci. (Paris) Life Sci.* **318,** 465–474.
6. Bechtold, N., Jaudeau, B., Jolivet, S., et al. (2000) The maternal chromosome set is the target of the T-DNA in the *in planta* transformation of *Arabidopsis thaliana*. *Genetics* **155,** 1875–1887.

7. Ye, G. N., Stone, D., Pang, S. Z., Creely, W., Gonzales, K., and Hinchee, M. (1999) *Arabidopsis* ovule is the target for *Agrobacterium in planta* vacuum infiltration transformation. *Plant J.* **19,** 249–257.
8. Desfeux, C., Clough, S. J., and Bent, A. F. (2000) Female reproductive tissues are the primary target of *Agrobacterium*-mediated transformation by the *Arabidopsis* floral-dip method. *Plant Physiol.* **123,** 895–904.
9. Koncz, C. and Schell, J. (1986) The promoter of the T_L-DNA gene 5 controls the tissue-specific expression of chimaeric genes carried by a novel type of *Agrobacterium* binary vector. *Mol. Gen. Genet.* **204,** 383–396.
10. Bent, A. F., Kunkel, B. N., Dahlbeck, D., et al. (1994) *RPS2* of *Arabidopsis thaliana:* A leucine-rich repeat class of plant disease resistance genes. *Science* **265,** 1856–1860.
11. Sambrook, J. and Russell, D. W. (2001) *Molecular Cloning: A Laboratory Manual.* 3rd Ed., Vol. 1., Cold Spring Harbor Laboratory Press, Cold Spring Harbor, NY.
12. Clough, S. J. and Bent, A. F. (1998), Floral dip: a simplified method for *Agrobacterium*-mediated transformation of *Arabidopsis thaliana. Plant J.* **16,** 735–743.
13. Luo, Z.-Q., Clemente, T. E., and Farrand, S. K. (2001) Construction of a derivative of *Agrobacterium tumefaciens* C58 that does not mutate to tetracycline resistance. *Mol. Plant Microbe Interact.* **14,** 98–103.
14. Johns, C. W. and Palmer, R. G. (1982) Floral development of a flower-structure mutant in soybeans, *Glycine max (L.)* Merr. (Leguminosae). *Amer. J. Bot.* **69,** 829–842.
15. Guard, A. T. (1931) Development of floral organs of the soy bean. *Bot. Gaz.* **91,** 97–102.
16. Olhoft, P. M. and Somers, D. A. (2001) L-cysteine increases *Agrobacterium*-mediated T-DNA delivery into soybean cotyledonary-node cells. *Plant Cell Rep.* **20,** 706–711.

7

Production of Transgenic Crops by the Floral-Dip Method

Ian S. Curtis

Summary

The application of floral dipping toward the production of transformed plants has been rather limited. However, this procedure has enabled the successful production of transformed *Medicago truncatula* plants (a model plant for legume genetics) at efficiencies higher than those obtained by tissue culture methods. Indeed, this simple system, without requiring any knowledge of plant tissue culture, has been a breakthrough in the production of the first transgenic radish plants. This root crop is of major importance in the Far East, and the development of such a gene transfer system in radish has enabled agronomically important germplasms to be produced. Although the radish is closely related to *Arabidopsis thaliana*, it appears the two plants have different mechanisms of T-DNA transfer using floral dip. This chapter describes the simple system that has been adopted in the routine production of transgenic radish.

Key Words: *Agrobacterium*-mediated transformation; floral dip; radish; *Raphanus sativus* L. var. *longipinnatus* Bailey; transgenic plants.

1. Introduction

The production of transgenic plants in tissue culture requires careful preparation of plant tissues, a procedure of transforming individual plant cells, and a screening system for selecting transformed plants. In some crops, such as a lettuce *(1)*, a well-established tissue culture system is available for the efficient production of transgenic plants at high transformation efficiencies (number of explants producing transformed shoots/total number of explants). However, in crops such as radish, it has been difficult to regenerate shoots from transformed cells of seedling explants. At present, there are very few species that can be routinely transformed in the absence of a tissue culture based regeneration sys-

From: *Methods in Molecular Biology, vol. 286: Transgenic Plants: Methods and Protocols*
Edited by: L. Peña © Humana Press Inc., Totowa, NJ

tem. One such plant, *Arabidopsis thaliana*, can be transformed by several *in planta* methods including vacuum infiltration *(2)*, transformation of germinating seeds *(3)*, and floral dipping *(4)* (*see* Chapter 6). These systems have contributed greatly toward the isolation and understanding the functions of many plant genes. In terms of *in planta* methods being applied to crop plants, there have been very few successful reports. The first successful report on the application of *in planta* transformation toward a crop was achieved in pakchoi (*Brassica rapa* L. ssp. *chinensis*) by the vacuum infiltration of an *Agrobacterium* suspension into flowering plants *(5)*. Despite confirming the transmission of the *bar* gene from the *Agrobacterium* into the progeny infiltrated plants, only two plants from 20,000 seeds sown yielded transformed plants (transformation efficiency of 0.0001%). However, in a separate study, the application of infiltration of seedlings or flowering plants could be used successfully in the production of transformed plants of the model legume, *Medicago truncatula (6)*. Here, the frequencies of the progenies producing transformed plants were variable but significantly higher compared to the pakchoi system (flower infiltration method, 4.7–76%; seedling infiltration, 2.9–27.6%) and also compared to *Arabidopsis* (0.1–3%) *(2,4)*. However, Southern blot analyses of these legume transformants revealed that these plants were a mixture of independent and sibling transformants (flower infiltration, 13–23% independent; seedling infiltration, 67–86%) and so the mechanism of transformation appears different in the two methods. In addition, as the flower infiltration method in *Medicago truncatula* produces a large number of sibling transformants and in *Arabidopsis* the plants are usually independent hemizygotes for the transgene, it appears that the two species have different mechanisms of transgene integration.

Here, a more recent *in planta* transformation method, floral-dipping, is described in the production of stably transformed plants of radish *(7)*. This procedure is simple, reproducible, and has been used successfully in the production of agronomically useful germplasms of this crop *(8)*.

2. Materials

2.1. Plant Material

1. Seeds of Korean variety "Jin Ju Dae Pyong" (Kyoungshin Seeds Co., Ltd., Seoul, South Korea).

2.2. Agrobacterium tumefaciens *Strain*

1. *Agrobacterium tumefaciens* strain AGL1 *(9)* with pCAMBIA3301 is used in all plant transformation work (*see* **Note 1**). The vector has both the selectable marker *bar* and reporter *gus*A genes located between T-DNA borders under the control of the CamV 35S promoter. The binary vector is transferred into AGL1 by electroporation.

2.3. Agrobacterium *Culture Media*

1. Yeast-Extract Peptone (YEP) medium: 10 g/L of tryptone (Difco, Detroit, MI), 10 g/L of yeast extract (Difco), 5 g/L of NaCl.
2. Kanamycin sulfate (Sigma, St. Louis, MO): 50 mg/mL of stock in water. Sterilize by filtration through a 0.2-µm membrane. Store at –20°C for 3–4 mo.
3. Rifampicin (Sigma): 4 mg/mL of stock dissolved in methanol. Sterilize by filtration; store at –20°C for 3–4 mo.
4. Liquid culture medium: YEP medium supplemented with 50 mg/L of kanamycin and 50 mg/L of rifampicin.
5. Agar-solidified medium: YEP medium with 14 g/L of agar (Bacto-agar, Difco), 50 mg/L of kanamycin, and 100 mg/L of rifampicin.

2.4. Other Solutions

1. GUS assay buffer: 10 m*M* $Na_2EDTA \cdot H_2O$, 0.1% Triton X-100, 0.1% 5-bromo-4-chloro-3-indolyl-β-D-glucuronide (X-Gluc), 0.1 *M* NaH_2PO_4, 0.5 *M* $K_3Fe(CN)_6$.
2. Floral dip inoculation medium (IM): 5% (w/v) sucrose, 0.05% (v/v) Silwet L-77 (Osi Specialities, Danbury, CT) dissolved in water, pH 5.2.

3. Methods

3.1. Preparation of Agrobacterium *for Floral Dipping*

1. From glycerol stocks, *A. tumefaciens* strain AGL1 carrying pCAMBIA3301 is streaked onto YEP agar-solidified medium containing antibiotics and cultured at 28°C for 2–3 d in the dark.
2. Using a platinum loop, a single bacterial colony is chosen to inoculate 10 mL of YEP liquid medium with selection contained in a 50 mL-capacity Falcon tube and incubated overnight on an orbital shaker at 180 rpm.
3. The culture is transferred to a conical flask containing 1 L of YEP medium with 50 mg/L of kanamycin and grown overnight as described.
4. Measure the optical density (OD) of the bacterial culture by spectrophotometry. An OD of 0.7–0.8 at a wavelength of 600 nm is sufficient for the transformation of radish.
5. The liquid culture is decanted into 300-mL-capacity sterile centrifuge tubes and centrifuged in a Beckman Centrifuge (20 min, 4800*g*, room temperature). Discard the supernatant by decanting into a container containing 1% domestic bleach (v/v) to kill residual bacteria.
6. Resuspend the pellet in the centrifuge tubes using 1 L of inoculation medium by gentle agitation (*see* **Note 2**).

3.2. Production of Radish Plants for Transformation

1. Seeds are sown in a soil mix of Vermiculite, Perlite, TKS2 soil-based compost, and peat (12:4:3:1; v/v) in a glasshouse under natural daylight supplemented with 61 $\mu E/m^2s$ daylight fluorescent illumination (26°C day/18°C night).

2. Plants at the four-leaf stage of development are transferred to (20-cm diameter, 30-cm high) pots containing Vermiculite, Perlite, TKS2 soil-based compost, and peat (2:1:1:2) to allow the plant to grow to maturity (one plant per pot).
3. After 7–14 d (six-leaf stage of development), plants are vernalized in a cold chamber (4 ± 2°C, 16-h photoperiod, 45 µE/m^{2}s daylight fluorescent tubes) for 10 d to initiate bolting. Cold-treated plants are transferred to the glasshouse and allowed to bolt (*see* **Note 3**).
4. Plants should be watered with a general-purpose liquid feed at fortnight intervals to ensure the plants grow healthy (*see* **Note 4**).

3.3. Floral Dipping Procedure

1. Prior to dipping, all siliques, flowers, and floral-buds showing petal color are removed from bolted plants using fine-pointed scissors (*see* **Note 5**).
2. Combine the suspended *Agrobacterium* cultures (suspended in IM; *see* **Subheading 3.1.**) into either a 1- or 2-L measuring cylinder or beaker and carefully invert the potted plant into the infection medium so that the inflorescence is completely submerged (*see* **Note 6**).
3. Gently mix the inflorescence of the plant in IM for 5 s and then remove the plant from the bacterial suspension and return to the upright position. Insert a clear polyethylene bag immediately over the dipped inflorescence, removing as much of the air inside the bag, and then tie just below the inflorescence.
4. Place the plants in a shaded area away from direct sunlight and allow them to remain overnight (*see* **Note 7**).
5. Remove the bag, and then allow the plants to flower. For successful seed production, each flower needs to be hand pollinated with a fine paint brush for 2–3 consecutive days at a time when the pollen is free flowing (around midday) (*see* **Note 8**).
6. Seeds should be harvested from siliques when they start to dry. Seeds should be incubated at 30°C for at least 10 d to aid ripening before they can be sown.

3.4. Selection of Transformed Plants

1. Seeds from floral-dipped plants are sown to soil and the origin of where the seeds were collected is noted (which plant dipped, location of bolt) (*see* **Note 9**).
2. Fifteen days after sowing, the plants are sprayed with 0.03% (v/v) Basta® at weekly intervals for 3 wk to select those plants that are resistant to the herbicide.
3. Plants that show no necrosis on the leaves after the third spraying have a functional *bar* gene. To confirm further the transgenicity of these herbicide-resistant plants, leaf pieces are excised and analyzed by GUS histochemical staining. GUS assays are performed by immersing tissues in assay buffer. After overnight incubation at 37°C, the tissues are immersed in 95% ethanol (1–2 h) and then stored in 70% ethanol for 2–5 d (*see also* Chapter 14).
4. From thousands of plants analyzed for GUS expression, plants that exhibit leaf necrosis (including wild-type) after herbicide spraying fail to show GUS expression and plants that are resistant to Basta® are all positive by GUS assays.

5. Herbicide-resistant plants are transferred to pots and allowed to grow to the six-leaf stage of development, at which time the plants are vernalized to promote bolting as described (*see* **Subheading 3.2., step 3,** and **Note 10**).
6. At the time of flowering, the plants must be hand pollinated to promote successful seed set. The seeds are harvested, ripened, and then sown to soil and the resulting plants are screened for transgene activity (T2 generation).
7. Seeds from these transformed T2 lines are collected and sown to determine which lines are homozygous (all plants are herbicide resistant) and heterozygous (usually 75% of plants are Basta® resistant) for the transgenes.

4. Notes

1. Experiments conducted in culture have shown that *A. tumefaciens* strain AGL1 is virulent toward explants of Korean cultivar "Jin Ju Dae Pyong" and so this bacterium was used in such floral-dip studies.
2. Suspending the bacterial pellet by vigorous shaking should be avoided, as it may harm the efficiency of the bacteria to transform the plant.
3. In earlier studies, radish plants of various stages of bolt development (primary bolt, single stem, 3–9 cm height; secondary bolt, 10–15 cm; tertiary bolt, 16–24 cm) were dipped into a suspension of *Agrobacterium* to determine whether plant development stage is a critical factor in the optimization of transformation efficiency. Plants that exhibit a single stem with many immature floral buds at the time of dipping are the most responsive in terms of the production of transformed seeds. Optimal transformation efficiency (1.4%) is achieved by dipping such plants into a bacterial suspension containing 5% sucrose and 0.05% Silwet L-77. Secondary bolted plants are also amenable to transformation but the yield of transformed seeds is considerably lower (0.2% optimum). Later developing bolted plants usually fail to produce transformed seeds.
4. A liquid fertilizer applied to cold-treated plants at 2-wk intervals helps to prevent leaf chlorosis, which is often seen approx 7 d after the plants are transferred back to the glasshouse. The feed also allows the plant to produce a strong thick stem from which many floral buds develop.
5. The removal of siliques, flowers, and floral buds with petal color at this stage helps to reduce the number of harvested nontransformed seeds. In addition, such wounding within the inflorescence aids the bacteria to infect the plant.
6. The presence of a surfactant in the IM is critical for the successful transformation of immature flowers of radish. Silwet L-77, a trisiloxane, in the presence of 5% sucrose is the most effective surfactant treatment in the production of transformed seeds compared to nonionic detergents such as Pluronic F-68 and Tween-20. The use of Silwet L-77 at 0.05% is a significantly improved treatment compared to concentrations used at 0.01 and 0.1%. In the absence of a surfactant, no transformed seeds are produced when inoculated with *Agrobacterium* in the presence of sucrose.
7. Keeping the infected plant in the shade helps prevent the bag from overheating, which would have a strong negative effect on plant transformation. Removal of

air in the bag allows the bacteria to move around the inflorescence and so increase the area capable of being infected by the bacterial agent.

8. The radish variety "Jin Ju dae Pyong" when grown in a glasshouse free of pollinating insects will not produce seeds owing to a sexual incompatibility problem. The use of a fine paintbrush to hand-pollinate the flowers is essential for the production of seeds and subsequently the production of transformed plants.
9. To be able to understand the mechanism of how *Agrobacterium* infects the inflorescence of radish, seeds from floral-dipped plants must be categorized in terms of where the seeds originated from the infected plants. Southern blot analyses have revealed that between 50 and 60% of all transformed plants were siblings for the transgene (had the same T-DNA insertion pattern) and that such populations of siblings originated from siliques from the same bolt or inflorescence stem. The case of *Arabidopsis*, which usually produces independently transformed seeds, suggests that radish and *Arabidopsis* follow different mechanisms of how the developing seeds are genetically transformed.
10. A few days prior to cold treatment, leaves from putative transformed plants are harvested for molecular analysis. Southern analysis is used to understand both the insertion pattern of the transgenes (independent or sibling transformant) and also to determine copy number (usually one or two copies).

Acknowledgments

I wish to thank Dr. Hong Gil Nam for inviting me to conduct these studies at the Pohang University of Science and Technology (POSTECH), Republic of Korea. This work was supported in part by the Korean Federation of Science and Technology Societies (KOFST).

References

1. Curtis, I. S., Power, J. B., Blackhall, N. W., de Laat, A. M. M., and Davey, M. R. (1994) Genotype-independent transformation of lettuce using *Agrobacterium tumefaciens*. *J. Exp. Bot.* **45,** 1441–1449.
2. Bechtold, N., Ellis, J., and Pelletier, G. (1993) *In planta Agrobacterium*-mediated gene transfer by infiltration of adult *Arabidopsis thaliana* plants. *C. R. Acad. Sci. (Paris) Life Sci.* **316,** 1194–1199.
3. Feldmann, K. A. and Marks, M. D. (1987) *Agrobacterium*-mediated transformation of germinating seeds of *Arabidopsis thaliana*: a non-tissue culture approach. *Mol. Gen. Genet.* **208,** 1–9.
4. Clough, S. J. and Bent, A. F. (1998) Floral dip: a simplified method for *Agrobacterium*-mediated transformation of *Arabidopsis thaliana*. *Plant J.* **16,** 735–743.
5. Qing, C. M., Fan, L., Lei, Y., et al. (2000) Transformation of Pakchoi (*Brassica rapa* L. ssp. *chinensis*) by *Agrobacterium* infiltration. *Mol. Breed.* **6,** 67–72.
6. Trieu, A. T., Burleigh, S. H., Kardailsky, I. V., et al. (2000) Transformation of *Medicago truncatula* via infiltration of seedlings or flowering plants with *Agrobacterium*. *Plant J.* **22,** 531–541.

7. Curtis, I. S. and Nam, H. G. (2001) Transgenic radish (*Raphanus sativus* L. var. *longipinnatus* Bailey) by floral-dip method—plant development and surfactant are important in optimizing transformation efficiency. *Trans. Res.* **10,** 363–371.
8. Curtis, I. S., Nam, H. G., Yun, J. Y., and Seo, K.-H. (2002) Expression of an antisense *GIGANTEA* (*GI*) gene fragment in transgenic radish causes delayed bolting and flowering. *Trans. Res.* **11,** 249–256.
9. Lazo, G. R., Stein, P. A., and Ludwig, R. A. (1991) A DNA transformation-competent *Arabidopsis* genomic library in *Agrobacterium*. *Bio/Technol.* **9,** 963–967.

8

Chloroplast Genetic Engineering to Improve Agronomic Traits

Henry Daniell, Oscar N. Ruiz, and Amit Dhingra

Summary

Major crop losses occur annually as a result of biotic and abiotic stresses. The ability to hyperexpress foreign proteins, single-step multigene engineering, lack of positive effect and gene silencing, vector sequences and pleiotropic effects have resulted in several hundred-fold more tolerance to the environmental stresses via chloroplast genetic engineering than nuclear genetic engineering. Maternal inheritance of chloroplast expressed transgenes renders the technology environmentally safe and promotes public acceptance. This review provides protocols for engineering agronomic traits like insect, herbicide and disease resistance; salt and drought tolerance; and phytoremediation via chloroplast genome.

Key Words: Disease resistance; drought tolerance; GM crops; herbicide resistance; maternal inheritance; pest resistance/management; phytoremediation; plastid transformation; transgene containment.

1. Introduction

In the postgenomic era, nuclear and chloroplast genomes have been the targets of genetic manipulation to enhance the agronomic traits or introduce value-added traits into crop plants. Among several approaches, engineering the chloroplast genome is emerging as a successful approach. Manipulation of the chloroplast genome has become routine in the model system tobacco and has been extended to other edible solanaceous crops (potato and tomato). Expression of the transgenes in this organelle offers unique advantages that render this technology safe and acceptable to the public. Gene containment is the most notable advantage that this technology offers because chloroplast genomes are

From: *Methods in Molecular Biology, vol. 286: Transgenic Plants: Methods and Protocols*
Edited by: L. Peña © Humana Press Inc., Totowa, NJ

Table 1
Biopharmaceutical Proteins Expressed Via the Chloroplast Genome

Biopharmaceutical proteins	Gene	Site of integration	Promoter	5'/3' regulatory elements	% tsp expression	Ref.
Elastin derived polymer	*EG121*	trn*I*/trn*A*	Prrn	*T7*gene*10*/*T*psbA	ND	***9***
Human somatotropin	hST	trn*V*/rps*12/7*	Prrn[a], PpsbA[b]	*T7*gene*10*[a] or psbA[b]/*T*rps*16*	7.0% [a] and 1.0% [b]	***18***
Cholera toxin	*CtxB*	trn*I*/trn*A*	Prrn	Ggagg/*T*psbA	4%	***19***
Antimicrobial peptide	MSI-*99*	trn*I*/trn*A*	Prrn	Ggagg/*T*psbA	Not tested	***14***
Insulin like growth factor	IGF-*1*	trn*I*/trn*A*	Prrn	*P*psb*A*/*T*psbA	33%	***20***
Interferon-α 5	INFα5	trn*I*/trn*A*	Prrn	*P*psb*A*/*T*psbA	ND	***21***
Interferon-α 2b	INFα2B	trn*I*/trn*A*	Prrn	*P*psb*A*/*T*psbA	19%	***22***
Human serum albumin	hsa	trn*I*/trn*A*	Prrn[a], PpsbA[b]	ggagg[a], psb*A*[b]/*T*psbA	0.02% [a], 11.1% [b]	***23***
Interferon-γ	IFN-γ	rbc*L*/acc*D*	*P*psb*A*	*P*psb*A*/*T*psbA	6%	***24***
Monoclonal antibodies	Guy's 13	trn*I*/trn*A*	Prrn	Ggagg/*T*psbA	ND	***25***
Anthrax protective antigen	Pag	trn*I*/trn*A*	Prrn	*P*psb*A*/*T*psbA	4–5%	***26***
Plague vaccine	Ca*F1*–Lcr*V*	trn*I*/trn*A*	Prrn	*P*psb*A*/*T*psbA	4.6%	***27***
Canine parvovirus vaccine	CPVVP2	trn*I*/trn*A*	Prrn	*P*psb*A*/*T*psbA	31.1%	***41***

Table 2
Agronomic Traits Engineered Via the Chloroplast Genome

Agronomic traits	Gene	Site of integration	Promoter	5'/3' Regulatory elements	Ref.
Insect resistance	*Cry1A*(c)	trn*V*/rps12/7	Prrn	rbc*L*/*T*rps*16*	***17***
Herbicide resistance	*CP4* (petunia)	rbc*L*/acc*D*	Prrn	ggagg/*T*psb*A*	***8***
Insect resistance	*Cry2Aa2*	rbc*L*/acc*D*	Prrn	ggagg (native)/*T*psb*A*	***10***
Herbicide resistance	bar	rbc*L*/acc*D*	Prrn	rbc*L*/*T*psbA	***15***
Insect resistance	*Cry2Aa2* operon	trn*I*/trn*A*	Prrn	Native 5-UTRs/*T*psb*A*	***11***
Disease resistance	*MSI*-99	trn*I*/trn*A*	Prrn	ggagg/*T*psbA	***14***
Drought tolerance	tps	trn*I*/trn*A*	Prrn	ggagg/*T*psbA	***16***
Phytoremediation	mer*A*[a]/mer*B*[b]	trn*I*/trn*A*	Prrn	ggagg[a, b]/*T*psbA	***12***
Salt tolerance	badh	trn*I*/trn*A*	Prrn	*T7*gene*10*/*T*rps*16*	***39,40***

maternally inherited in most plant species *(1,2)*. In addition, chloroplast engineering overcomes the challenges of low-level expression, gene silencing, position effect, and multistep engineering of multiple genes, which are current limitations of nuclear transformation *(3,4)*. Chloroplast transformation has been used to express several biopharmaceutical proteins (**Table 1**) and engineer several agronomic traits (**Table 2**) that are detailed in this chapter. In addition, several reporter and selectable marker genes have been expressed via the chloroplast genome (**Table 3**).

1.1. Genome Organization and Principle of Chloroplast Transformation

The chloroplast genome is usually a circular molecule that is self replicating and varies in size from 120–220 kb among different plant species *(5)*. It is predominantly present as a single molecule, but recent cytogenomic analysis has revealed that structural organization of chloroplast DNA is highly dynamic. It is arranged in both linear and circular conformation with one to four copies of the genome *(6)*. A typical plant cell contains approx 100 chloroplasts and each chloroplast further harbors approximately 100 copies of the same genome. This implies that a single gene is represented by at least 10,000 copies in a single plant cell. Single chloroplast genomes of most plant species possess two inverted repeat regions and thus the copy number of the genes encoded by this region is approx 20,000. Therefore, it is quite appealing to introduce a transgene into the chloroplast genome and obtain high levels of expression, taking advantage of the high copy number. Indeed several sites of insertion, including some in the inverted repeat region, have been used for expression of foreign genes into the chloroplast genome (*see* **Tables 1** and **2**). Site-specific integration of transgenes into the chloroplast genome differs from random nuclear integration because chloroplast transgenes are integrated via homologous recombination. Chloroplast transformation vectors are thus designed with homologous flanking sequences on either side of the transgene and introduced into the chloroplast via particle bombardment or using protoplasts by polyethylene glycol (PEG) treatment. Transformation is accomplished by integration of the transgene into a few genome copies initially followed by approx 15–20 cell divisions under selection pressure, thereby yielding a homogeneous population of plastid genomes. If the gene is introduced in the inverted repeat (IR) region, integration in one inverted repeat is followed by the phenomenon of copy correction that duplicates the introduced transgene into the other inverted repeat as well (**Fig. 1**). Chloroplast transformation vectors may also carry an origin of replication that facilitates replication of the plasmid inside the chloroplast, thereby increasing the templates to be presented for homologous recombination and consequently enhancing the probability of transgene integration *(7–9)*.

Table 3
Reporter and Selectable Marker Genes Expressed Via the Chloroplast Genome

Reporter/selectable genes	Organism	Ref.
cat—chloramphenicol acetyl transferase	a. Cucumber etioplasts^T b. Cultured tobacco cells^T	a. ***28*** b. ***7***
uid*A*—β-glucuronidase	Wheat leaves and calli^T	***29***
aad*A*—aminoglycoside adenylyl transferase	a. Chlamydomonas^S b. Tobacco^S	a. ***30*** b. ***31***
npt*II*—neomycin phosphotransferase	Tobacco^S	***32***
aph*A-6*—aminoglycoside phosphotransferase	a. Chlamydomonas^S b. Tobacco^S	a. ***33*** b. ***34***
gfp—green fluorescent protein	a. Tobacco and Arabidopsis^T b. Potato^S	a. ***35*** b. ***36***
aad*A*–gfp—fusion protein	Tobacco^S and rice^T	***37***
badh—betaine aldehyde dehydrogenase	Tobacco^S	***38***
hphs—hygromycin phosphotransferase (synthetic)	Tobacco^S	***42***

T, Transient expression; S, stable integration.

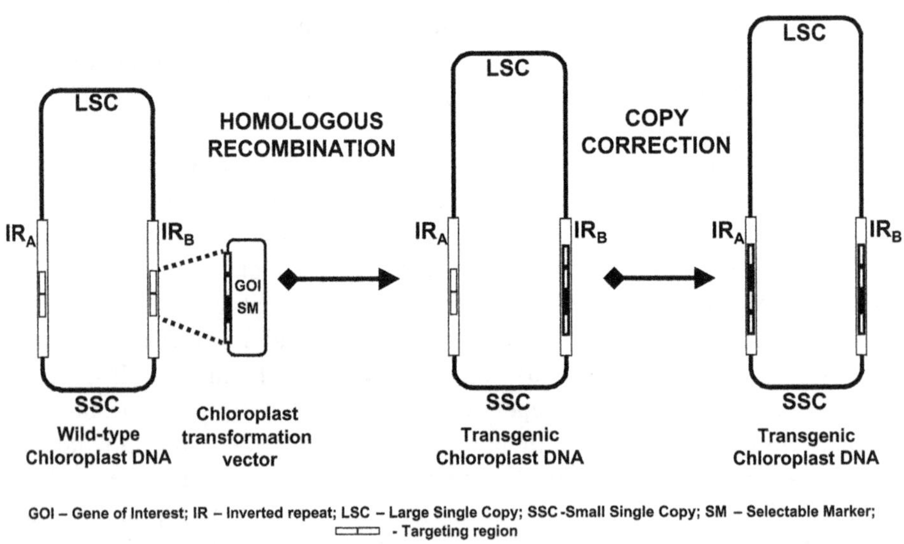

Fig. 1. Schematic representation of chloroplast transformation showing the phenomenon of homologous recombination and copy correction.

2. Materials

2.1. Chloroplast Transformation Vector Construction

1. Total cellular DNA from tobacco.
2. Primers designed to land on the chloroplast genome for the amplification of flanking sequences and the requisite regulatory elements such as promoters and 5'- and 3'-untranslated regions.
3. Pfu-based DNA polymerase and deoxynucleotide triphosphates (dNTPs) for polymerase chain reaction (PCR).
4. DNA modifying enzymes such as T4 DNA polymerase for producing blunt-ends, calf intestinal alkaline phosphatases to remove 5'- and 3'-phosphoryl groups from nucleic acids, and T4 DNA ligase to form phosphodiester bonds.
5. PCR cloning kit.

2.2. Tissue Culture and Particle Bombardment

1. Media: Murashige and Skoog (MS) salt mixture and Phytagar (Invitrogen, Carlsbad, CA), sucrose, myoinositol, benzylaminopurine, naphthaleneacetic acid (Sigma, St. Louis, MO).
2. Particle bombardment: particle gun—PDS 1000 He, microcarriers—gold or tungsten particles, macrocarriers, macrocarrier holders, rupture disks—1100 psi (Bio-Rad, Hercules, CA), calcium chloride—biotech grade, spermidine free base (Sigma). Working solutions of $CaCl_2$ and spermidine free base as well as the handling of consumables for gene bombardment not listed in this work are prepared as explained in Chapter 4.

2.3. Preparation of Tobacco Tissue Culture Media

1. MS medium: MS medium is used at several stages: for seed germination, plant propagation, and root induction after the second round of regeneration on selection medium. MS medium is prepared by mixing a 4.3-g packet of MS salts (Invitrogen) and 30 g of sucrose in a 1-L beaker and by adjusting the volume to 900 mL with distilled water. The pH is then adjusted to 5.8 with 1 *N* KOH. Finally, the volume is made up to 1000 mL. The solution is placed in a 2-L flask and 6 g of Phytagar is added at a final concentration of 0.6%. The solution is autoclaved for 20 min at 121 psi and allowed to cool to 40°C before any antibiotics are added to the medium. The growth medium is poured into deep Petri dishes, jars, or Magenta boxes.
2. RMOP medium: RMOP medium is used for shoot induction from tobacco leaves after they have been bombarded. This media contains phytohormones involved in the regeneration of shoots. The media is prepared by adding a 4.3-g packet of MS salts (Invitrogen Corp., NY), 30 g of sucrose, 100 mg of *myo*-inositol, 1 mL of a 100 mg/mL thiamine-HCl solution, 1 mL of a 100 mg/mL 6-benzyl-aminopurine (BAP) solution, and 100 μL of a 100 mg/mL naphthalene acetic acid (NAA) solution in a 1-L beaker. The volume is adjusted to 900 mL and the pH is adjusted to 5.8 with 1 *N* KOH. The volume is brought to 1000 mL with sterile distilled water or MilliQ grade water and Phytagar is added to the medium at a final concentration of 0.6%. The medium is autoclaved for 20 min at 121 psi in a 2-L flask (*see* **Note 1**).

Finally, when the temperature of the medium drops below 40°C, antibiotics may be added if the medium is to be used for selection. The medium is then poured into deep Petri dishes (100-mm × 25-mm).

3. Spectinomycin solution: The spectinomycin solution is prepared by dissolving 1 g of spectinomycin into 10 mL of distilled water at a final concentration of 100 mg/mL. Then, the solution is filter-sterilized under aseptic conditions under a laminar flow hood and stored at –20°C.

2.4. Preparation of Gold Particles Suspension

1. The gold particles to be used for bombardment may be prepared in advance, at least 1 d prior to bombardment.
2. Weigh out 50 mg of gold particles (0.6 μm) into a 1.5-mL centrifuge tube.
3. Add 1 mL of 100% ethanol (molecular biology grade), and vortex-mix for 3 min.
4. Pellet the gold particles by centrifuging the tube for 2–5 min at maximum speed in a benchtop microcentrifuge and then discard the supernatant.
5. Add 1 mL of 70% ethanol and vortex-mix for 2 min.
6. Incubate the tube for 15 min at room temperature. Mix the contents of the tube about three times during the incubation.
7. Centrifuge the tube at maximum speed for 3 min, and then discard the supernatant.
8. Add 1 mL of distilled water and vortex-mix for 1 min or until the particles are completely suspended.
9. Allow the particles to settle down for 1 min at room temperature and then centrifuge the tube at maximum speed for 2 min; discard the supernatant.
10. Repeat **steps 8** and **9** two additional times.
11. Add 50% (v/v) glycerol to the gold particles at a final concentration of 60 mg/mL.
12. Store the gold particles at –20°C until ready to use.

2.5. Working Solutions for the Plant Bioassays

1. 0.1 *M* Cacodylate buffer, pH 7.4: 2.5% glutaraldehyde, 2% paraformaldehyde, and 5 m*M* $CaCl_2$.
2. 0.05 *M* Glycine prepared in 1X phosphate-buffered saline (PBS).
3. 2% Glutaraldehyde diluted in 1X PBS.
4. 10 m*M* Phenylmercuric acetate (PMA) working solution diluted in dimethyl sulfoxide.
5. 0.5–5 m*M* Glyphosate.
6. 3% or 6% PEG (mol. wt. 8000, Sigma).

3. Methods

3.1. Amplification and Cloning of Flanking Sequences

Flanking sequences are required for homologous recombination. It is desirable, to amplify these with a polymerase that has proofreading properties. In our laboratory we use Pfu Turbo (Stratagene, La Jolla, CA). Set up the PCR as follows: DNA: 100–200 ng; buffer: 1X; dNTPs: 200–300 μ*M*; primer 1: 15–20 pmol; primer 2: 15–20 pmol; enzyme: 1–2.5 U. Make up the total volume to 50 μL with sterile Milli Q (Millipore, Bedford, MA) grade water. The PCR is carried out by denaturing the PCR mix at 94°C for 5 min followed by 30 cycles of denaturation at 94°C for 30 s, annealing at 55–60°C for 30 s, and extension at 72°C for 2–4 min as per the size of the expected PCR product. This is followed by an 8 to 10-min extension at 72°C. Treat the amplified DNA fragment with *Taq* polymerase in the presence of dATP to add A-overhangs that facilitate direct cloning of PCR products into pCR 2.1 cloning vector provided with the TOPO cloning kit (Invitrogen). For the addition of A-overhangs incubate the PCR product with 200 m*M* dATP and *Taq* DNA polymerase at 72°C for 10 min. For cloning the DNA fragment with modified ends into pCR 2.1 vector follow the manufacturer's protocol (Invitrogen). The cloned fragment containing the flanking sequences is derived from the PCR 2.1 vector by digesting with appropriate restriction enzymes and is then blunt-ended using T4 DNA polymerase following manufacturer's protocol (NEB). In brief, resuspend the DNA in 1X T4 DNA polymerase reaction buffer supplemented with 100 m*M* dNTPs. Add 1 U of T4 DNA polymerase per microgram of DNA and incubate for 15 min at 12°C. Stop the reaction by adding ethylenediaminetetraacetic acid (EDTA) to a final concentration of 10 m*M* and heating to 75°C for 20 min. The blunt-ended fragment is then ligated with the help of T4 DNA ligase to *Pvu*II-digested pBluescript II KS dephosphorylated with calf intestinal alkaline phosphatase (CIAP) as per the manufacturer's instructions (Promega, Madison, WI). For dephosphorylation, purify the digested plasmid DNA to be dephosphorylated by ethanol precipitation and resuspend the pellet in 40 μL of 10 m*M* Tris-HCl, pH 8.0. Set up the following reaction: DNA (up to 10 pmol of 5'-ends): 40 μL; CIAP 10X reaction buffer: 5 μL; diluted CIAP (0.01 U/mL): up to 5 μL. Make up the total volume to 50 μL with sterile Milli Q grade water (Millipore). Incubate the reaction at 37°C for 30 min and add another aliquot of diluted CIAP (equivalent to the amount used earlier) and continue incubation at 37°C for an additional 30 min. Finally add 300 μL of CIAP stop buffer. Extract with phenol–chloroform and ethanol precipitate the DNA by adding 0.5 vol 7.5 *M* ammonium acetate, pH 5.5, and 2 vol of 100% ethanol to the final aqueous phase.

3.1.1. Constructing the Chloroplast-Specific Expression Cassette

The basic chloroplast specific expression cassette is comprised of a promoter, selectable marker, and 5'/3' regulatory sequences to enhance the efficiency of transcription and translation of the gene (**Fig. 2**). The chloroplast-specific promoters and regulatory elements are amplified from the total cellular DNA using

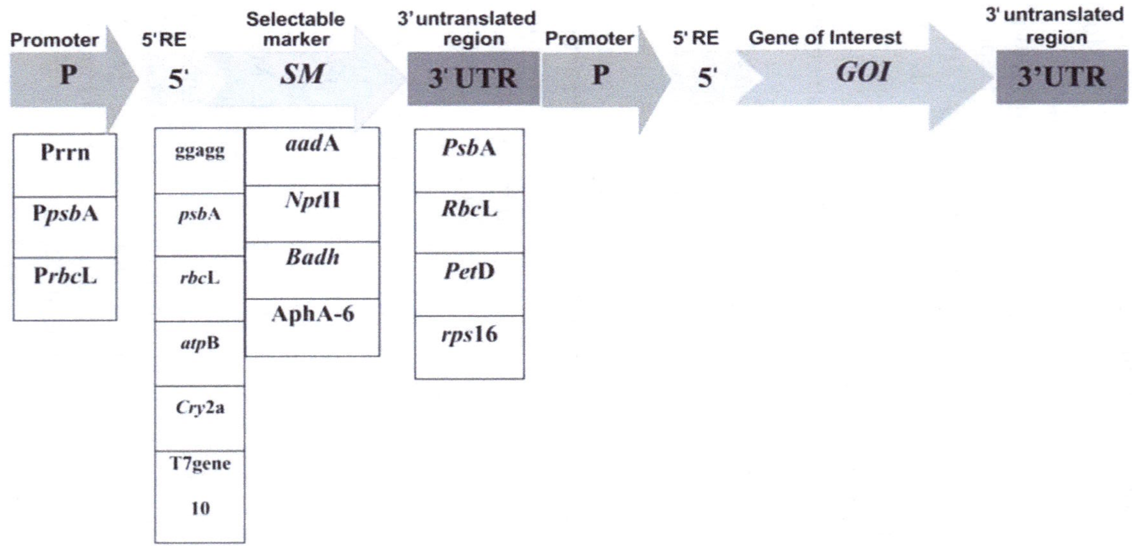

Fig. 2. Schematic representation of the chloroplast-specific expression cassette. For a list of regulatory elements and genes of interest used for chloroplast transformation, refer to **Tables 1** and **2**.

specific primers designed on the basis of the sequence information available for the chloroplast genome of tobacco (Accession no. NC_001879). The chloroplast specific expression cassette is cloned into a unique site in between the flanking sequences. Approximately 1 kb of homologous flanking regions is adequate to facilitate recombination. Care must be taken not to interrupt any genes while inserting the expression cassette. The site of insertion is therefore the intergenic spacer region.

3.2. Preparation for Bombardment

3.2.1. Preparation of Tobacco Tissues for Bombardment

The leaf material for particle bombardment is obtained from wild-type tobacco plants, which are generated in two ways. Plants may be generated from seeds by germination of seeds in MS medium. The seeds are germinated in a Petri dish and then individual seedlings are moved to jars or Magenta boxes containing MS medium. The time between seed germination and plant growth with the right size leaves is about 2 mo. The leaves to be used in bombardment need to be green, with no damage or defects, and with an approximate size of 2 in. × 1 in. Another way to generate the plant tissue is by propagating nodal sections. In this method, nodal segments of aseptically grown plants are transferred to jars containing MS medium. This system decreases the time to obtain leaves of adequate size (*see* **Note 2**).

3.2.2. Preparation of Consumables for Bombardment

Consumables to be used during bombardment should be prepared in advance. Autoclave the macrocarrier holders, stopping screens, filter paper, and Kimwipes. Sterilize the macrocarriers and rupture disks by submerging them in 100% ethanol for 10 min. Place the macrocarriers and rupture disks over autoclaved Kimwipes and dry them under the laminar flow hood.

3.3. Particle Bombardment of Tobacco Leaves

3.3.1. DNA Coating of the Gold Particles (see **Note 3**)

1. Vortex-mix the previously prepared gold particles that were stored at –20°C until they are completely resuspended.
2. Pipet out 50 μL of the gold particle suspension into a 1.5-mL microcentrifuge tube.
3. Add 10 μL of plasmid DNA that has a concentration of 1 μg/μL, and vortex-mix for 5 s.
4. Add 50 μL of freshly prepared filter-sterilized 2.5 *M* $CaCl_2$, and vortex-mix for 5 s.
5. Add 20 μL of 0.1 *M* spermidine free base and vortex-mix for 5 s.
6. Vortex the mixture for 20 min at 4°C.

7. Add 200 μL of room temperature absolute ethanol to the mixture, vortex for 5 s, and then centrifuge the mixture for 30 s at 800*g*. Remove the supernatant and repeat **step 7** four times.
8. After the final step, resuspend the pellet in 30 μL of 100% ethanol.
9. Keep the particles on ice until ready to use.

3.3.2. Tobacco Leaf Samples for Bombardment (see **Note 4**)

1. Take five fully expanded and undamaged green leaves from a young wild-type tobacco plant growing in a jar or a Magenta box.
2. Place an autoclaved Whatman filter disk on RMOP medium and a leaf on it with its adaxial side facing the medium. The abaxial side is the one that is to be bombarded (*see* **Note 5**).
3. Cover the Petri dish and repeat for all the leaves.

3.3.3. Macrocarrier Loading (see **Note 4**)

1. The macrocarrier is placed inside the macrocarrier holder with its concave side facing outward. Use the macrocarrier insertion tool to push the macrocarrier in place inside the holder.
2. Resuspend the gold particles completely by vortex-mixing and pipetting to eliminate any clumps. Aliquot 5 μL of the gold particles coated with the plasmid DNA (microcarrier) and spread it out in the center of the macrocarrier.
3. Repeat **steps 1** and **2** with additional macrocarriers based on the number of leaves to be bombarded.
4. DNA-coated gold particles remaining after loading all the macrocarriers may be used for a second application on the macrocarriers, already containing the gold particles. Alternatively, remaining gold particles coated with DNA could be run on agarose gels to test DNA binding efficiency.

3.3.4. Particle Bombardment (see **Notes 4** and **6**)

The particle bombardment procedure explained in Chapter 4 may be followed. Exceptions to this procedure are indicated as follows:

1. Using the adjustment handle of the helium regulator, rotate it clockwise until the pressure is set at 1350 psi.
2. Place a rupture disk (1100 psi) into the rupture disk retaining cap and rotate it into position in the gene gun tightly.

3.4. Tobacco Tissue Regeneration and Selection (see **Note 7**)

1. Incubate bombarded leaves in the dark for 48 h at 27°C.
2. After 48 h, cut the leaves into small pieces of about 25 mm^2 each with the help of a sterile scalpel blade.
3. Place the small pieces on agar containing RMOP medium (100 mm × 25 mm in a Petri dish) supplemented with suitable selection agent; this is called the first round

of selection. Place the bombarded (abaxial) side of the leaf in direct contact with the selection medium. RMOP medium induces shoot formation.

4. Incubate the leaf tissues at 27°C in a photoperiod of 16 h light and 8 h dark.
5. First putative transgenic shoots may be obtained within 4–6 wk after the bombarded leaf sections are placed on the selection medium.
6. Once the putative transgenic shoot obtained in the first round of selection starts developing leaves that are approx 1 cm^2 in size, small sections of these are placed on new RMOP medium supplemented with spectinomycin for second round of selection.
7. For the second round of selection, cut the leaf into pieces of about 4 mm^2 and place them on RMOP medium containing suitable antibiotic as the selection agent. Again, ensure that the abaxial side of the leaf is in direct contact with the medium.
8. These leaf sections will produce transgenic shoots in about 3–4 wk. Once the shoot develops into a small plantlet, it is detached from the leaf tissue and transferred to MS agar medium supplemented with spectinomycin. This step is termed as the third round of selection where root formation occurs.
9. After about 4–6 wk in the rooting medium, the plant is completely developed and ready to be moved to soil in pots.
10. To transfer the plant to soil, gently pull out the plant from the agar and wash the roots thoroughly with tap water to remove any agar attached to the roots.
11. Grow plants in high nutrient soil at appropriate temperature 16-h/8-h light/dark photoperiod in a growth chamber or a greenhouse.
12. Collect seeds from T_0 generation transgenic plants.

3.5. Molecular and Biochemical Analyses of Transformed Plants

3.5.1. PCR Analysis

PCR analysis is used to screen the transgenic plants and distinguish true chloroplast transgenic plants from mutants or nuclear transgenic plants. Site-specific chloroplast integration of the transgene cassette is determined by using a set of primers of which one anneals to the native chloroplast genome and the other anneals within the transgene cassette. Mutant and nuclear transgenic plants are not expected to produce a PCR product with these primers. Chloroplast-specific integration of the transgene cassette can be confirmed further by using a set of primers that anneal to flanking sequences used for homologous recombination. The size of the PCR product should then depend on size of the transgene cassette. This initial screening is very important for eliminating the mutants and nuclear transgenic plants from the transgenic plant population.

1. Isolate total DNA from wild-type and transgenic plants using DNeasy™ Plant Mini Kit (Qiagen, Valencia, CA) and use this DNA as a template for PCR reactions.

2. To a 50-μL PCR reaction add: 1 μL of 100 ng/μL genomic DNA, 5 μL of 10X PCR reaction buffer, 1 μL each of 10 μ*M* forward and reverse primers, 200–250 μ*M* each of dNTPs, 0.5 μL (2.5 U) *Taq* DNA polymerase, and sterile distilled water.
3. Perform the PCR under following reaction conditions: denaturation for 5 min at 94°C, followed by 30 cycles of denaturation at 94°C for 1 minute, annealing at 60–65°C for 1 min, extension at 72°C for 1 min. The time of extension can be varied depending on the size of the amplicon. Usually 1 min/kb is the prescribed elongation time for most DNA polymerases. This is followed by a final step of 7–10 min of elongation at 72°C.
4. Detect PCR amplification products in agarose gels by electrophoresis and stained with ethidium bromide (EtBr).

3.5.2. Southern Blot Analysis

Southern blot analysis is performed to determine the copy number and degree of homoplasmy of the introduced transgene. A single plant cell harbors approx 10,000 copies of the chloroplast genome. Initially, the transgene cassette integrates into a few of these genomes and under selection pressure its copy number gradually increases. After three rounds of selection, the untransformed genome copies are replaced by the transgenic genome copies, leading to complete homoplasmy. This can be ascertained by Southern analysis of the total DNA isolated from plants generated after the third round of selection. The Southern blot is probed with radiolabeled flanking sequences used for homologous recombination. The transgenic plants possess higher molecular weight plastid genome that is distinguishable on the autoradiogram from the smaller molecular weight fragment representing the untransformed plastid genome. If the transgenic plants are heteroplasmic, a wild-type fragment is visible along with the larger transgenic fragment. Absence of the wild-type fragment confirms the establishment of homoplasmy (*see* **Note 8**).

3.5.3. Bioassays

Bioassays assess the effectiveness or properties or functionality of the introduced transgene(s). In addition, transgenic plants may be exposed to more rigorous challenges than what they would normally encounter in the real environment. Bioassays provide an accurate estimate on the new capabilities conferred to the plant by the integrated transgene(s).

3.5.3.1. Insect Resistance

Cry2Aa2 is one of the insecticidal proteins produced by the bacterium *Bacillus thuringensis*. Genetically modified plants with insecticidal proteins have shown significantly increased resistance against insects. Cry2Aa2 protein is encoded by an operon and has been expressed via chloroplast genome as a single gene ***(10)*** or as an operon ***(11)*** that contains the *cry2Aa2* gene and the ORF 2, which is a chaperone that folds the Cry2Aa2 protein into crystals. DeCosa et al. ***(11)*** demonstrated the formation of cuboidal crystals when the complete cry2Aa2 operon was expressed. In addition to the ORF 2 protein, crystal formation was also facilitated by hyperexpression of the insecticidal protein through chloroplast genetic engineering. Chloroplast transgenic plants showed that Cry2Aa2 accumulated up to 46.1% of total protein and this is the highest reported level of foreign protein expression in transgenic plants. Transgenic plants expressing the single *cry2Aa2* gene or the complete cry2Aa2 operon showed high insecticidal activity when compared with wild type tobacco plants. When nontransgenic control tobacco leaves were fed to the tobacco budworm (*Heliothis virescens*), cotton bollworm (*Helicoverpa zea*) and beet armyworm (*Spodoptera exigua*), the leaf pieces were completely consumed after 24 h. When the tobacco budworm was fed with a leaf from transgenic plants expressing the single gene, it died after 5 d, whereas the insect fed with the leaf from the plant expressing the operon died in 3 d (**Fig. 3**). Similar results were obtained when the assays were repeated with the cotton bollworm and the beet armyworm. This proved that the hyper expression of the *cry2Aa2* gene through chloroplast engineering of the *cry2Aa2* gene can confer 100% resistance to insects that feed on transgenic plants. Most importantly, chloroplast transgenic plants killed insects that were 40,000-fold resistant to insecticidal proteins. It was also observed that the old senescent leaves contained very high levels of the insecticidal protein, in spite of high protease activity. This could be attributed to chaperone-assisted crystal formation that prevented proteolytic degradation of the protein and allowed its higher accumulation. Indeed electron microscopic analysis revealed the presence of cuboidal crystals of the insecticidal protein in mature and old chloroplast transgenic leaves expressing the cry2Aa2 operon (**Fig. 4**).

3.5.3.1.1. Insect Bioassay

1. Cut 2-cm^2 leaf disks from transgenic and wild-type plants.
2. Place the leaf segments over distilled water-soaked cardboard lids in 50 × 12 mm plastic Petri dishes with tight-fitting lids.
3. Add 5–10 neonate insects (*Heliothis virescens, Helicoverpa zea, Spodoptera exigua*) per sample, with three replicates per treatment.
4. Examine the insect mortality daily for a period of 5 d.
5. Repeat the experiment at least three times, but preferably four to five times.
6. As a control, include a leaf disk from another chloroplast transgenic plant harboring only the basic transformation cassette devoid of the *cry2Aa2* gene. This would confirm that the observed results are actually because of the insecticidal protein and not some other extraneous factors.

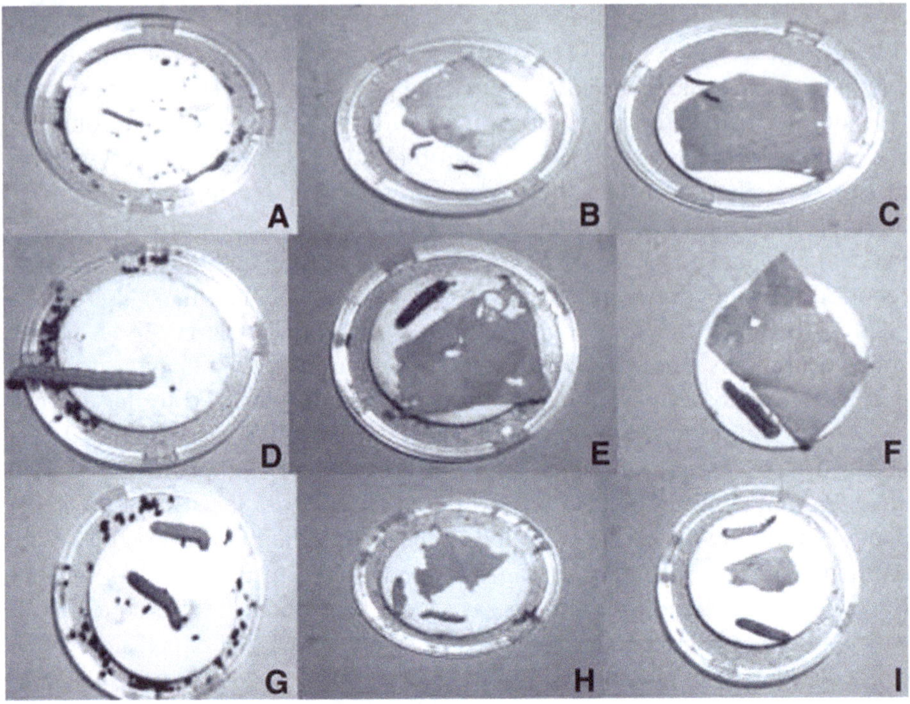

Fig. 3. Insect bioassays. **(A, D, G)** Untransformed tobacco leaves; **(B, E, H)** single gene-derived cry2Aa2 transformed leaves; **(C, F, I)** operon-derived cry2Aa2 transformed leaves. **(A–C)** Bioassays with *Heliothis virescens* (tobacco budworm); **(D–F)** bioassays with *Helicoverpa zea* (cotton bollworm); **(G–I)** bioassays with *Spodoptera exigua* (beet armyworm). For each replicate samples from the same leaf were used.

3.5.3.1.2. Transmission Electron Microscopy and Immunogold Labeling

1. Cut the transformed and untransformed leaf into 1- to 3-mm^2 sections.
2. Fix sections in 0.1 *M* cacodylate buffer, pH 7.4, for 15 min under vacuum and 12 h at 4°C.
3. Rinse the samples twice in 0.1 *M* cacodylate Savant (Savant, Holbrook, NY) pH 7.4, after setting fixation.
4. Dehydrate fixed samples through a graded ethanol series to 95%, then implant in LRW resins at 60°C for 24 h.
5. Cut ultrathin sections using a Leica Ultracut T ultramicrotome and collect section onto nickel grids
6. Incubate sections in 0.05 *M* glycine prepared in PBS for 15 min to inactivate residual aldehyde groups.

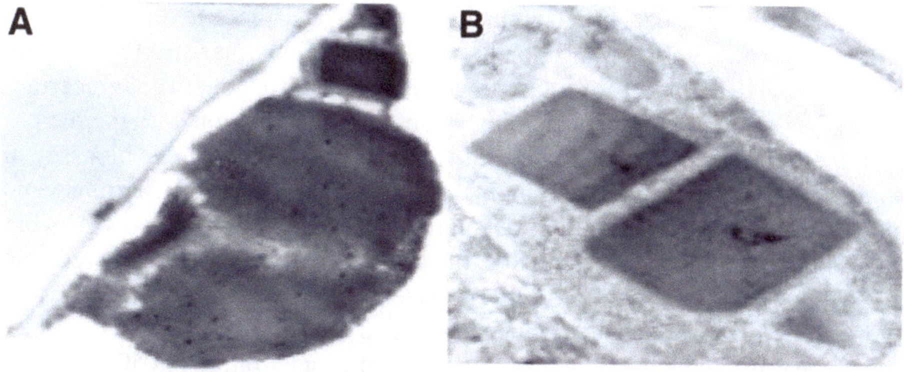

Fig. 4. Transmission electron micrographs. (**A**) Detection of Cry2A protein by immunogold labeling using Cry2A antibody. (**B**) Accumulation of folded Cry2A protein as cuboidal crystals in transgenic chloroplasts.

7. Place grids onto drops of blocking solution (PBS containing 2% nonfat dry milk) and incubate for 30 min.
8. Incubate sections for 1 h in rabbit anti-Cry2A polyclonal antibody (dilution range: 1:1000–1:25,000 in blocking solution).
9. Wash sections with blocking solution six times for 5 min each.
10. Incubate sections for 2 h with a goat anti-rabbit IgG secondary antibody conjugate to 10-nm gold diluted 1:40 in blocking solution.
11. Wash sections six times for 5 min each in blocking solution and three times for 5 min each with PBS. Fix sections in 2% glutaraldehyde for 5 min.
12. Wash fixed section in PBS three times for 5 min each, then in distilled water five times for 2 min each.
13. Stain sections using uranyl acetate and lead citrate and observe samples under transmission electron microscope at 60 kV.

3.5.3.2. Phytoremediation

Mercury and its most toxic form, organomercurials, present a serious hazard to the environment and ecosystems. Chemical and physical remediation procedures as well as bacterial bioremediation methods have proven ineffective because of the high cost and environmental concerns. As an alternative, phytoremediation has been proposed as a system for safe and cost-effective remediation of toxic chemicals in the environment. Mercury and organomercurials mainly target the chloroplasts. For this reason, it is advantageous to use chloroplast genetic engineering to increase resistance to mercury and organomercurials and at the same time detoxify the highly toxic organomercurials and metal mercury forms present in the contaminated environment *(12)*.

To achieve this, two bacterial enzymes that confer resistance to different forms of mercury known as mercuric ion reductase and organomercurial lyase were overexpressed in the chloroplast through chloroplast genetic engineering. When the chloroplast transgenic plants containing the operon with the mercuric ion reductase and organomercurial lyase were tested through a bioassay in which the extremely toxic organomercurial PMA was used, the transgenic plants were substantially more resistant than wild type tobacco plants growing under the same conditions. The 16-d-old tobacco plants (seedlings) were able to grow well in soil containing PMA concentrations of 50 μ*M*, 100 μ*M*, and even survived at the highest concentration of 200 μ*M*. On the other hand, wild-type tobacco plants struggled to survive at concentrations of 50 μ*M* PMA (**Fig. 5**). When nuclear transgenic tobacco seeds containing the *merA* and *merB* genes were germinated in medium containing PMA, they were resistant only to concentrations of 5 μ*M* PMA *(13)*.Twenty-four-day-old chloroplast transgenic plants treated with 100, 200, 300, and 400 μ*M* PMA showed an increase in the total dry weight when compared with wild-type growing at the same concentrations. On the other hand, the total dry weight of wild-type plants progressively decreased with each increase in PMA concentration from 0 to 400 μ*M*. Chlorophyll content of the leaf is an indication of the chloroplast structural and physiological integrity. When 15-mm diameter leaf discs from wild-type and transgenic plants were grown for 10 d in a concentration of 10 μ*M* PMA, the chlorophyll concentration of the transgenic plants increased, whereas the wild-type was reduced. These bioassays show the efficiency and activity of the chloroplast expressed enzymes and establishes that chloroplast genetic engineering can be used for phyto-remediation. This is the first report of the use of chloroplast genetic engineering for phytoremediation.

3.5.3.2.1. Organomercurial Bioassay

1. Surface-sterilize tobacco seeds in 7% sodium hypochlorite containing 0.1% Tween-20. Vortex-mix the tube for 5 min, then wash the seeds five times with distilled water. Finally dry the seeds in a SpeedVac at medium temperature.
2. Germinate the transgenic sterilized seeds in a plate containing one-half strength MS medium with suitable selection agent and 0.3% Phytagar. Adjust the pH to 5.7 with 1 *N* KOH. The wild-type seeds are germinated in the same media but lacking selection agent.
3. Incubate plates with seeds at 4°C for 3 d, and then transfer to a growth chamber in a 16-h light photoperiod at a temperature of 24°C, humidity of 75–90%, and a photon flux density of 750 μE/m^2.
4. Transfer seedling to soil (Sand–Davis Mix, 50:50) approx 10 d after germination and maintain in the greenhouse at 22°C using a 16-h light photoperiod.
5. Five replicate pots each containing a single seedling (wild-type and transgenic) is used for the assay.

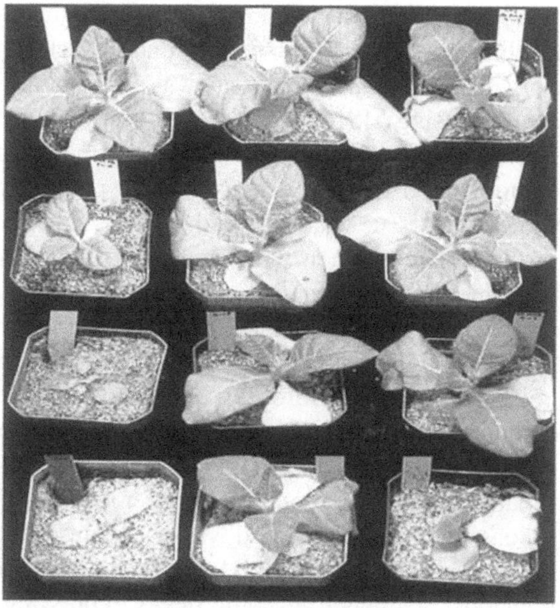

Fig. 5. Effect of PMA concentration on the growth of wild-type and transgenic tobacco lines. Plants were treated with 200 mL of Hoagland's nutrient solution supplemented with 0, 50, 100, and 200 μ*M* PMA. Photographs were taken 14 d after treatment. WT, Negative control Petit Havana; 5A, pLDR-MerAB transgenic line; 9, pLDR-MerAB-3'-UTR transgenic line.

6. Water the pots twice a week with one-half-strength Hoagland solution.
7. Six days after the initial transfer of seedlings into soil, apply three different concentrations of PMA to the pots containing wild-type and transgenic plants, in three replicates for each of the concentrations.
8. Prepare a 10 m*M* PMA working solution.
9. Add 100 mL of one-half-strength Hoagland solution containing concentrations of 50, 100, and 200 μ*M* PMA. Control receives the same treatment but without PMA.
10. Grow plants under conditions as explained in **step 4** for at least 14 d. Then take picture and assess total biomass by measuring the length of the root and shoot as well as root and shoot total dry weight.

3.5.3.3. Disease Resistance

Helical structured antimicrobial peptides (AMPs) are expressed as protective agents against pathogens in many organisms. We have expressed MSI-99, an analog of magainin 2 in transgenic chloroplasts *(14)*. This AMP confers protection against prokaryotic organisms because of the high specificity for negatively charged phospholipids, which are mostly found in bacteria and less abundant in eukaryotic organisms. *In planta* bioassay was performed with chloroplast transgenic plants expressing MSI-99. The leaves were inoculated with the phytopathogen *P. syringae* pv *tabaci* and the absence of necrosis around the inoculation area demonstrated increased resistance to pathogen colonization and infection. No necrotic tissue was observed in transgenic plants even when 8×10^5 cells were inoculated. When wild-type plants were inoculated with 8×10^3 cells of the same phytopathogen (a much lower number of cells than in the cells used for transgenic plants), a large necrotic area was observed (**Fig. 6C,D**). This suggests that high levels of AMP are expressed by the chloroplast and that this is released from the chloroplast during pathogen infection. Studies of bacterial population at the site of inoculation 4 d after inoculation showed that wild-type plants had a cell population of 13,750 ± 750 colony-forming units (cfu) compared to the lower count in transgenic plants of 4650 ± 125 cfu. When similar bioassays were performed with the plant pathogen, the yeast *Colletotrichum destructivum* in nontransformed controls, the plant developed anthracnose lesions, whereas transgenic plants expressing MSI-99 did not develop any lesions (**Fig. 6A,B**). This study shows that chloroplast genetic engineering can be used to confer high-level resistance to phytopathogenic organisms in plants.

3.5.3.3.1. In Planta Bioassay

1. Grow a culture of *P. syringae* pv *tabaci* overnight (ATCC 17914) in liquid nutrient broth. Alternatively, grow *C. destructivum* (ATCC 42492) in Czapek yeast autolysate agar at 24°C.
2. Centrifuge the *P. syringae* pv *tabaci* culture and resuspend the pellet in 50 mL of 0.01 *M* phosphate buffer. Make dilutions of the suspension in phosphate buffer. Obtain *C. destructivum* by flooding the agar plate with 9 mL of distilled water and remove the spores aseptically. Dilute the inoculum to a final density of approx 1×10^6 spores/mL.
3. Prepare the leaf by scraping the leaf with fine-grain sandpaper to an area of 7 mm.
4. Add 10 μL of 8×10^5, 8×10^4, 8×10^3, and 8×10^2 cell culture of *P. syringae* pv *tabaci* to each prepared area in transgenic and wild-type plants. To inoculate *C. destructivum*, place eight drops containing 10 μL of the diluted inoculum.
5. Take photograph 5 d after inoculation.
6. In another assay use 25 mL of 8×10^5, 8×10^4, 8×10^3, and 8×10^2 cell culture of *P. syringae* pv *tabaci* and inject it into the leaf of wild-type and transgenic tobacco plants by using a needle with a syringe.
7. Take photograph 5 d after inoculation.

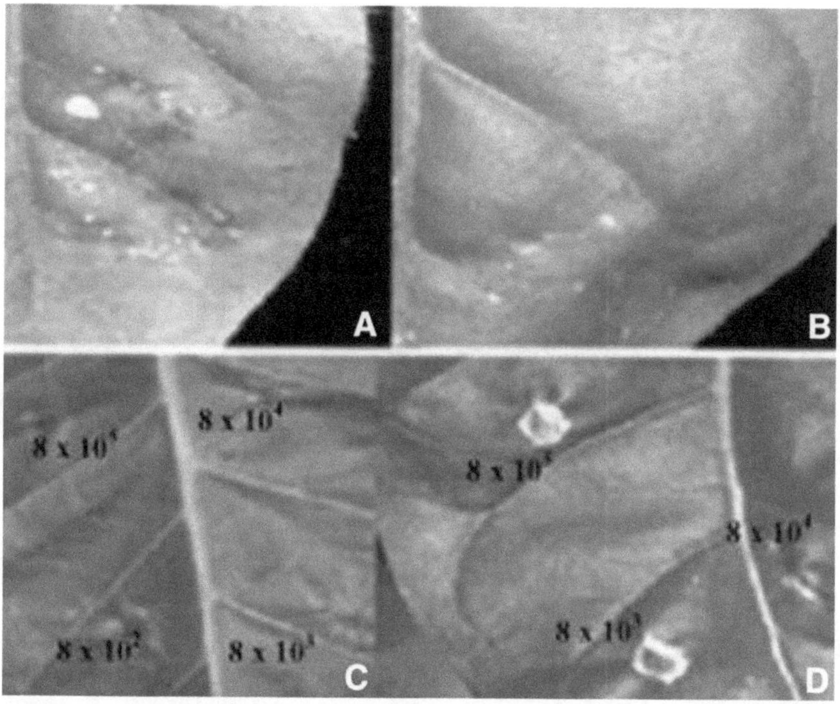

Fig. 6. *In planta* bioassays for disease resistance. **(A, B)** Fungal disease resistance. Leaves were inoculated on the adaxial surface with eight drops of 10 μL each of the culture containing 1 × 10^6 spores/mL of the fungal pathogen *Colletotrichum destructivum.* **(A)** Wild-type leaf. **(B)** transgenic leaf. **(C, D)** Bacterial disease resistance: 8 × 10^5, 8 × 10^4, 8 × 10^3, and 8 × 10^2 cell cultures of bacterial pathogen *Pseudomonas syringae* pv *tabaci* were added to a 7-mm scraped area in transgenic and nontransgenic tobacco lines **(C, D)**. Photos were taken 5 d after inoculation.

8. Collect leaf disks containing the inoculated area from transgenic and wild-type plants, 4 d after inoculation.
9. Grind the samples in 300 μL of 10 m*M* $MgCl_2$.
10. Transfer the homogenates to 5 mL of PO_4 buffer.
11. Plate dilutions of the samples in *Pseudomonas* Agar F (Difco, Detroit, MI) for 48 h at 28°C.
12. Enumerate the colonies.

3.5.3.4. Herbicide Resistance

Glyphosate is a broad-spectrum herbicide that kills majority of grasses and broad-leaf weeds. Glyphosate acts by competitive inhibition of the 5-enol-

Fig. 7. Herbicide resistance assay. Chloroplast transgenic and wild-type plants 18-wk-old were sprayed with 5 m*M* glyphosate solution. (**A**) Chloroplast transgenic line. (**B**) Wild-type control.

pyruvyl shikimate-3-phosphate enzyme (EPSPS). This disrupts the aromatic amino acid biosynthetic pathway that occurs only in plants and microorganisms. Because this potent herbicide lacks selectivity, for successful weed control in crop plantations, the crops have to be genetically modified to resist glyphosate. Usually, the targets for most herbicides, including glyphosate, are amino acids and fatty acid biosynthetic pathways found in the chloroplast. We have shown that the hyperexpression of the petunia EPSPS (which is highly sensitive to glyphosate) through chloroplast transformation conferred resistance to high levels of glyphosate and the transgene was maternally inherited ***(8)***. When concentrations of up to 5 m*M* glyphosate were sprayed on chloroplast transgenic plants expressing the petunia EPSPS, they survived without any detrimental symptoms (**Fig. 7**). Untransformed tobacco plants were highly susceptible to glyphosate, dying 7 d after exposure to 0.5 m*M* glyphosate. More recently, tobacco plants were transformed via chloroplast genetic engineering with more resistant forms of EPSPS, including AroE (*Bacillus*) and CP4 (*Agrobacterium*, ***15***).

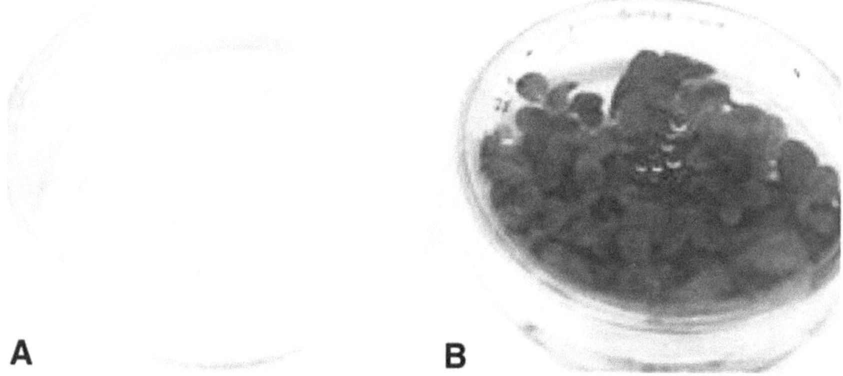

Fig. 8. Maternal inheritance of transgenes. **(A)** Wild type and **(B)** chloroplast transgenic seeds expressing the *EPSPS* gene were germinated on MSO medium supplemented with 500 mg/L of spectinomycin.

3.5.3.4.1. Glyphosate Tolerance Test

1. Spray wild-type and transgenic plants growing in soil with equal volumes of different concentrations of glyphosate (0.5–5 m*M*).
2. Take picture each week after initial exposure to glyphosate.

3.5.3.5. MATERNAL INHERITANCE

Transgenes integrated into chloroplast genomes are, in general, inherited maternally. This is evident when transgenic seeds (as shown in **Fig. 8**) are germinated on MSO basal medium containing 500 μg/mL of spectinomycin. There should be no detrimental effect of the selection agent in transgenic seedlings, whereas untransformed seedlings will be affected. In **Fig. 8B**, all transgenic seedlings carry the spectinomycin resistance trait and show maternal inheritance without any Mendelian segregation of introduced transgenes. If further confirmation is necessary, then pollen from chloroplast transgenic lines may be used to fertilize wild-type untransformed plants; progeny should not carry the trait if the transgenes were inherited maternally.

3.5.3.6. DROUGHT TOLERANCE

Environmental stress factors such as drought, salinity, or freezing are hazardous to plants mostly because of their sessile way of life. Osmoprotectants are produced in plants, yeast, and other organisms, and this confers resistance to several factors including drought. The *TPS1* gene from yeast encodes the trehalose phosphate synthase, an enzyme that produces the osmoprotectant trehalose.

Fig. 9. Comparison of nuclear and chloroplast transgenic lines to illustrate pleiotropic effect. **1**, Wild-type tobacco; **2–5**, T_0 nuclear transgenic lines expressing *tps1* (**2** has the lowest expression levels compared to **5**); **6**, T_1 chloroplast transgenic line expressing *tps1*; **7**, wild-type tobacco plant.

Attempts to confer resistance to drought by expressing this enzyme via nuclear transformation have proven ineffective because of adverse pleotropic effects even at very low levels of trehalose accumulation. We reported the hyperexpression of the trehalose phosphate synthase and the increased accumulation of trehalose in chloroplasts of transgenic plants *(16)*. When TPS1 was expressed through chloroplast transformation, no pleotropic effects were detected and the plant was as healthy as wild-type controls (**Fig. 9**). Drought tolerance bioassays in which transgenic and wild-type seeds were germinated in MS medium containing concentrations of 3–6% PEG showed that the chloroplast transgenic plants producing high levels of trehalose germinated, grew, maintained green color, and remained healthy (**Fig. 10**). Wild-type seeds germinated under similar conditions showed severe dehydration, loss of chlorophyll (chlorosis), and retarded growth that finally ended in the death of the seedlings. Loss of chlorophyll in the nontransgenic plants reveals that drought affects thylakoid membrane stability.

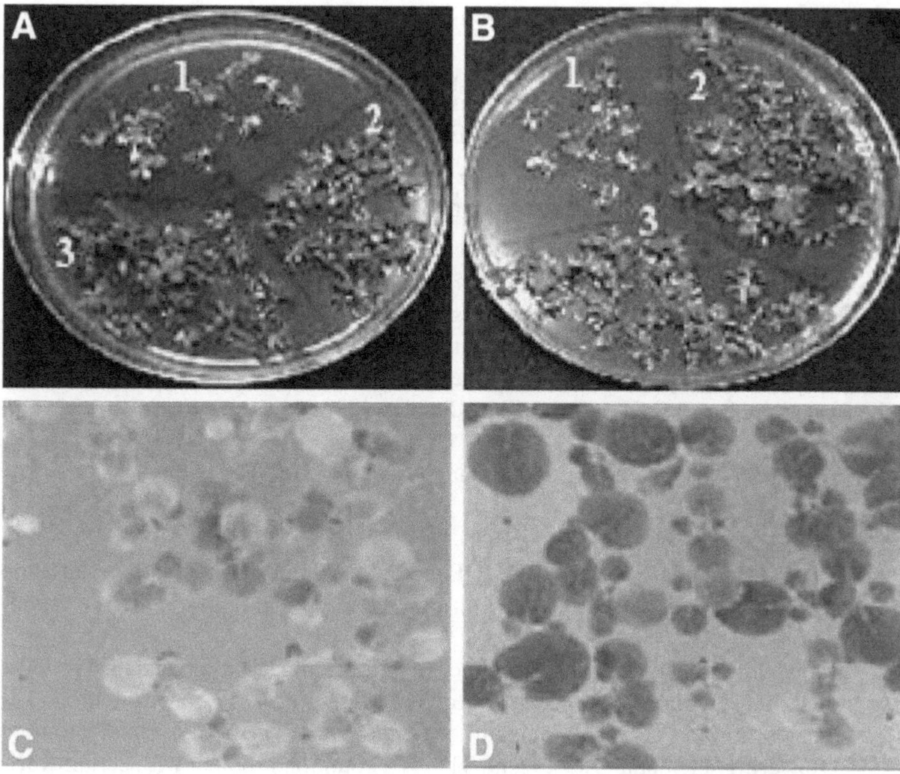

Fig. 10. Drought tolerance assays. **(A, B)** Dehydration–rehydration assay. Three-week-old seedlings were dried for 7 h and rehydrated in MS medium for 48 h. *1*, Untransformed; *2,3*, T_1 and T_2 chloroplast transgenic lines. **(C, D)** PEG growth assay. Seedlings 4-wk-old were grown on MS medium with 6% PEG. **(C)** Untransformed. **(D)** T2 chloroplast transgenic line.

Production of trehalose in the chloroplast of transgenic plants conferred membrane stability. In another assay, when seedlings from transgenic and wild-type tobacco plants were dried for 7 h, and showed dehydration symptoms, but when the seedlings were rehydrated in MS medium for 48 h, all chloroplast transgenic plants accumulating trehalose recovered and grew well. The wild-type controls became bleached and died (**Fig. 10**). In addition, when potted transgenic and wild-type plants were not watered for 24 d and were then rehydrated for 24 h, the transgenic plant recovered while the control plant did not recover. These results show that expression of the enzyme trehalose phosphate synthase via chloroplast genetic engineering confers resistance to drought.

3.5.3.6.1. Drought Tolerance Bioassays

For the PEG bioassay:

1. Germinate sterilized chloroplast transgenic and wild-type tobacco seeds in MS medium plates containing 3 or 6% PEG (mol wt 8000, Sigma).
2. Take a photograph 4 wk after plating the seeds.

For the dehydration/rehydration assay:

1. Germinate chloroplast transgenic and wild-type tobacco seeds on agarose with or without 500 μg/mL of spectinomycin, respectively.
2. Take 3-wk-old seedlings from transgenic and wild-type and air-dry for 7 h at room temperature in 50% relative humidity.
3. Rehydrate for 48 h by introducing the seedlings root into MS medium.
4. Place rehydrated seedlings in MS media plate and allow them to grow for several days.
5. Compare wild-type and transgenic plants. Take picture.

4. Notes

1. It is important not to autoclave RMOP medium for more than 30 min because the phytohormones added to the medium may break down, inhibiting tissue regeneration.
2. It is of importance to note the number of times a plant is propagated. We have noted that if a plant is propagated more than five times, the transformation efficiency of the leaf decreases. In addition, if the plant to be used for bombardment is flowering, the plant is senescent. This is a common problem with certain varieties such as Petit Havana. We have noted that this also decreases the transformation efficiency of the plant.
3. The order of adding gold, DNA, $CaCl_2$, and spermidine is essential for the proper coating of the gold particles. $CaCl_2$ should be prepared fresh and the DNA-coated gold particles have to be used within 2 h.
4. The preparation of leaf tissues and macrocarriers and the bombardment has to take place under aseptic conditions under a laminar flow hood. Before each addition of gold particles into the macrocarrier, make sure to vortex-mix the particles for at least 30 s to resuspend, and use immediately. It is essential to avoid clumps of gold particle when loading the macrocarriers because this will damage the leaf tissue during bombardment and decrease transformation efficiency.
5. The side of the leaf to be bombarded has to be the abaxial side because it does not contain the waxy cuticle found in the upper side of the leaf. This allows for better penetration of the gold particles and increases transformation efficiency. It is essential to cut the leaf as close to the time of bombardment as possible, as this helps in decreasing the activation of nucleases and proteases (often detected in detached leaves) that could affect the transformation process.
6. The microcarrier launch assembly has to be placed in level one (L1 = 3 cm), and the target plate shelf has to be placed in level four (L4 = 12 cm). The stopping screen support has to be placed in between the spacer rings (one under and one

over the stopping screen). This setup is essential for efficient chloroplast transformation.

7. When placing the leaf pieces on RMOP agar medium containing the selection agent, make sure to leave enough space in between pieces (we recommend about five pieces per plate). This will allow the full expansion of the leaf segments. The first round of selection is an appropriate time to screen for integration of transgenes into the chloroplast genome by PCR. When moving the transgenic plants to soil from the third round of selection, it is essential to remove any agar attached to the roots. This will decrease the possibility of fungal and bacterial contamination when the plants are grown in soil.
8. The presence of the foreign gene into the nuclear or mitochondrial genome can be detected by using the foreign gene as a DNA probe and prolonged exposure of blots on films.

References

1. Daniell, H. (2002) Molecular strategies for gene containment in transgenic crops. *Nat. Biotechnol.* **20,** 581–586.
2. Daniell, H. and Parkinson, C. L. (2003) Jumping genes and containment. *Nat. Biotechnol.* **21,** 374–375.
3. Daniell, H. and Dhingra, A. (2002) Multigene engineering: dawn of an exciting new era in biotechnology. *Curr. Opin. Biotechnol.* **13,** 136–141.
4. Daniell, H., Khan, M. S, and Alison, L. (2002) Milestones in chloroplast genetic engineering: an environmentally friendly era in biotechnology. *Trends Plant Sci.* **7,** 84–91.
5. Palmer, J. D. (1985) Comparative organization of chloroplast genomes. *Annu. Rev. Genet.* **19,** 325–354.
6. Lilly, J. W., Havey, M. J., Jackson, S. A., and Jiang, J. M. (2001) Cytogenomic analyses reveal the structural plasticity of the chloroplast genome in higher plants. *Plant Cell* **13,** 245–254.
7. Daniell, H., Vivekananda, J., Nielsen, B. L., Ye, G. N., Tewari, K. K., and Sanford, J. C. (1990) Transient foreign gene expression in chloroplasts of cultured tobacco cells after biolistic delivery of chloroplast vectors. *Proc. Natl. Acad. Sci. USA* **87,** 88–92.
8. Daniell, H., Datta, R., Varma, S., Gray, S., and Lee, S. B. (1998) Containment of herbicide resistance through genetic engineering of the chloroplast genome. *Nat. Biotechnol.* **16,** 345–348.
9. Guda, C., Lee, S. B., and Daniell, H. (2000) Stable expression of biodegradable protein based polymer in tobacco chloroplasts. *Plant Cell Rep.* **19,** 257–262.
10. Kota, M., Daniell, H., Varma, S., Garczynski, S. F., Gould, F., and William, M. J. (1999) Overexpression of the *Bacillus thuringiensis* (Bt) Cry2Aa2 protein in chloroplasts confers resistance to plants against susceptible and Bt-resistant insects. *Proc. Natl. Acad. Sci. USA* **96,** 1840–1845.
11. DeCosa, B., Moar, W., Lee, S. B., Miller, M., and Daniell, H. (2001) Overexpression of the *Bt* Cry2Aa2 operon in chloroplasts leads to formation of insecticidal crystals. *Nat. Biotechnol.* **19,** 71–74.

12. Ruiz, O. N., Hussein, H., Terry, N., and Daniell, H. (2003) Phytoremediation of organomercurial compounds via chloroplast genetic engineering. *Plant Physiol.* **132,** 1344–1352.
13. Bizily, S. P., Rugh, C .L., and Meagher, R. B. (2000) Phytodetoxification of hazardous organomercurials by genetically engineered plants. *Nat. Biotechnol.* **18,** 213–217.
14. DeGray, G., Kanniah, R., Franzine, S., John, S., and Daniell, H. (2001) Expression of an antimicrobial peptide via the chloroplast genome to control phytopathogenic bacteria and fungi. *Plant Physiol.* **127,** 852–862.
15. Iamtham, S. and Day, A. (2000) Removal of antibiotic resistance genes from transgenic tobacco plastids. *Nat. Biotechnol.* **18,** 1172–1176.
16. Lee, S. B., Kwon, H. B., Kwon, S. J., et al. (2003) Accumulation of trehalose within transgenic chloroplasts confers drought tolerance. *Mol. Breed.* **11,** 1–13.
17. McBride, K. E., Svab, Z., Schaaf, D. J., Hogan, P. S., Stalker, D. M., and Maliga, P. (1995) Amplification of a chimeric *Bacillus* gene in chloroplasts leads to an extraordinary level of an insecticidal protein in tobacco. *Biotechnology* **13,** 362–365.
18. Staub, J. M., Garcia, B., Graves, J., etal. (2000) High-yield production of a human therapeutic protein in tobacco chloroplasts. *Nat. Biotechnol.* **18,** 333–338.
19. Daniell, H., Lee, S. B., Panchal, T., and Wiebe, P. O. (2001) Expression of the native cholera toxin B subunit gene and assembly as functional oligomers in transgenic tobacco chloroplasts. *J. Mol. Biol.* **311,** 1001–1009.
20. Ruiz, G. (2002) Optimization of codon composition and regulatory elements for expression of the human IGF-1 in transgenic chloroplasts. MS thesis, University of Central Florida, Orlando, FL.
21. Torres, M. (2002) Expression of interferon α5 in transgenic chloroplasts of tobacco. MS thesis, University of Central Florida, Orlando, FL.
22. Falconer, R. (2002) Expression of interferon α2b in transgenic chloroplasts of a low-nicotine tobacco. MS thesis, University of Central Florida, Orlando, FL.
23. Fernandez-San Millan, A., Mingo-Castel, A., and Daniell, H. (2003) A chloroplast transgenic approach to hyper-express and purify human serum albumin, a protein highly susceptible to proteolytic degradation. *Plant Biotechnol. J.* **1,** 71–79.
24. Leelavathi S. and Reddy V. S. (2003) Chloroplast expression of His-tagged GUS-fusions: a general strategy to overproduce and purify foreign proteins using transplastomic plants as bioreactors. *Mol. Breed.* **11,** 49–58.
25. Daniell, H., Dhingra, A., and San-Milan, A. F. (2001) *12th International Congress on Photosynthesis*, Vol. S40-04 1-6, CSIRO, Brisbane, Australia.
26. Daniell, H., Watson, J., Koya, V., and Leppla, S. (2004) Expression of *Bacillus anthracis* protective antigen in transgenic chloroplasts of tobacco, a non-food/feed crop. *Vaccine*, in press.
27. Singleton, M. L. (2003) Expression of CaF1 and LcrV as a fusion protein for a vaccine against *Yersinia pestis* via chloroplast genetic engineering. MS thesis, University of Central Florida, Orlando, FL.

28. Daniell, H. and McFadden, B. A. (1987) Uptake and expression of bacterial and cyanobacterial genes by isolated cucumber etioplasts. *Proc. Natl. Acad. Sci. USA* **84,** 6349–6353.
29. Daniell, H., Krishnan, M., and McFadden, B. F. (1991) Transient expression of beta-glucuronidase in different cellular compartments following biolistic delivery of foreign DNA into wheat leaves and calli. *Plant Cell Rep.* **9,** 615–619.
30. Goldschmidt Clermont, M. (1991) Transgenic expression of aminoglycoside adenine transferase in the chloroplast—a selectable marker for site-directed transformation of *Chlamydomonas*. *Nucleic Acids Res.* **19,** 4083–4089.
31. Svab, Z. and Maliga, P. (1993) High-frequency plastid transformation in tobacco by selection for a chimeric *aadA* gene. *Proc. Natl. Acad. Sci. USA* **90,** 913–917.
32. Carrer, H., Hockenberry, T. N., Svab, Z., and Maliga, P. (1993) Kanamycin resistance as a selectable marker for plastid transformation in tobacco. *Mol. Gen. Genet.* **241,** 49–56.
33. Bateman, J. M. and Purton, S. (2000) Tools for chloroplast transformation in *Chlamydomonas*: expression vectors and a new dominant selectable marker. *Mol. Gen. Genet.* **263,** 404–410.
34. Huang, F. C., Klaus, S. M. J., Herz, S., Zou, Z., Koop, H. U., and Golds, T. J. (2002) Efficient plastid transformation in tobacco using the *aphA-6* gene and kanamycin selection. *Mol. Genet. Genom.* **268,** 19–27.
35. Hibberd, J. M., Linley, P. J., Khan, M. S., and Gray, J. C. (1998) Transient expression of green fluorescent protein in various plastid types following microprojectile bombardment. *Plant J.* **16,** 627–632.
36. Sidorov, V. A., Kasten, D., Pang, S. Z., Hajdukiewicz, P. T. J., Staub, J. M., and Nehra, N. S. (1999) Stable chloroplast transformation in potato: use of green fluorescent protein as a plastid marker. *Plant J.* **19,** 209–216.
37. Khan, M. S. and Maliga, P. (1999) Fluorescent antibiotic resistance marker for tracking plastid transformation in higher plants. *Nat. Biotechnol.* **17,** 910–915.
38. Daniell, H., Muthukumar, B., and Lee, S. B. (2001) Marker free transgenic plants: engineering the chloroplast genome without the use of antibiotic selection. *Curr. Genet.* **39,** 109–116.
39. Devine, A. L. and Daniell, H. (2004) Plastids, in *Annual Plant Reviews*, Vol. 13, (Moller, E., ed.) Blackwell, Oxford, UK, Chapter 10, pp. 283–323.
40. Dhingra, A. and Daniell, H. (2004) Chloroplast genetic engineering via organogenesis or somatic embryogenesis, in *Arabidopsis Protocols*, 2nd Ed., in press.
41. Molina, A., Hervas-Stubbs, S., Daniell, H., Mingo-Castel, A. M., and Veramendi, J., (2004) High-yield expression of a viral peptide animal vaccine in transgenic tobacco chloroplasts. *Plant Biotechnol. J.* **2,** 141–153.
42. Dhingra, A. (2000) Analysis of components involved in regulation of plastid gene expression, plastome organization and its manipulation with emphasis on rice (*Oryza sativa* L.). PhD thesis, University of Delhi South Campus, New Dehli, India.

III

Regeneration

9

Organogenesis From Transformed Tomato Explants

Anne Frary and Joyce Van Eck

Summary

Tomato was one of the first crops for which a genetic transformation system was reported involving regeneration by organogenesis from *Agrobacterium*-transformed explants. Since the initial reports, various factors have been studied that affect the efficiency of tomato transformation and the technique has been useful for the isolation and identification of many genes involved in plant disease resistance, morphology and development. In this method, cotyledon explants from in vitro-grown seedlings are precultured overnight on a tobacco suspension feeder layer. The explants are then inoculated with *Agrobacterium* and returned to the feeder layer for a 2-d period of cocultivation. After cocultivation, the explants are transferred to an MS-based selective regeneration medium containing zeatin. Regenerated shoots are then rooted on a separate selective medium. This protocol has been used with several tomato cultivars and routinely yields transformation efficiencies of 10–15%.

Key Words: *Agrobacterium tumefaciens*; biotechnology; gene transfer; genetic engineering; genetic transformation; *Lycopersicon esculentum*; morphogenesis; regeneration; transgenic plants.

1. Introduction

Organogenesis from transformed explants is a very common method for generating transgenic plants. This technique is often favored because, unlike embryogenesis, morphogenesis from organ explants has been developed for many different plant species. Thus, established organogenesis systems are often easily adapted into genetic transformation protocols. In addition, regeneration from pieces of nonmeristematic transformed tissue may present a lower risk of chimerism than regeneration from plant meristems. The technique was first popularized by Horsch et al. *(1)*, who described a simplified method for *Agro-*

From: *Methods in Molecular Biology, vol. 286: Transgenic Plants: Methods and Protocols*
Edited by: L. Peña © Humana Press Inc., Totowa, NJ

bacterium-mediated transformation of plants. Since then, organogenesis has become the preferred method for the regeneration of transgenic individuals in many species including tobacco, petunia, tomato, potato, cauliflower, squash, cotton, chrysanthemum, sunflower, and apple. This chapter uses *Agrobacterium tumefaciens*-mediated transformation of tomato, *Lycopersicon esculentum*, to illustrate the methods for organogenesis from transformed explants.

The earliest reports of *Agrobacterium*-mediated transformation of tomato were by Horsch et al. ***(1)*** and McCormick et al. ***(2)*** nearly 20 yr ago. Other methods for introducing foreign DNA into the crop, including microinjection ***(3)***, particle bombardment ***(4,5)*** (*see* Chapter 4), electroporation, and polyethylene glycol (PEG)-mediated transformation of protoplasts ***(6,7)*** (*see* Chapter 5), have also been described. However, none of these techniques has proven to be as popular as the *Agrobacterium*-mediated method, which is favored for its practicality, effectiveness, and efficiency.

The ability to genetically engineer tomato has been of great value because of its agronomic and economic importance and its usefulness as a model system. Tomato was one of the first crop species with a molecular genetic map ***(8)*** and, more recently, is the subject of new areas of study including functional genomics, proteomics, and metabolomics. Genetic transformation of tomato played an integral role in the map-based cloning of the first disease resistance ***(9)*** and quantitative trait ***(10)*** loci isolated in plants and in the identification of genes for many other important agronomic, morphological, and developmental traits (e.g., *see* **refs. *11–13***). *Agrobacterium*-mediated transformation of tomato with an antisense polygalacturonase construct was also used to develop the first commercial transgenic plant product, the Flavr Savr tomato ***(14)***. In addition, the technique has been used to introduce very large fragments of DNA (up to 150 kb) into the tomato genome ***(15)***.

Since the early reports of transformation, several groups have described various factors that affect the efficiency of tomato transformation. Conditions that influence tomato transformation include the choice and age of explants ***(16–18)***, the length of preculture ***(18,19)***, the strain and concentration of the *Agrobacterium tumefaciens* culture used for cocultivation ***(17,19)***, the length of cocultivation and medium used ***(17)***, the use of a petunia or tobacco suspension culture feeder layer ***(17–20)***, the orientation (adaxial side up vs abaxial side up) of cotyledon explants ***(21)***, the gelling agent ***(21)***, the plate sealant ***(21)***, and the frequency of transfer to fresh selective medium ***(21)***. However, it is important to note that conditions that result in an efficient transformation system for one genotype, do not always translate into an efficient system for other genotypes ***(18)***.

Our work required the development of an efficient transformation system for several different tomato lines for complementation analyses; testing of new vector systems; and studying value-added traits, promoter efficacy, and functional genomics. Based on information from reports in the literature, we developed a standard tomato transformation protocol that has been routinely used by several research groups for transformation of various *L. esculentum* freshmarket cultivars including Moneymaker, Yellow Pear, Rio Grande, Momor, the processing line E6203, and Micro-Tom which was developed for the ornamental market. Average transformation efficiencies (the percent of explants that give rise to transformed plantlets) using this protocol ranged from 10 to 15%.

2. Materials

2.1. Tissue Culture Media

2.1.1. Stocks

1. 1X Murashige and Skoog (MS) basal salts mixture powder. If MS salts in powdered form are not available, MS major salt, minor salt and iron stocks can be prepared and stored at 4°C for several months. MS major salt stock (10X): 19 g/L of KNO_3, 16.5 g/L of NH_4NO_3, 4.4 g/L $CaCl_2 \cdot 2H_2O$, 3.7 g/L $MgSO_4 \cdot 7H_2O$, 1.7 g/L of KH_2PO_4. MS minor salt stock (100X): 0.62 g/L of H_3BO_3, 2.23 g/L $MnSO_4 \cdot 4H_2O$, 0.86 g/L of $ZnSO_4 \cdot 7H_2O$, 0.083 g/L of KI, 0.025 g/L of $Na_2MoO_4 \cdot 2H_2O$,1 mL of $CuSO_4 \cdot 5H_2O$ stock (2.5 mg/mL), 1 mL $CoCl_2 \cdot 6H_2O$ stock (2.5 mg/mL). MS iron stock (200X): 8.6 g/L of ethylenediaminetetraacetic acid (EDTA) ferric-sodium salt, light sensitive; store in brown bottle.
2. 0.4 mg/mL of thiamine-HCl, dissolve in H_2O; store at 4°C for up to 1 mo.
3. 0.5 mg/mL of pyridoxine-HCl, dissolve in H_2O; store at –20°C.
4. 0.5 mg/mL of nicotinic acid, dissolve in H_2O; store at –20°C.
5. 1 mg/mL of 2,4-dichlorophenoxyacetic acid (2,4-D), dissolve in H_2O, store at 4°C for up to 1 mo.
6. 1 mg/mL of kinetin, dissolve in a few drops of 1 *M* HCl; store at –20°C.
7. 2 mg/mL of glycine, dissolve in H_2O; store at –20°C.
8. 0.25 mg/mL of folic acid, dissolve in H_2O; store at –20°C.
9. 0.5 mg/mL of D-biotin, dissolve in H_2O, store at –20°C.
10. 1000X Nitsch vitamin stock: 2 g/L of glycine, 10 g/L of nicotinic acid, 0.5 g/L of pyridoxine HCl, 0.5 g/L of thiamine-HCl, 0.5 g/L of folic acid, 40 mg/L of D-biotin; adjust pH to 7.0 to clear solution, and store at –20°C.
11. 1 mg/mL of zeatin, dissolve in a few drops of 1 *M* HCl, filter-sterilize, store at –20°C.
12. Appropriate selective agent stock for vector, dissolve as necessary, filter sterilize, store at –20°C. (This protocol has been successfully used with kanamycin, hygromycin, and bialaphos as selective agents. All of these compounds are toxic and should be handled with appropriate care.)
13. 300 mg/L of timentin (a mixture of ticarcillin disodium and potassium clavulanate) or 500 mg/L of carbenicillin, dissolve in H_2O, filter sterilize, store at –20°C. These compounds are toxic and should be handled with appropriate care.

2.1.2. Media

1. 1/2 MSO medium: 1/2X MS salts; 100 mg/L of myoinositol, 2 mg/L of thiamine-HCl, 0.5 mg/L of pyridoxine-HCl, 0.5 mg/L of nicotinic acid, 1% and sucrose, 0.8% agar, pH 5.8. Autoclave and store at room temperature for up to 1 mo.
2. KC Biological MS (KCMS) medium: 1X MS salts, 100 mg/L of myoinositol, 1.3 mg/L of thiamine-HCl, 0.2 mg/L of 2,4-D, 200 mg/L of KH_2PO_4, 0.1 mg/L of kinetin, and 3% sucrose. For solid medium, add 0.52% Agargel, pH 5.5. Autoclave and store at room temperature for up to 1 mo.
3. Luria Bertani (LB) medium: 10 g/L of Bacto-tryptone, 5 g/L of yeast extract, 10 g/L of NaCl, 1.5% Bacto-agar Difco (BD, Franklin Lakes, NJ). Autoclave, cool to 55°C, add appropriate filter-sterilized selection agent, and store at 4°C (length of time depends on selection agent).
4. Yeast extract medium (YE): 400 mg/L of yeast extract, 10 g/L of mannitol, 100 mg/L of NaCl, 200 mg/L of $MgSO_4 \cdot 7H_2O$, 500 mg/L of KH_2PO_4. Autoclave and store at room temperature for up to 3 mo.
5. MS liquid medium: 1X MS salts, 100 mg/l myoinositol, 2 mg/L glycine, 0.5 mg/L of nicotinic acid, 0.5 mg/L of pyridoxine-HCl, 0.4 mg/L of thiamine-HCl, 0.25 mg/L of folic acid, 0.05 mg/L of D-biotin, 3% sucrose, pH 5.6, autoclave, store at room temperature for up to 3 mo.
6. 2Z medium: 1X MS salts, 100 mg/L of myoinositol, 1X Nitsch vitamins, 2% sucrose, 0.52% Agargel, pH 6.0. Autoclave; cool to 55°C; and add filter-sterilized stocks of vector-selective agent, timentin, or carbenicillin and 2 mg/L of zeatin. Store at room temperature for up to 1 wk.
7. 1Z medium: same composition as 2Z except zeatin is reduced to 1 mg/L. Store at room temperature for up to 1 wk.
8. Selective rooting medium: 1X MS salts, 1X Nitsch vitamins, 3% sucrose, 0.8% Bacto-agar, pH 6.0. Autoclave, cool to 55°C, and add filter-sterilized stocks of vector selective agent and timentin or carbenicillin. Store at room temperature for up to 1 mo.

2.2. Preparation of Plant Material

1. Seeds from *L. esculentum* line(s) of choice.
2. 20% Household bleach (1.05% sodium hypochlorite) plus 0.1% Tween-20.
3. Sterile distilled water.
4. Solid 1/2 MSO medium in Magenta (GA7) boxes.

2.3. Preculture

1. Tobacco suspension culture (approx 7 d after subculture) grown in liquid KCMS medium.
2. Solid KCMS medium in 100 × 15 mm petri dishes.
3. Sterile Whatman filter paper (7-cm circles).
4. Tomato seedlings (6–8 d old) obtained from **step 3.1.**
5. Sterile Petri dishes or paper towels.
6. Sterile distilled water.

2.4. Preparation of Agrobacterium tumefaciens

1. *A. tumefaciens* strain containing tumor-inducing (Ti) plasmid harbouring the gene of interest.
2. LB medium supplemented with appropriate selection agent in 100 × 15 mm Petri dishes.
3. YM liquid medium.
4. MS liquid medium.

2.5. Infection and Cocultivation

1. Precultured tomato explants prepared as described in **step 3.2.**
2. *A. tumefaciens* culture prepared as described in **step 3.3.**
3. Sterile Magenta boxes or other wide-mouthed containers.
4. Sterile paper towels.

2.6. Selective Plant Regeneration

1. Selective 2Z medium in 100 × 15 mm Petri dishes.
2. Selective 1Z medium in 100 × 20 mm Petri dishes and Magenta boxes.
3. Micropore filter tape (3M Corporation, St. Paul, MN).
4. Selective rooting medium in Magenta boxes.

2.7. Transfer to Soil

1. Four-inch pots.
2. Sterile potting mix.
3. Clear plastic bags or containers.

3. Methods

3.1. Preparation of Plant Material

1. Surface sterilize seeds in 20% household bleach for 10 min and rinse three times with sterile distilled water (*see* **Note 1**).
2. Culture on 1/2 MSO medium in Magenta boxes (25 seeds/box). Maintain at 24 ± 2°C, under a 16-h photoperiod (cool white fluorescent lights, 60–100 E/m^2/s for 6–8 d depending on the tomato line being used. Cotyledons should be expanded and seedlings should be used before first true leaves emerge.

3.2. Preculture

3.2.1. Preparation of Feeder Plates

1. One day prior to explant preparation, pipet approx 2 mL of a 7-d-old tobacco suspension culture onto solid KCMS medium in Petri dishes. (*See* **Note 2** for maintenance of the tobacco suspension.)
2. Seal plates with Parafilm and incubate overnight in the dark at 24°C ± 2°C.
3. Cover the suspension culture with a 7-cm circle of sterile Whatman filter paper.

3.2.2. Preparation of Explants

1. Remove seedlings to damp sterile paper towel or Petri dish containing sterile water and excise cotyledons (*see* **Note 3**).
2. Cut both ends of cotyledon to remove the tip and petiole. If cotyledon sections are longer than 1 cm, cut them in half (*see* **Note 4**).
3. Place cotyledon explants on feeder layer plates prepared the previous day as described in **Subheading 3.2.1.** At this stage as many as 80 explants can be cultured on a single plate; however, fewer should be used if contamination may be a problem.
4. Seal plates with Parafilm and culture for 1 d at 24 ± 2°C, under a 16-h photoperiod.

3.3. Preparation of A. tumefaciens

1. Streak *A. tumefaciens* strain onto fresh plate of LB medium containing the appropriate selection agent and incubate for 48 h at 28°C.
2. Transfer four well-formed colonies to a flask containing 50 mL of YM liquid medium supplemented with the appropriate selection agent and maintain in a shaking incubator at 28°C until an OD600 of 0.4–0.6 is reached (usually overnight).
3. Centrifuge the cells at 8000*g* for 10 min at 20°C.
4. Resuspend pellet in 50 mL of MS liquid medium.

3.4. Infection and Cocultivation

1. Transfer cotyledon explants prepared as described in **step 3.2** to 25 mL of prepared *Agrobacterium* suspension in a sterile Magenta box or similar wide-mouthed container and incubate for 5 min.
2. Remove bacterial suspension with a sterile pipet.
3. Blot explants on sterile paper towels and place with the adaxial sides down (that is, upside down) on the original feeder plates (*see* **Note 5**).
4. Maintain at 19–25°C in the dark for 48 h of cocultivation (*see* **Note 6**).

3.5. Selective Plant Regeneration

1. Transfer cotyledon explants to plates containing selective 2Z medium with the adaxial sides facing up (*see* **Note 7**). A total of 25 explants are cultured on each plate at this stage.
2. Seal plates with micropore tape and maintain at 24 ± 2°C under a 16-h photoperiod of cool white fluorescent lights (*see* **Note 8**).
3. After 3 wk, transfer the cultures to plates containing 1Z medium (*see* **Note 9**).
4. Transfer explants to fresh medium at 3-wk intervals (*see* **Note 10**). Discard explants that are completely brown or bleached and trim off dead tissue from regenerating explants. Approximately 10 explants are cultured on each plate at this stage—the exact number depends on the size of the callus and regenerating tissue.
5. When shoots begin to regenerate from the callus, transfer cultures to 1Z medium in Magenta boxes.

6. Excise shoots from callus when they are approx 2 cm tall and transfer to selective rooting medium in Magenta boxes (*see* **Note 11**).
7. Maintain plants at 24 ± 2°C under a 16-h photoperiod of cool white fluorescent lights.

3.6. Transfer to Soil

1. When the selected transgenic lines have well-formed root systems, they can be transferred to soil (*see* **Note 12**).
2. Remove a plant from its culture vessel and gently wash the medium from the roots. Use tepid water.
3. Transfer each plant to a 4-in. pot containing a sterile potting mix.
4. Cover each plant with a clear plastic container or a plastic bag secured to the pot to provide a humid environment (*see* **Note 13**). Transfer to a growth chamber or to a shaded area of a greenhouse. Do not place in direct sunlight (*see* **Note 14**).
5. After 1 wk, gradually lift the plastic container each day during the next week. If a plastic bag is used, cut a small hole each day during the course of the next week, then remove the bag.

3.7. Timetable for Tomato Transformation

This timetable is based on a 6-d germination period. Adjust the times to the germination period of your tomato seeds.

D 1. Sterilize seeds and transfer to 1/2 MSO medium (*see* **Subheading 3.1.**).

D 5. Streak *Agrobacterium* onto LB selective medium (*see* **Subheading 3.3.1.**).

D 6. Prepare feeder layer plates (*see* **Subheading 3.2.1.**).

D 7. Prepare cotyledon explants (*see* **Subheading 3.2.2.**).
Inoculate liquid overnight culture (*see* **Subheading 3.3.2.**).

D 8. Prepare *Agrobacterium* suspension for infection (*see* **Subheadings 3.3.3.** and **3.3.4.**).
Infect and cocultivate cotyledon explants (*see* **Subheading 3.4.**).

D 10. Transfer cotyledon explants to selective 2Z medium (*see* **Subheadings 3.5.1.** and **3.5.2.**).

D 31. Transfer cultures to 1Z medium (*see* **Subheading 3.5.3.**). Continue to transfer to fresh 1Z medium at 3-wk intervals (*see* **Subheading 3.5.4.**).

4. Notes

1. If contamination is a problem, the seeds can be soaked for 2 min in 70% ethanol before the bleach treatment.
2. The tobacco suspension culture (NT1) is maintained in liquid KCMS medium on a rotary shaker and is subcultured weekly by adding 2 mL of the old culture to 48 mL of fresh medium.
3. Remove only as many seedlings as can be prepared in a few minutes, as they wilt quickly.

4. Hypocotyls may also be used but they regenerate more slowly and tend to produce more nontransformed shoots.
5. Explant orientation at this step is only for convenience.
6. Culture at lower temperatures has been associated with higher transformation efficiencies.
7. Explant orientation at this step has a significant effect on transformation efficiency. Explants placed adaxial side up curl into the medium, make better contact with it, and produce more transformed shoots.
8. Micropore tape gives higher transformation efficiency than Parafilm.
9. Zeatin level is reduced at this step to conserve resources. The reduction does not have a negative effect on transformation efficiency.
10. Transfer is necessary every 3 wk as plates sealed with micropore tape dry out quickly. More frequent transfers do not increase transformation rate.
11. Do not transfer shoots without meristems because they will not develop meristems in rooting medium.
12. To ensure that the plants are free of *Agrobacterium* before transfer to soil, shoots can be rooted twice on medium containing timentin or carbenicillin and then once on medium lacking this antibiotic. Plantlets that show *Agrobacterium* contamination at this stage should be destroyed by autoclaving. Alternatively, a PCR assay with *Agrobacterium*-specific primers can be done to ensure that plants are not contaminated.
13. Cover each plant with either the plastic container or a bag immediately after transfer to soil. The plants wilt quickly after removal from the culture containers. For plastic containers, Magenta boxes can be used or even clear plastic bottles that have the top of the bottle removed.
14. If placed in direct sunlight, heat will build up under the plastic containers or bags and kill the plants. If a growth chamber or shaded area of a greenhouse is not available, then maintenance in a lab setting under lights will be sufficient for the first 2 wk. After that period, they can be moved to a greenhouse.

References

1. Horsch, R. B., Fry, J. B., Hoffmann, N. L., et al. (1985) A simple and general method for transferring genes into plants. *Science* **227,** 1229–1231.
2. McCormick, S., Niedermeyer, J., Fry, J., Barnason, A., Horsch, R., and Fraley, R. (1986) Leaf disc transformation of cultivated tomato (*Lycopersicon esculentum*) using *Agrobacterium tumefaciens*. *Plant Cell Rep.* **5,** 81–84.
3. Toyoda, H., Matsuda, Y., Utsumi, R., and Ouchi, S. (1988) Intranuclear microinjection for transformation of tomato callus cultures. *Plant Cell Rep.* **7,** 293—296.
4. Xu, Y., Yu, H., and Hall, T .C. (1994) Rice triosephosphate isomerase gene 5' sequence directs glucuronidase activity in transgenic tobacco but requires an intron for expression in rice. *Plant Physiol.* **106,** 459–467.
5. Van Eck, J. M., Blowers, A. D., and Earle, E. D. (1995) Stable transformation of tomato cell cultures after bombardment with plasmid and YAC DNA. *Plant Cell Rep.* **14,** 299–304.

6. Nakata, K., Tanaka, H., Yano, K., and Takagi, M. (1992) An effective transformation system for *Lycopersicon peruvianum* by electroporation. *Jpn. J. Breed.* **42,** 487–495.
7. Koornneef, M., Hanhart, C., Jongsma, M., et al. (1986) Breeding of a tomato genotype readily accessible to genetic manipulation. *Plant Sci.* **45,** 201–208.
8. Tanksley, S. D., Ganal, M. W., Prince, J. P., et al. (1992) High density molecular linkage maps of the tomato and potato genomes. *Genetics* **132,** 1141–1160.
9. Martin, G. B., Frary, A., Wu, T., et al. (1994) A member of the tomato *Pto* gene family confers sensitivity to fenthion resulting in rapid cell death. *Plant Cell* **6,** 1543–1552.
10. Frary, A., Nesbitt, T. C., Frary, A., et al. (2000) *fw2.2*: a quantitative trait locus key to the evolution of tomato fruit size. *Science* **289,** 85–88.
11. Jones, D. A., Thomas, C. M., Hammond-Kosack, K. E., Balint-Kurti, P. J., and Jones, J. D. (1994) Isolation of the tomato *Cf-9* gene for resistance to *Cladosporium fulvum* by transposon tagging. *Science* **266,** 789–793.
12. Milligan, S. B., Bodeau, J., Yaghoobi, J., Kaloshian, I., Zabel, P., and Williamson, V. M. (1998) The root knot nematode resistance gene *Mi* from tomato is a member of the leucine zipper, nucleotide-binding leucine-rich repeat family of plant genes. *Plant Cell* **10,** 1307–1319.
13. Liu, J., Van Eck, J., Cong, B., and Tanksley, S. D. (2002) A new class of regulatory genes underlying the cause of pear-shaped tomato fruit. *Proc. Natl. Acad. Sci. USA* **99,** 13,302–13,306.
14. Sheehy, R. E., Kramer, M., and Hiatt, W. R. (1988) Reduction of polygalacturonase activity in tomato fruit by antisense RNA. *Proc. Nat. Acad. Sci. USA* **85,** 8805–8809.
15. Frary, A. and Hamilton, C. M. (2001) Efficiency and stability of high molecular weight DNA transformation: an analysis in tomato. *Transgenic Res.* **10,** 121–132.
16. Chyi, Y. S. and Phillips, G. C. (1987) High efficiency *Agrobacterium*-mediated transformation of *Lycopersicon* based on conditions favorable for regeneration. *Plant Cell Rep.* **6,** 105–108.
17. Fillatti, J. J., Kiser, J., Rose, R., and Comai, L. (1987) Efficient transfer of a glyphosate tolerance gene into tomato using a binary *Agrobacterium tumefaciens* vector. *Biotechnology* **5,** 726–730.
18. Hamza, S. and Chupeau, Y. (1993) Re-evaluation of conditions for plant regeneration and *Agrobacterium*-mediated transformation from tomato (*Lycopersicon esculentum*). *J. Exp. Bot.* **44,** 1837–1845.
19. Davis, M. E., Miller, A. R., and Lineberger, R. D. (1991) Temporal competence for transformation of *Lycopersicon esculentum* (L. Mill.) cotyledons by *Agrobacterium tumefaciens*: relation to wound-healing and soluble plant factors. *J. Exp. Bot.* **42,** 359–364.
20. van Roekel, J. S., Damm, B., Melchers, L. S., and Hoekema, A. (1993) Factors influencing transformation frequency of tomato (*Lycopersicon esculentum*). *Plant Cell Rep.* **12,** 644–647.

21. Frary, A. and Earle, E. D. (1996) An examination of factors affecting the efficiency of *Agrobacterium*-mediated transformation of tomato. *Plant Cell Rep.* **16,** 235–240.

10

Genetic Transformation of Conifers Utilizing Somatic Embryogenesis

Krystyna Klimaszewska, Robert G. Rutledge, and Armand Séguin

Summary

Over the last 5 yr, the production of transgenic conifers has been greatly facilitated by the ability to transform somatic embryonal tissues (somatic embryos) via cocultivation with *Agrobacterium tumefaciens*. This has allowed us to develop protocols for the genetic transformation of several spruce species. Furthermore, these procedures can produce an average of 20 independent transgenic lines (translines) per gram fresh mass of embryonal tissue, providing for the first time the magnitude-of-scale required for implementing large-scale functional genomics studies in conifers. Combined with efficient regeneration of transgenic trees via somatic embryos, the potential for genetic engineering of conifers has been demonstrated by stable reporter gene expression (GUS or GFP) resulting from single insert T-DNA integration events.

Key Words: *Agrobacterium* cocultivation; conifers; genetic transformation; *Picea*; somatic embryogenesis; spruce.

1. Introduction

Somatic embryogenesis (SE) depicts an asexual process that leads to the formation of an embryo from somatic cells either *in planta* through apomixis or in vitro through tissue culture. Somatic embryos are capable of germinating and producing plants in a manner very similar to their zygotic counterparts. Since its discovery in Norway spruce in 1985, SE has become a method of choice for clonal propagation of conifers because of its high and sustained productivity and the amenability of embryogenic cultures to cryogenic storage. Production of conifer somatic seedlings involves several steps: (a) induction of embryonal tissue (rapidly proliferating cultures of early-stage somatic embryos); (b) estab-

From: *Methods in Molecular Biology, vol. 286: Transgenic Plants: Methods and Protocols*
Edited by: L. Peña © Humana Press Inc., Totowa, NJ

lishment and proliferation of embryonal lines in culture via periodic subculturing, from which large amounts of embryonal tissue can be produced; (c) maturation of somatic embryos via transfer of embryonal tissue onto medium containing abscisic acid (ABA); and (d) germination of mature somatic embryos and transfer of the resulting somatic seedlings into soil. Conifer SE has become a primary enabling technology for genetic engineering because embryonal tissue is amenable to genetic transformation via cocultivation with *A. tumefaciens*. Combined with rapid selection of transgenic embryonal tissues and highly efficient seedling regeneration via somatic embryo maturation (**Fig. 1**), the production of transgenic spruce has now become routine. To illustrate this process, we will describe first the initiation and maintenance of embryonal tissues from three spruce species (*Picea mariana*, *P. glauca*, and *P. abies*), followed by the production of transgenic embryonal tissues, regeneration of the transgenic trees and molecular characterization of transgenic tissues.

2. Materials

2.1. Plant Material

1. Immature cones or mature seeds of spruce spp.
2. Spruce embryogenic cultures.

2.2. Tissue Culture

2.2.1. Stock Solutions

1. 10X MLV stock (modified Litvay's medium *[1]*): 8.21 g/L of NH_4NO_3, 9.5 g/L of KNO_3, 9.25 g/L of $MgSO_4 \cdot 7H_2O$, 1.7 g/L of KH_2PO_4 (monobasic), 0.11 g/L of $CaCl_2 \cdot 2H_2O$, 100 mL MLV micronutrient stock (100X), 100 mL of MLV vitamin stock (100X), 1 g/L of *myo*-inositol, 0.4 g/L of iron-ethylenediaminetetraacetic acid (Fe EDTA). Store frozen in 100-mL aliquots for further use.
2. 100X MLV micronutrient stock: 0.415 g/L of KI, 3.1 g/L of H_3BO_3, 2.1 g/L of $MnSO_4 \cdot H_2O$, 4.3 g/L of $ZnSO_4 \cdot 7H_2O$, 0.125 g/L of $Na_2MoO_4 \cdot 2H_2O$, 0.05 g/L of $CuSO_4 \cdot 5H_2O$, 0.013 g/ of $CoCl_2 \cdot 6H_2O$. Store frozen in 100-mL aliquots for further use.
3. 100X MLV vitamin stock: 0.05 g/L of nicotinic acid, 0.01 g/L of pyridoxine-HCl, 0.01 g/L of thiamine-HCl. Store frozen in 100-mL aliquots for further use.
4. 6-Benzylaminopurine (BA), 0.5 mg/mL stock solution.
5. 2,4-Dichlorophenoxyacetic acid (2,4-D), 1 mg/mL stock solution.
6. Glutamine, 25 mg/mL stock solution.
7. 10 m*M* (±)-*cis*, *trans*-ABA, 2.64 mg/mL stock solution.

2.2.2. Media

1. MLV: 1X MLV salts with vitamins, 1 g/L of casein hydrolysate (casamino acids), 20 g/L of sucrose, 2.2 mL/L of 2,4-D (1 mg/mL stock solution), 2.2 mL/L of BA (0.5 mg/mL stock solution), pH 5.7. Autoclave and add 20 mL of filter

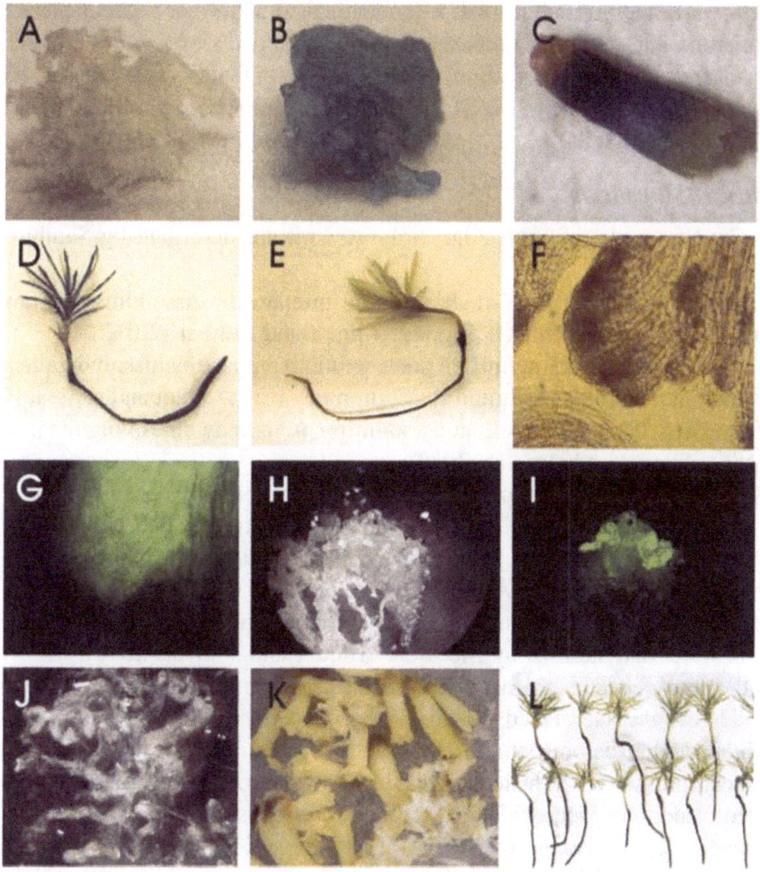

Fig. 1. Regeneration of transgenic spruce via somatic embryogenesis. Control (A) and stably transformed embryonal tissue (B) with the *gus* gene, magnification x2. GUS assay of a transgenic somatic embryo (C), magnification x16. Regeneration of a control somatic seedling (D) and transgenic somatic seedling (E), magnification x1.4. Closeup of a GFP transformed embryogenic suspension culture with normal light (F) and with blue light (G), magnification x150. Growing embryogenic cluster of transformed cells with normal light (H) and with blue light (I), magnification x20. Complete sequence of somatic embryogenesis in spruce with embryogenic cell suspension (J), development of somatic embryo (K), and germinating somatic seedlings (L).

sterilized glutamine (25 mg/mL stock solution) to cooled medium. If semisolid medium is required, add 4 g/L of gellan gum (Phytagel, Sigma, St. Louis, MO) before adjusting pH and before autoclaving.

MLV maturation medium (MLVM): identical to MLV but replace 2,4-D and BA with ABA at 60 μM and add 60 g/L of sucrose and 6 g/L of gellan gum.

3. MLV germination medium MLVG: identical to MLV but without plant growth regulators and with 6 g/L gellan gum.
4. *Agrobacterium* growth medium (YEP): 10 g/L of Bacto yeast extract, 10 g/L of Bacto-peptone, 5 g/L of NaCl, pH 7.2. After autoclaving, add 2 mL of filter-sterilized 1 *M* MgS04.

2.3. Other Chemicals

1. Standard reagents for molecular biology work are described by Sambrook and Russell *(2)*.
2. Acetosyringone: 200 *mM* stock solution, prepare by dissolving the powder in dimethyl sulfoxide (DMSO, Sigma). Aliquot and store at -20°C.
3. Kanamycin sulfate: 25 mg/mL of stock solution, prepare by dissolving the powder in water and sterilize by filtration (22 lim pore size). Aliquot and store at -20°C.
4. Rifampicin: 25 or 50 mg/mL stock solution, prepare by dissolving the powder in DMSO. Aliquot and store at -20°C.
5. Gentamicin sulfate: 20 mg/mL stock solution, prepare by dissolving the powder in water and sterilize by filtration. Aliquot and store at -20°C.
6. Timentin: 250 mg/mL stock solution; prepare by dissolving the powder in water and sterilize by filtration. Aliquot and store at -20°C.
7. Cefotaxime sodium: 250 mg/mL stock solution; prepare by dissolving the powder in water and sterilize by filtration. Aliquot and store at -20°C.
8. All the above cited chemicals are unstable at high temperatures. Thus, they should be added to the autoclaved medium once it has cooled to approx 60°C. Swirl to mix thoroughly before pouring the medium in Petri dishes.
9. p-glucuronidase (GUS) histochemical buffer: 0.5 mg/mL of X-GLUC (5-bromo-4-chloro-3-indolyl glucuronide), 100 mM sodium phosphate, pH 7.0, 0.5 *mM* ferrocyanide, 0.5 *mM* ferricyanide, 0.5% (v/v) Triton X-100, 1 *mM* ethylenediamine tetraacetic acid (EDTA).
10. 4-Methyl umbelliferyl glucuronide (MUG) base buffer: 150 mM phosphate, 0.1% Triton X-100, 0.1% Na sarcosine, 10 mM EDTA, 50 lig/mL of RNaseA, 25 mM sodium metabisulfite, pH 7.0.

2.4. Culture Conditions

Embryonal tissues are maintained at approx 23°C in the dark in an environmentally controlled growth chamber. Maturation of somatic embryos is done under indirect light (20-30 lLimol m^s) and germination of plantlets is conducted at approx 23°C under a 16-h photoperiod (90-110 lLimol m^s, TRUE-LITE®).

2.5. Transformation Vectors and A. tumefaciens Strain

Agrobacterium tumefaciens strain C58/pMP90 *(3)* is used in our laboratory for all the genetic transformation experiments in conifers. The *Agrobacterium* strain also contains a binary vector with a selectable marker (kanamycin or hygromycin). We have used successfully two types of binary vectors, pBINPLUS *(4)* and the pCAMBIA series *(see* Website: http://www.cambia.org).

For a detailed list of *Agrobacterium* binary vectors and technical details about these vectors, the reader should consult Hellens et al. *(5)*.

3. Methods

The methods described in the subheadings that follow outline (a) the initiation of somatic embryogenesis (somatic embryonal tissue), (b) consideration about the plasmid vectors and the transformation of *A. tumefaciens*, (c) the transformation of embryonal cultures, (d) the characterization of transgenic cell colonies, and (e) plant regeneration and establishment of transgenic trees in a greenhouse.

3.1. Somatic Embryogenesis: Initiation and Proliferation of Embryonal Tissues

1. Immature or mature seeds of *P. glauca*, *P. mariana*, and *P. abies* are surface-disinfected according to the published protocols *(6)* and the whole megagametophytes (containing the zygotic embryo) or isolated mature zygotic embryos are excised from the seeds. The explants are placed onto MLV medium.
2. Once the embryonal tissue appears either directly from the zygotic embryo or from the micropylar end of a megagametophyte (usually after 6–10 wk of culture) it is separated from the explant and placed onto fresh medium of the same composition for proliferation. The embryogenic cell line is considered established if it grows in a consistent and vigorous manner following multiple subcultures, conducted every two weeks.
3. The same medium is used for proliferation of embryonal tissue, cocultivation with *A. tumefaciens* and selection of transformed tissues, except that the cocultivation medium is supplemented with 50 μ*M* acetosyringone. Unless stated otherwise, all cultures are kept in the dark at 23°C.

3.2. Introduction of Binary Vectors into Agrobacterium

Agrobacterium strains are distinguished by their antibiotic resistance, carried either chromosomally or by a tumor-inducing (Ti) plasmid. For genetic transformation, the *Agrobacterium* strain generally contains a binary vector that will confer another antibiotic resistance for bacterial propagation of the binary vector, as well as antibiotic selection of transgenic tissues.

1. This procedure was adapted from An et al. *(7)*. A single *Agrobacterium* colony is grown in 2 mL of YEP media at 28°C overnight.
2. Then, 50 mL of YEP is inoculated with 2.0 mL of the overnight culture and grown to $OD_{600\ nm}$ of 0.5.
3. Following centrifugation for 5 min at 2700*g*, the cell pellet is resuspended into 10 mL of 0.15 *M* NaCl, followed by centrifugation for 5 min at 2700*g*, and the cells are resuspended into 1.0 mL of ice-cold 20 m*M* $CaCl_2$.
4. Then, 1 μg of DNA is mixed with 200 μL of cells in a 1.5-mL Eppendorf tube and incubated for 30 min on ice. This mixture is then frozen for 1 min in liquid

nitrogen and the tube is placed into a 37°C water bath until the cell mixture has thawed.

5. Then, 1 mL of YEP medium is added and the cells are incubated at 28°C for 2–4 h with slow shaking.
6. Following centrifugation for 1 min the cells are resuspended in 100 µL of YEP medium, followed by plating onto semisolid YEP medium containing 100 µg/mL of rifampicin, 50 µg/mL of kanamycin sulfate, and 20 µg/mL of gentamicin sulfate and the plate is incubated at 28°C. Transformed colonies appear within 2–3 d.

3.3. Agrobacterium *Transformation of Embryonal Tissue*

3.3.1. Cocultivation of Embryonal Tissues With A. tumefaciens

1. Day 1. A single colony of *A. tumefaciens* containing the desired binary plasmid is inoculated into YEP medium containing 50 or 100 mg/mL of rifampicin, 50 mg/mL of kanamycin sulfate, and 20 mg/mL of gentamicin sulfate, and incubated overnight at 28°C with shaking (250 rpm).
2. Day 2. Fifty microliters of bacterial suspension is transferred into 50 mL of YEP medium containing the same antibiotics and grown overnight under the conditions described in **step 1**.
3. Day 3. The bacterial cells are pelleted by centrifugation and resuspended in the MLV medium to an optical density of $OD_{600\ nm}$ 0.6.
4. Embryonal tissue is collected in a sterile test tube, and suspended in a liquid MLV medium at a ratio of 100 mg in 1.0 mL of medium to which an equal volume of bacterial suspension is added. This results in a final ratio of 50 mg of embryonal tissue per milliliter of *Agrobacterium* suspension (OD 0.3).
5. Finally, a sterile solution of acetosyringone is added to a final concentration of 50 µ*M*.
6. For the first hour the cocultivation is carried out in the liquid medium in the dark at 23°C on a shaker (110 rpm). Subsequently, 1.5 mL of suspension containing 75 mg of embryonal tissue is poured over a 70-mm filter paper disk (Whatman no. 2) that is placed in a Büchner funnel and a short low-pressure pulse is applied to drain excess liquid medium.
7. Each filter paper, all covered with a layer of embryonal tissue, is then placed onto a 90 × 15-mm Petri dish containing semisolid MLV medium (*see* **Subheading 2.2.2.**) containing 50 µ*M* acetosyringone and incubated at 23°C in the dark for 2 d.

3.3.2. Inhibition of A. tumefaciens *Growth*

1. Day 5. Filter papers are either transferred onto fresh semisolid medium of the same composition but supplemented with 400 mg/L of timentin, or the cells are collected and washed in a liquid MLV medium.
2. The washing step is often necessary to reduce the density of *A. tumefaciens* cells that multiply during the cocultivation and that may overgrow the plant cell culture. Usually, this has an adverse effect on the embryonal culture viability.

To collect embryonal tissues for washing, the filter paper is placed adjacent to the inner wall of the Erlenmeyer flask and 10 mL of the liquid medium is poured from a pipet over the surface of the filter paper until all or most of the cells are dislodged and collected in the flask. Tissues from more than one filter paper representing the same experimental treatment may be collected in one flask for washing.

3. An additional 10 mL of the medium may be added to reduce further the density of bacterial cells. The cells are then collected on a fresh filter paper using a Büchner funnel, and the filter paper is placed onto a fresh medium of the same composition and cultured as described above (*see* **Note 1**).

3.3.3. Selection of Transformed Cell Colonies, Proliferation of Transgenic Lines

1. Day 8. The filter papers are examined under the stereomicroscope and tissue colonies showing signs of growth are picked and subcultured individually onto fresh medium containing 400 mg/L of timentin and 25 mg/L of kanamycin for *P. abies*, or 35 mg/L of kanamycin for *P. mariana* and *P. glauca*. Note that each cluster of growing embryonal tissue has been produced by an individual transformation event and thus each represents a distinct transgenic line or "transline" (*see* **Note 2**).
2. Day 15. Subculture each clump of embryonal tissue onto fresh medium containing the same concentration of kanamycin, but with 300 mg/L of cefotaxime instead of timentin.
3. Continue subculturing every 2 wk. Transformed tissues should show vigorous growth, whereas the nontransformed tissues initially show dramatic reductions in growth rate, and eventually die, usually by the third subculture (after 6–8 wk on kanamycin-containing medium).
4. The translines are considered established after 16 wk of continuous growth on a medium with kanamycin (and cefotaxime); thereafter, the subcultures are performed on medium without kanamycin and cefotaxime (*see* **Note 3**).

3.4. Molecular Characterization of Spruce Translines

Following antibiotic selection and once sufficient embryonal tissue has grown, various techniques can be used for molecular characterization of each putative transline. A primary objective is to provide supporting evidence that each embryonal transline is truly transgenic, although growth in the presence of antibiotics has in our hands proven to be a reliable indicator that embryonal tissues are transgenic. Second, the level of transgene expression can be an important consideration for the selection of those translines from which somatic embryos will be matured and somatic seedlings produced. Finally, the same techniques can be used to follow both spatial and temporal transgene expressions in somatic embryos and seedlings.

Protein and nucleic acids extraction procedures for plants usually start with the grinding of the material in liquid nitrogen. For all the methods described in the following subheadings, traditional grinding using a mortar and pestle in liquid nitrogen works fine. However, specific grinding tools are available commercially and we have successfully used the FastPrep® System from Q-BIOgene (*see* Website: http://www.qbiogene.com/fastprep/index.shtml). This system is a benchtop homogenizer that works at high speed and is very efficient for rapid lysis of various samples (from tissue culture material to needles). The samples to analyze are placed in 2.0-mL resistant tubes containing lysing matrix particles (ceramic beads) and the extraction buffer. A short pulse (30–40 s) of vigorous shaking of the tubes will cause the particles to impact the sample from many directions in the extraction buffer. This operation will result in the release of nucleic acids and proteins into the homogenization buffer and the sample can then be recovered from the lysate after centrifugation. Under the next subheadings we will provide some of the techniques used for the molecular characterization of the translines. Unless otherwise stated, all molecular methods are carried out according to Sambrook and Russell *(2)*.

3.4.1. Histochemical Staining for GUS Activity

1. Assuming that the transformation vector expresses the *gus* gene, histological staining with X-Gluc is an effective and rapid method for visualizing *gus* transgene activity. We use a procedure based on that described by Jefferson *(8)* in which tissues are completely submerged into GUS histochemical buffer and incubated in the dark at 37°C for 24 h, although room temperature can also produce equivalent results and allows microscopic visualization of staining over time (*see also* Chapter 14).
2. The speed of color development and staining intensity can provide a qualitative indication of the level of *gus* transgene expression. In addition, the pattern of staining can be used to evaluate tissue specificity of the promoter used to drive *gus* gene expression. Embryonal tissue, somatic embryos, developing buds, and needles have all been found to produce very low levels of endogenous GUS activity, levels that are well below the detection level of X-Gluc histochemical staining. Caution must be exercised, however, for tissues taken from plants in soil, as artifactual staining can be produced by infecting fungi and bacteria (our unpublished results).

3.4.2. Quantitative Fluorescent Assay for GUS Activity

1. A fluorescence-based assay using MUG is commonly used to provide a quantitative measurement of *gus* transgene activity, which, although it requires some effort to set up, is still significantly easier to conduct than transgene messenger RNA (mRNA) accumulation assays (*see* **Subheading 3.4.4.**). The approach involves preparation of a tissue extract followed by a GUS enzymatic assay in which a small amount of tissue extract is mixed with the MUG substrate that produces a

fluorescent molecule, methyl umbelliferone (MU), on cleavage by GUS *(8,9)* (*see also* Chapter 14). The amount of GUS activity present in the sample is thus determined by taking samples of the reaction mixture over time, and determining the concentration of MU via fluorescence using a fluorometer. GUS activity (generally expressed as picomoles of MU produced per minute) is then normalized against the concentration of total protein in the sample.

2. However, we have found that normalization to DNA concentration can be more effective, in that the determination of DNA concentration via SYBR Green I fluorescent is much easier to conduct and more accurate than that of protein determination. In addition, this allows GUS activity to be expressed upon a "per cell" basis, as opposed to protein content of the tissue, which better reflects *gus* transgene activity (*see* **refs.** *8* and ***10*** for a more detailed discussion of this point).
3. The following procedures were developed for large-scale processing and analysis of transgenic tissues, in which samples are processed at room temperature and enzymatic protein and DNA assays are conducted in microtiter plates. For additional details and illustration of this protocol, the reader is referred to Rutledge and Côté *(10)*, in which the difficulties of assaying GUS activity from transgenic tissues of woody plants are also examined in detail.
4. Extracts are prepared by macerating tissues for 90 s at speed 6.0 using a FastPrep® System from Q-BIOgene in 2.0-mL screw-cap microfuge tubes containing 100 mg of glass beads (0.5 mm, Biospec Products, Bartlesville, OK), a single ceramic bead (5 mm, Bio101) and 1.0 mL of MUG base buffer. Base buffer can be supplemented further with 100 mg/mL polyvinyl polypyrrolidone (PVPP) for tissues containing phenolics and/or tannins. However, addition of PVPP was found to be unnecessary for embryonal tissues and somatic embryos of spruce. Extracts are then clarified by centrifugation at room temperature and 150 μL aliquots placed into the wells of a 96-well microtiter plate also at room temperature, from which samples are taken for enzymatic, protein and DNA determinations, using a multichannel pipet.
5. Protein concentrations are determined using the Bio-Rad Bradford Protein Assay kit (Bio-Rad, Hercules, CA) by mixing 5 μL of extract with 200 μL of diluted reaction mix in a microtiter plate, which is read in a standard microtiter plate spectrophotometer along with diluted BSA protein standards according to the manufacturer's instructions. Care must be taken to ensure that protein concentrations are within the linear range of the assay.
6. DNA concentration determinations are conducted by mixing 5 μL of extract or DNA standard with 200 μL of a 1/3000 dilution of SYBR Green I (Molecular Probes, Eugene, OR) in Tris-ethylenediaminetetraacetic acid (TE) (10 m*M* Tris, 1 m*M* EDTA, pH 7.0). The DNA concentrations in the tissues extracts are then calculated by comparison to the fluorescence produced by a dilution series of a DNA standard (Lambda DNA, Roche Molecular Systems, Alameda, CA) prepared in extraction buffer. In our laboratory, fluorescence is measured using a Fluorolite1000 microtiter-plate reader (Dynatech Laboratories, Chantilly, VA; filter set: 485 nm excitation–BP22 and 530 nm emission–BP30).

7. Microtiter-plate MUG assays are conducted in duplicate for each extract, using a 200-µL reaction volume containing 5 µL of extract and 1.0 m*M* MUG (Sigma) in base buffer lacking RNaseA. The microtiter plate is covered and incubated at 37°C in an air-circulating oven. Then, 20-µL aliquots are taken from the reactions using a multichannel pipette every 10 min for a total of 60 min, and immediately mixed with 180 µL of stop buffer (0.2 *M* Na_2CO_3) in the wells of another microtiter plate. MU concentrations are determined based on a standard curve derived from six MU standards placed into this microtiter plate, and fluorescence is determined using a Fluorolite1000 microtiter-plate reader (Dynatech Laboratories; filter set: 365 nm excitation—BP15 and 450 nm emission-BP65). GUS enzymatic rates are then calculated by averaging the slope of MU production from each of the duplicate reactions and reported as pmol MU/minute/milligram of protein or nmol MU/minute/microgram of DNA (*see* **ref. *10*** for additional details of this procedure).

3.4.3. PCR Analysis of Genomic DNA

Confirming the integration of the transgene into genomic DNA is another method for molecular characterization of translines. In many cases, the presence of the gene used for antibiotic selection is the primary target, although other segments of the transformation vector (such as the transgene itself) are also used.

1. Genomic DNA is extracted from embryogenic tissue using Qiagen Genomic-tips (Qiagen, Valencia, CA) and the Qiagen protocol for plants. As mentioned previously, homogenization of plant tissues is done using the FastPrep® System.
2. Polymerase chain reaction (PCR) analysis is usually conducted on DNA samples from individual transgenic lines, non-transgenic lines, and from a negative control that contained no DNA. PCR reaction mixtures (50 µL) contained 1.5 m*M* $MgCl_2$, 100 µ*M* of each dNTP, 1 µ*M* of each primer, 1X of the supplied buffer, 2.5 U of *Taq* DNA polymerase (Life Technologies Inc., Rockville, MD) and 100 ng of template DNA. Thermocycler parameters were as follows: 10 min preheat at 95°C; 25 cycles of 94°C for 1 min, 58°C for 1 min, and 72°C for 1 min. We have performed several PCR amplifications specific to the neomycin phosphotransferase (*nptII*) gene using genomic DNA from various transgenic spruce species. The use of those *nptII* gene specific primers (5' CTGGCCACGACGGGCG TTCCTTG 3' and 5' GAATCGGGAGCGGCGATACCGTAAA 3') resulted in the amplification of a 545-bp DNA fragment. The products of the PCR reactions were analyzed by electrophoresis on a 1% agarose gel and stained with ethidium bromide.

3.4.4. RNA Extraction and Transcript Accumulation Analysis

In many cases, in particular for transformation vectors lacking the *gus* gene, it is necessary to quantify transgene activity in terms of mRNA transcript accumulation. Standard molecular techniques are used for Northern or real-time PCR analysis. The procedure of RNA extraction has been based on either the method described by Chang et al. ***(11)***, or the RNAeasy Plant Mini kit (Qiagen). Tissue homogenization using the FastPrep® System as described for tissue extraction for GUS assay has also been found to be an effective alternative to grinding tissue in liquid nitrogen and has been found to be a very rapid and effective method for extracting RNA from embryonal tissue and somatic embryos of spruce (*see* **Note 4**).

3.5. Plant Regeneration and Growth in a Greenhouse

The maturation of non-transformed and transgenic somatic embryos, germination, and growth in a potting mix are performed according to the published protocol ***(12)***.

3.5.1. Somatic Embryo Maturation

1. The tissue pieces are first suspended in a liquid MLV medium without plant growth regulators in a centrifuge test tube and are vigorously shaken to dissociate the pieces into fine cell suspension.
2. Then, 3 mL (or less) of the cell suspension is transferred onto a filter paper disk in the Büchner funnel and a short, low-pressure pulse is applied to drain the liquid and anchor the cells to the filter paper.
3. Subsequently, the filter paper with the cells is placed on MLVM medium. The cultures are placed under low-intensity light, 16-h photoperiod, at 23°C for 7–8 wk.

3.5.2. Somatic Embryo Germination and Conversion to Plantlets

1. The cotyledonary and morphologically normal somatic embryos are collected individually under the stereomicroscope and placed horizontally on a MLVG medium in Petri dishes for germination and plantlet conversion.
2. The somatic embryos are kept for the first 7–10 d under low light intensity and then placed under higher light intensity until both a root and an epicotyl develop.

3.5.3. Potting and Acclimatization of Plantlets

1. Plantlets that show both a growing root and an epicotyl with several needles may be planted in a substrate (1:1 peat–perlite mix) in small containers and kept in a mist chamber for 12–14 d under greenhouse conditions.
2. After this period, the humidity is gradually reduced and the plantlets can be grown outside the mist chamber.

4. Notes

1. Alternatively, cells are collected manually with a spatula and washed in 15 mL of MLV. The cells are then collected on a filter paper by filtration using the Büchner funnel. Cefotaxime at 300 mg/L also can be used as a bacteriostatic agent.

2. High density of cultured cells after cocultivation with *A. tumefaciens* can potentially create a problem with respect to the recovery of chimeric translines. If the transformed cells are proliferating in close proximity to each other then there is a possibility that two independently transformed cell colonies will "fuse" and create a chimeric transline. Thus, it is important to make frequent microscopic observations of postcocultivation cultures and separate rapidly growing colonies to prevent them from "fusing."
3. Our results (unpublished data) indicate that the selection period can be shorter, only 10–12 wk, without creating a problem of escapes or recurring *Agrobacterium* growth. However, the latter requires further research.
4. It should be noted, however, that although it has also been very successful for extracting RNA from flushing buds, we have not been successful in using the RNAeasy kit for extracting RNA from mature needles.

Acknowledgments

Our invaluable team of collaborators, C. Côté, D. Lachance, C. Levasseur, M.-J. Morency, G. Pelletier, and D. Stewart are gratefully acknowledged. A special thanks goes to C. Levasseur for providing the photographs. This research was supported by grants from the National Biotechnology Strategy of Canada.

References

1. Litvay, J. D., Verma, D. C., and Johnson, M. A. (1985) Influence of loblolly pine (*Pinus taeda* L.) culture medium and its components on growth and somatic embryogenesis of the wild carrot (*Daucus carota* L.). *Plant Cell Rep.* **4,** 325–328.
2. Sambrook, J. and Russell, D. W. (2001) *Molecular Cloning: A Laboratory Manual*, Cold Spring Harbor Laboratory Press, Cold Spring Harbor, NY.
3. Koncz, C., and Schell, J. (1986) The promoter of TL-DNA gene 5 controls the tissue-specific expression of chimaeric genes carried by a novel type of *Agrobacterium* binary vector. *Mol. Gen. Genet.* **204,** 383–396.
4. van Engelen, F. A., Molthoff, J. W., Conner, A. J., Nap, J.-P., Pereira, A., and Stiekema, W. J. (1995) pBINPLUS: an improved plant transformation vector based on pBIN19. *Trans. Res.* **4,** 288–290.
5. Hellens, R., Mullineaux, P., and Klee, H. (2000) A guide to *Agrobacterium* binary Ti vectors. *Trends Plant Sci.* **5,** 446–451.
6. Cheliak, W. M. and Klimaszewska, K. (1991) Genetic variation in somatic embryogenic response in open-pollinated families of black spruce. *Theor. Appl. Genet.* **82,** 185–190.
7. An, G., Ebert, P. R., Mitra, A. and Ha, S. B. (1988) *Binary Vectors in Plant Molecular Biology Manual*, Vol. A3 (Gelvin, S. B. and Schilperoort, R. A., eds.), Kluwer Academic, Dordrecht, The Netherlands, pp. 1–19.
8. Jefferson, R. A. (1987) Assaying chimeric genes in plants: The GUS gene fusion system. *Plant Mol. Biol. Rep.* **5,** 387–405.
9. Gallagher, S. R. (1992) *GUS Protocols: Using the GUS Gene as a Reporter of Gene Expression*, Academic Press, San Diego, CA, p. 221.

10. Côté, C. and Rutledge, R. G. (2003) An improved MUG fluorescent assay for the determination of GUS activity within transgenic tissue of woody plants. *Plant Cell Rep.* **21,** 619–624.
11. Chang, S., Puryear, J., and Cairney, J. (1993) A simple and efficient method for isolating RNA from pine trees. *Plant Mol. Biol. Rep.* **11,** 113–116.
12. Klimaszewska, K., Lachance, D., Pelletier, G., Lelu, M.-A., and Séguin, A. (2001) Regeneration of transgenic *Picea glauca*, *P. mariana*, and *P. abies* after cocultivation of embryogenic tissue with *Agrobacterium tumefaciens*. *In Vitro Cell. Dev. Biol. Plant* **37,** 748–755.

11

Regeneration of Transgenic Cassava From Transformed Embryogenic Tissues

Peng Zhang and Johanna Puonti-Kaerlas

Summary

Production of transgenic plants is gradually becoming routine in cassava biotechnology. Green cotyledons of maturing somatic embryos (somatic cotyledons for short) and friable embryogenic suspensions (FES) are the target tissues for transformation by *Agrobacterium* or biolistics. Putative transgenic shoots develop from transformed somatic cotyledons via shoot organogenesis or from FES via somatic embryogenesis under selection. Maturation of transgenic somatic embryos is induced by transfer to maturation medium with reduced concentrations of selective agents. Mature somatic embryos can also develop directly from FES cells under selection. Transgenic plants are regenerated by the elongation of transgenic shootlets from organogenesis experiments and by the germination of or shoot development from transgenic mature embryos cultured without selection. β-Glucuronidase (GUS) assays and rooting tests can be used to screen for escapes from selection, which improves the regeneration rate of truly transgenic plants.

Key Words: Friable embryogenic callus; genetic transformation; *Manihot esculenta* Crantz; plant regeneration; somatic embryogenesis.

1. Introduction

Somatic embryogenesis forms the basis for all current transformation methods of cassava. Somatic embryogenesis is the production of embryolike structures from somatic cells. Somatic embryos are independent bipolar structures that are not vascularly attached to their tissues of origin and that can develop and germinate into plantlets through developmental steps that correspond to those of the zygotic embryos. Primary somatic embryos can be induced to produce secondary somatic embryos by further subculturing on auxin-containing medium. By constant subculturing of somatic embryos, a cyclic embryogenesis system

From: *Methods in Molecular Biology, vol. 286: Transgenic Plants: Methods and Protocols*
Edited by: L. Peña © Humana Press Inc., Totowa, NJ

can be established. Cyclic embryogenesis is the most routinely used method of *de novo* plant regeneration of cassava in vitro and provides constant source of material for transformation. Plant regeneration from somatic embryos can be established via either direct germination ***(1–4)*** or shoot organogenesis ***(5,6)***.

However, somatic embryogenesis *per se* appears not to be fully compatible with the available selection systems, and there are only few reports on regeneration of transgenic cassava plants directly from somatic embryos. Currently, shoot organogenesis from cotyledons of somatic embryos ***(7–9)*** and embryogenic suspensions derived from friable embryogenic callus (FEC) ***(10,11)*** are the two most frequently used systems for production of transgenic cassava plants.

In the organogenesis system, development of shoot primordia is induced directly on cytokinin-containing media from cotyledon explants of germinating somatic embryos. A cycling system in which secondary somatic embryos are induced on cotyledon explants from maturing somatic embryos is used to provide a constant source of regeneration-competent explant material. In some cultivars, for example, TMS60444, a fraction of somatic embryos can also produce a new tissue type which consists mainly of small globular embryolike structures. When isolated, it produces a highly friable embryogenic callus (FEC). Once pure FEC is obtained, it can be transferred easily to liquid medium to establish a rapidly proliferating embryogenic suspension. By transfer to solid medium supplemented with auxin, development of maturing embryos can be induced ***(4)***.

FEC and the shoot organogenesis systems are compatible with both biolistic and *Agrobacterium*-mediated gene transfer and with several selectable markers. Antibiotic selection using hygromycin ***(8,9,12)*** and positive selection using mannose ***(9,12)*** (*see* Chapter 16) can be used for both shoot organogenesis and FEC-based methods, and paromomycin, kanamycin, or visual selection using firefly luciferase as a screenable marker or a combination of antibiotic selection and luciferase screening ***(10,11,13,14)*** can also be used to produce transgenic plants from FEC. Combining β-glucuronidase (GUS) assays with hygromycin or mannose selection allows rapid and easy visual preselection of fully transgenic suspensions before the regeneration step, resulting in 100% selection efficiency and reducing the length of time required for liquid culture before plant regeneration, and rooting assays can also be used to eliminate escapes from selection when plants are regenerated from somatic cotyledon explants ***(8,9)***.

2. Materials

2.1. Plant Material

1. Shoot cultures of cassava cultivars MCol22 and TMS60444.
2. Four-week-old somatic embryos of cassava cultivar MCol22.
3. Three-month-old friable embryogenic suspension (FES) of cultivar TMS60444.

2.2. Cassava Tissue Culture Media

2.2.1. Stocks

1. 1X Murashige and Skoog (MS) basal salt mixture including vitamins (powder, Duchefa Biochemie, Haarlem, The Netherlands) ***(15)***.
2. 1X Schenk and Hildebrandt (SH) basal salt mixture (powder, Duchefa Biochemie) ***(16)***.
3. 1X Gresshoff and Doy (GD) basal salt mixture including vitamins (Duchefa Biochemie).
4. 1000X MS vitamins ***(15)***; dissolve in distilled water and store at –20°C in aliquots.
5. 2 m*M* $CuSO_4$; store at 4°C.
6. 1 mg/mL of 6-benzylaminopurine (BAP), dissolve in 1 *N* NaOH (stable at 4°C for up to 3 mo).
7. 12 mg/mL of picloram; dissolve in 1 *N* NaOH and store at –20°C in aliquots.

2.2.2. Media

1. Basic shoot culture medium (CBM): 1X MS salts with vitamins, 2 μ*M* $CuSO_4$, 2% sucrose, 0.3% Gelrite, pH 5.8, autoclave.
2. Axillary bud enlargement medium (CAM): 1X MS salts with vitamins, 2 μ*M* $CuSO_4$, 10 mg/L of BAP, 2% sucrose, 0.3% Gelrite, pH 5.8, autoclave.
3. Somatic embryo induction medium (CIM): 1X MS salts with vitamins, 2 μ*M* $CuSO_4$, 12 mg/L of picloram, 2% sucrose, 0.3% Gelrite, pH 5.8, autoclave.
4. Embryo maturation medium (CMM): 1X MS salts with vitamins, 2 μ*M* $CuSO_4$, 0.1 mg/L of BAP, 2% sucrose, 0.3% Gelrite, pH 5.8, autoclave.
5. Friable embryogenic callus medium (GD): 1X GD salts with vitamins, 12 mg/L of picloram, 2% sucrose, pH 5.8, autoclave.
6. Suspension culture medium (SH): 1X SH salts, 1X MS vitamins, 12 mg/L of picloram, 6% sucrose, pH 5.8, autoclave.
7. Shoot elongation medium (CEM): 1X MS salts with vitamins, 2 μ*M* $CuSO_4$, 0.4 mg/L of BAP, 2% sucrose, 0.3% Gelrite, pH 5.8, autoclave.
8. Somatic embryo emerging medium (MSN): 1X MS salts with vitamins, 1 mg/L of α-Naphthaleneacetic acid (NAA), 2% sucrose, 0.3% Gelrite, pH 5.8, autoclave.
9. *Agrobacterium* growth medium (YEB): 1 g/L of Bacto yeast extract, 5 g/L of Bacto beef extract, 5 g/L of Bacto peptone, 5 g/L of sucrose, pH 7.2. After autoclaving add 2 mL of filter sterilized 1 *M* $MgSO_4$.

2.3. Other Chemicals

1. Acetosyringone: 200 m*M* stock solution; prepare by dissolving the powder in dimethyl sulfoxide (DMSO, Sigma, St. Louis, MO) and store at –20°C in aliquots.
2. Carbenicillin: 250 mg/mL stock solution; prepare by dissolving the powder in water and sterilize by filtration (22 μm pore size). Store at –20°C in aliquots.
3. Hygromycin: 25 mg/mL; dissolve in water and sterilize by filtration. Store at –20°C in aliquots.

Fig. 1. Schematic representation of the T-DNA region of the binary vector pHMG. RB, Right border; LB left border; p35S CaMV 35S promoter; 35! and Nos!, CaMV 35S and *Agrobacterium* nopaline gene terminator, respectively. For gene abbreviations, *see* text.

4. Rifampicin: 25 mg/mL stock solution, prepare by dissolving the powder in 0.1 *M* HCl and sterilize by filtration. Store at –20°C in aliquots.
5. Spectinomycin: 50 mg/mL stock solution, prepare by dissolving the powder in water and sterilize by filtration. Store at –20°C in aliquots.

All the above mentioned chemicals are unstable at high temperatures. Thus, they should be added to the autoclaved medium first after it has cooled to 60°C. Swirl to mix thoroughly before pouring the medium to containers.

2.4. Culture Conditions

All plant material is cultured at 26°C under a 16/8-h photoperiod (90–110 $\mu mol/m^2s$, TRUE-LITE™) in an environmentally controlled growth chamber unless otherwise stated. The embryogenic suspension cultures are cultured on a gyratory shaker (108 rpm) at 28°C under continuous light (approx 50 $\mu mol/m^2s$) and subcultured at 3-d intervals. The cultures are sieved through a metallic net (ϕ 500 μm) every 4 wk to obtain the fraction consisting of embryogenic units ranging in size from 250–500 μm.

2.5. Agrobacterium tumefaciens *Strain and Example Plasmid Used*

Agrobacterium tumefaciens strain LBA4404 ***(18)*** carrying the binary vector pHMG (**Fig. 1**) is used here as an example for genetic transformation experiments. The plasmid pHMG harbors an intron-interrupted hygromycin phosphotransferase gene (*hpt*) and an intron-interrupted *uidA*. Both genes are driven by the CaMV 35S promoters. Another selectable marker gene, phosphomannose isomerase (*pmi*), is also present in the T-DNA region. In the following procedures, the *hpt* is used as selectable marker. For protocols using other selectable markers, *see* **refs.** ***10–14***.

2.6. GUS Assays

GUS assays are performed by placing tissues in an assay buffer (10 m*M* $Na_2EDTA \cdot H_2O$, 0.1% Triton X-100, 0.3% 5-bromo-4-chloro-3-indoldyl β-D-glucuronide [X-Gluc], 0.1 *M* NaH_2PO_4, 0.5 *M* $K_3Fe[CN]_6$) ***(19)***. After 3–6 h incubation at 37°C, the tissues are washed several times with 96% ethanol and stored in 96% ethanol.

3. Methods

The methods described in the following subheadings outline (a) Induction of primary somatic embryos and cyclic embryogenic cultures, (b) Induction of FEC and embryogenic suspensions, (c) *Agrobacterium*-mediated FES transformation, and (d) Transformation of somatic cotyledons (*see* **Note 1**).

3.1. Induction of Primary Somatic Embryos and Cyclic Embryogenic Cultures

3.1.1. Primary Embryogenesis from Axillary Buds, Immature Leaves, and Leaf Lobes

1. Cut nodal explants (young stem cuttings with axillary buds) from 4-wk-old in vitro plants of MCol22 or TMS60444 and place them horizontally on CAM medium for 6 d.
2. Remove the enlarged axillary buds from the nodal explants with a syringe needle and transfer onto plates containing somatic embryo induction medium CIM.
3. Alternatively, isolate apical meristems, shoot tips and 1–6 mm long immature leaf lobes from the shoot tips of MCol22 or TMS60444 and culture on CIM medium (*see* **Note 2**).
4. Incubate the plates at 26°C in the dark or a continuous weak light (approx 10 $\mu mol/m^2s$) (*see* **Note 3**). After 2 wk, check the emerging embryos on the explants. Once new globular embryos appear (**Fig. 2A**), transfer them onto fresh CIM medium (*see* **Note 4**).

3.1.2. Establishment Cyclic Embryogenic Cultures

1. Subculture somatic embryos/embryo clusters on CIM medium every second week. Somatic embryos can be multiplied via secondary embryogenesis in this way (*see* **Note 5**).
2. Alternatively, after two or three cycles, harvest and transfer somatic embryos (**Fig. 2B**) onto CMM at 26°C with a 16-h photoperiod to induce embryo maturation and production of green cotyledons.
3. Collect 2 wk-old green cotyledons (**Fig. 2C**), cut into 0.25-cm^2 pieces with a scalpel and place on CIM medium to induce secondary embryogenesis. Approximately 50 pieces can be placed per Petri dish.
4. After 2 wk, transfer the new somatic embryos/embryo clusters that develop from cut ends of the cotyledons onto CIM medium for establishing cyclic embryogenic cultures.

3.2. Induction of FEC and Embryogenic Suspensions

1. Transfer somatic embryo clusters of TMS60444 onto GD medium and culture at 26°C in the dark.
2. Subculture the cultures at 4-wk intervals. After two or three cycles, check the cultures under a microscope.
3. Separate the FEC (a light yellow, highly friable callus with numerous spherical compact embryogenic units with dense cytoplasm) from the surface of embryogenic structures and place onto plates containing 25 mL of GD medium at 26°C under a 16-h photoperiod (*see* **Note 6**).
4. After 2 wk, select pure FEC under a stereomicroscope to eliminate undesirable tissues, such as nonembryogenic friable calli and somatic embryos, and transfer to fresh GD medium.

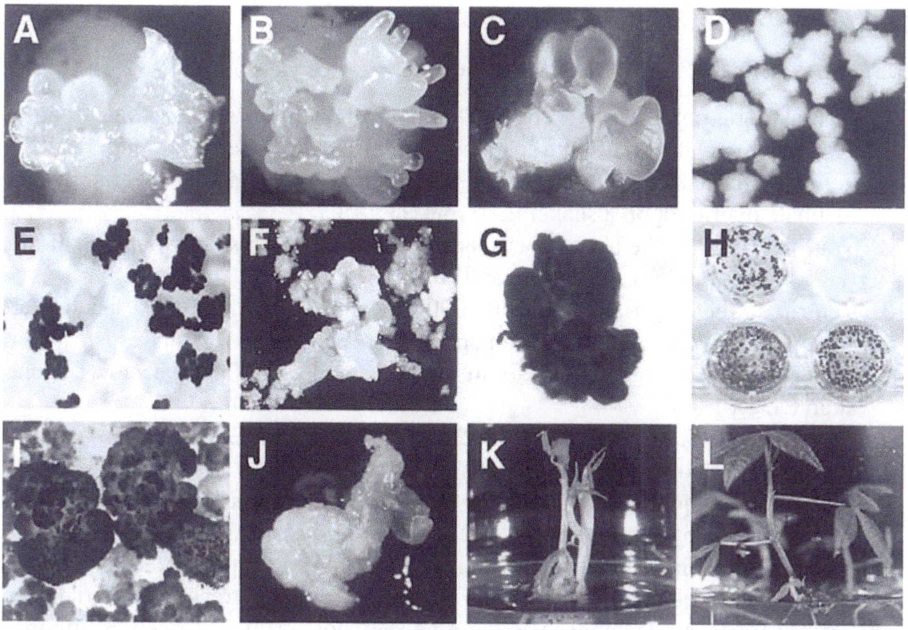

Fig. 2. Regeneration of transgenic cassava via somatic embryogenesis and shoot organogenesis. **(A)** Primary embryos. **(B)** Secondary embryos. **(C)** Cotyledon stage somatic embryos. **(D)** Friable embryogenic suspensions. **(E)** GUS assay of transformed embryogenic suspensions after 3-wk selection. **(F)** Formation of somatic embryos from embryogenic suspension cells. G, GUS assay of transgenic somatic embryos. **(H)** GUS assay of transgenic suspension lines; the well on the right of the top row is negative control. **(I)** Close-up of GUS stained transgenic suspension cultures. J, Hygromycin resistant shoot primordia developed from somatic cotyledon. **(K)** Shoots regenerated from transgenic somatic embryos. **(L)** Shoot cultures of transgenic cassava plants.

5. Subculture the FEC cultures on GD medium at 3-wk intervals (*see* **Note 7**).
6. Transfer approx 0.5 g of FEC in 30 mL of SH medium in 190-mL jars and culture on a gyratory shaker (108 rpm) at 28°C under continuous light (approx 50 µmol/m^2/s).
7. Remove the old medium with a pipet and replace with 30 mL of fresh SH medium every third day.
8. After 3 wk, sieve the cultures through a metallic net (ϕ 500 µm) to obtain the fraction consisting of embryogenic units ranging in size from 250-500 µm (*see* **Note 8**). Subculture aliquots of 1 mL settled cell volume (SCV) of embryogenic suspension cells (**Fig. 2D**) in a fresh jar with 30 mL of fresh medium. Repeat the filtering every 3 wk.

3.3. *Agrobacterium*-Mediated FES Transformation

3.3.1. Inoculation With *A. tumefaciens*

1. Pick a single colony of *Agrobacterium* harboring the plasmid pHMG from an agar plate and inoculate into a round-bottom plastic tube containing 5 mL of liquid YEB medium with 25 mg/L rifampicin and 100 mg/L of spectinomycin. Incubate overnight on a shaker (240 rpm) at 28°C.
2. Transfer 25 µL of the bacterial suspension to 50 mL of fresh medium in 250-mL flasks and culture for 12–20 h to an OD_{600} 0.5–1.0.
3. Centrifuge the bacterial solution at 6000 rpm at 4°C for 10 min, resuspend in an equal volume of liquid MS medium, pH 5.3, and centrifuge again. Resuspend the bacterial pellet in liquid MS medium supplemented with 200 µ*M* acetosyringone at an OD_{600} 1.0.
4. Grow the *Agrobacterium* culture for 2 h in 190-mL jars at 28°C (80 rpm) and use for inoculation of explants.
5. Transfer an aliquot of 2 mL of SCV of embryogenic cassava suspensions by a sterile pipette to jars containing 10 mL of bacterial suspension.
6. After 45 min, remove excess bacterial suspension with a pipet. Spread inoculated tissues using a pipet onto sterile filter papers (ϕ 9 cm) and transfer to Petri dishes with solidified SH medium supplemented with 100 µ*M* acetosyringone. Cocultivate at 25°C for 3 d.

3.3.2. Selection and Establishment of Transformed Suspensions

1. Collect the transformed tissues from the filter paper with a pair of tweezers and transfer to jars containing 30 mL of SH medium.
2. Wash the tissues by pipetting up and down several times. Draw off the liquid and repeat the wash two times.
3. To assess transformation efficiency, put several suspension clusters into GUS buffer. Incubate at 37°C for 6 h, wash several times with 96% ethanol, and then store in 96% ethanol. Successful T-DNA transfer is indicated by the dark blue precipitant in the tissues.
4. Transfer the rest of the FEC into fresh jars containing 30 mL of SH medium with 12.5 mg/L of hygromycin and 500 mg/L of carbenicillin. Culture for 3 d on a shaker at 137 rpm.
5. Replace the culture medium with SH medium containing 25 mg/L of hygromycin and 500 mg/L of carbenicillin. Culture on a shaker at 108 rpm. Refresh the medium at 3-d intervals. Include both positive and negative controls to monitor the culture conditions and selection efficiency (*see* **Note 9**).
6. After 2 wk of cultivation in selective medium, the antibiotic-resistant suspension cells develop into yellowish, friable embryogenic clusters indistinguishable from control FEC grown under nonselective conditions. Stain a small part of cell clusters in GUS assay buffer for 6 h at 37°C. Transgenic cells are easily distinguished by the dark blue precipitate (**Fig. 2E**) from a background of white and brown-colored dead tissues.

3.3.3. Somatic Embryogenesis From Transformed Suspension Cultures

1. Spread the suspensions, which have grown under selection for 2–3 wk to Petri dishes containing 25 mL of MSN medium with 10 mg/L of hygromycin. Culture at 26°C under a 16 h photoperiod to allow embryo development.
2. After 2–4 wk, resistant somatic embryos (**Fig. 2F**) are formed. Use several embryos from each embryo cluster for GUS assay. Blue stain is detectable in transgenic lines (**Fig. 2G**). Transfer GUS-positive embryo lines individually onto CMM containing 12.5 mg/L of hygromycin for cotyledon emergence.
3. Alternatively, transfer transformed suspensions to GD solid medium with 10 mg/L of hygromycin to produce hygromycin-resistant FECs in 2–3 wk. Then pick up FECs individually and culture in 30 mL of SH medium supplemented 10 mg/L of hygromycin to establish transgenic suspension lines. Use a small fraction of the transgenic suspension lines for the GUS assay. Transgenic suspension lines show dark blue staining (**Fig. 2H**). The independent transgenic suspension lines can be used directly for molecular analysis or transfer to MSN medium for embryo development as described in **steps 1** and **2** (*see* **Note 10**).

3.4. Shoot Organogenesis From Transformed Somatic Cotyledons

1. Cut the cotyledons from 2-wk-old somatic embryos into 0.25-cm^2 pieces with a scalpel and place on CBM medium to prevent the explants from drying.
2. Inoculate the cotyledon pieces with *Agrobacterium* in jars containing 20–30 mL of bacterial suspension from **step 4** of **Subheading 3.3.1.** and place on a shaker at 80 rpm.
3. After 45 min, remove the bacterial suspension with a pipet and transfer the cotyledon pieces onto plates containing COM supplemented with 100 μ*M* acetosyringone and cocultivate at 25°C for 3 d.
4. After cocultivation, growing bacteria can be visible at the edges of the cotyledon explants on the medium. Wash the inoculated explants with sterile water in jars three times. Then wash two times with liquid MS medium supplemented with 500 mg/L carbenicillin. Place the explants on sterile filter paper to draw off excess liquid.
5. Put several cotyledon pieces into GUS buffer for transient assay. Incubate at 37°C for 6 h, wash several times with 96% ethanol, and store in 96% ethanol. Successful T-DNA transfer is indicated by the dark blue precipitant in the cotyledon pieces. Transfer the rest of the explants onto regeneration and selection medium as follows.
6. Place the inoculated somatic cotyledon pieces onto plates containing COM medium supplemented with 4 mg/L of $AgNO_3$, 10 mg/L of hygromycin, and 500 mg/L of carbenicillin in the dark at 26°C for 1 wk.
7. Transfer the explants to fresh COM medium with 20 mg/L of hygromycin and 500 mg/L of carbenicillin under the same culture conditions.
8. After 2 wk, cut off emerging shoot primordia that appear on the explants (**Fig. 2J**) and place on CEM medium supplemented with 10 mg/L of hygromycin for another 2 wk.

9. Transfer elongated shootlets onto CBM medium for further growth. Simultaneously, pick up one leaf from the shoots and place in GUS staining buffer for 6 h. Transgenic lines will show dark blue staining.
10. When the shoots are approx 1–2 cm in height, they can be used for rooting assays (*see* **Subheading 3.5., steps 2** and **3**) to eliminate escapes from the selection (*see* **Note 11**).
11. Transgenic plant lines are ready for molecular analysis, such as polymerase chain reaction (PCR).

3.5. Plant Regeneration From Transformed Somatic Embryos

1. Transfer maturing/mature somatic embryos from FES transformation experiments onto CEM medium in jars for germination. Generally it takes 2–4 wk for new shoots (**Fig. 2K**) to develop on the cotyledonary stage embryos.
2. Transfer the developing shoots on CBM for further growth (**Fig. 2L**). After 3 wk, excise the shoots (approx 1 cm long) developing from axillary buds of the stem cuttings and transfer to CBM medium supplemented with 8 mg/L of hygromycin for a rooting screen. Use shoots of wild-type plants as a negative control.
3. Check the shoots after 1 wk. Transgenic lines can root normally, whereas control shoots fail to produce any roots.
4. The plant lines are ready for molecular analysis.

4. Notes

1. All operations are performed under sterile conditions under a horizontal laminar flow hood using sterile materials.
2. Primary somatic embryogenesis of cassava is an explant-dependent event. Therefore, careful selection of explants should be considered. Meristems from axillary buds or apical shoots are the best candidates. Immature leaves smaller than 6 mm can be used as well, but the compact embryolike structures that are formed on them rarely develop to maturing somatic embryos. Care should be taken to place the explants with the upper side (meristem tip, adaxial side of the leaf) up on the medium to ensure efficient embryo production.
3. Our recent study has shown that weak light (approx 10 to 20 $\mu mol/m^2/s$) can enhance the frequency of somatic embryogenesis in most of tested cultivars.
4. To maintain the vigorous growth of somatic embryos, somatic embryos and embryo clusters have to be carefully separated from callus and other nonmorphogenic tissues and transferred onto fresh CIM every second week. It is advisable to subculture small embryo clusters rather than individual embryos, as the former appears to grow and develop better than single isolated embryos.
5. It is important to remove all nonmorphogenic tissues regularly to ensure the production of good quality embryos.
6. The formation of FEC in cassava is cultivar dependent, but works reliably with, for example, TMS60444. Because the production of FEC and plant regeneration from FEC takes a relatively long time, the risk of mutations should be minimized. Therefore, we suggest using FES less than age 6 mo for transformation.

7. All other tissues, including compact callus should not be transferred.
8. For handling the suspensions, a 25-mL pipet with ϕ3.0 mm opening (or a plastic pipet with the tip cut off) is most convenient. The suspensions should be pipetted up and down a few times to break the larger embryogenic clusters into smaller ones. Washing the filter a few times with culture medium also increases the yield of suitable embryogenic units. After filtering the cultures are left to stand either in a test tube or a culture jar to allow the embryogenic clusters to settle at the bottom. When transferring aliquots, care should be taken not to transfer the cell fraction consisting of long tubiform highly vacuolated cells that collects on top of the embryogenic fraction.
9. Stepwise selection protocols are strongly recommended to allow recovery of transformed cells during earlier stage of selection.
10. By using the GUS assays, and selecting the fractions with the highest rate of blue stain, fully transgenic suspension lines can be produced. Regeneration from such cultures will avoid production of nontransgenic escapes.
11. The selection pressure used for organogenesis is not very tight. Therefore, in addition to truly transgenic plants, escapes will be also regenerated. Rooting assays allow the elimination of these escapes. Furthermore, shoots emerging from cotyledon explants can also be chimeric. By repeated GUS assays combined with node culture of the originally chimeric transgenics, fully transgenic plants can be produced.

Acknowledgments

We thank Prof. Dr. Wilhelm Gruissem and Ingo Potrykus for their support on the cassava projects. Cassava cultivars were provided by CIAT (Centro Interancional de Agricultura Tropical, Cali, Columbia) and IITA (International Institute of Tropical Agriculture, Ibadan, Nigeria). This project was funded by the Centre for International Agriculture (ZIL), Zürich, The Rockefeller Foundation and the Swiss Federal Institute of Technology, Zürich.

References

1. Stamp, J. A. and Henshaw, G. G. (1987) Somatic embryogenesis from clonal leaf tissue of cassava. *Ann. Bot.* **59,** 445–450.
2. Mathews, H., Schopke, C., Carcamo, R., Chavarroaga, P., Fauquet, C., and Beachy R. N. (1993) Improvement of somatic embryogenesis and plant recovery in cassava. *Plant Cell Rep.* **12,** 328–333.
3. Li, H. Q., Huang, Y. W., Liang, C. Y., and Guo, J. Y. (1995) Improvement of plant regeneration from cyclic somatic embryos in cassava, in *Cassava Biotechnology Network* (Ed.) *Proc. Second Int. Scientific Meeting*, Bogor, Indonesia, CIAT Working Document 150, pp. 289–302.
4. Taylor, N. J., Edwards, M., Kiernan, R. J., Davey, C. D. M., Blakesley, D., and Henshaw G. G. (1996) Development of friable embryogenic callus and embryogenic suspension culture systems in cassava (*Manihot esculenta* Crantz). *Nat. Biotech.* **14,** 726–730.

5. Li, H. Q., Huang, Y. W., Liang, C. Y., et al. (1998) Regeneration of cassava plants via shoot organogenesis. *Plant Cell Rep.* **17,** 410–414.
6. Zhang, P., Phansiri, S., and Puonti-Kaerlas, J. (2001) Improvement of cassava shoot organogenesis by the use of silver nitrate in vitro. *Plant Cell Tiss. Org. Cult.* **67,** 47–54.
7. Li, H. Q., Sautter, C., Potrykus, I., and Puonti-Kaerlas, J. (1996) Genetic transformation of cassava (*Manihot esculenta* Crantz). *Nat. Biotech.* **14,** 736–740.
8. Zhang, P., Legris, G., Coulin, P., and Puonti-Kaerlas, J. (2000) Production of stably transformed cassava plants via particle bombardment. *Plant Cell Rep.* **19,** 939–945.
9. Zhang, P., Potrykus, I., and Puonti-Kaerlas, J. (2000) Efficient production of transgenic cassava using negative and positive selection. *Transgenic Res.* **9,** 405–415.
10. Schöpke, C., Taylor, N., Carcamo, R., et al. (1996) Regeneration of transgenic cassava plants (*Manihot esculenta* Crantz) from microbombarded embryogenic suspension cultures. *Nat. Biotech.* **14,** 731–735.
11. González, A. E., Schöpke, C., Taylor, N. J., Beachy, R. N., and Fauquet, C. M. (1998) Regeneration of transgenic cassava plants (*Manihot esculenta* Crantz) through *Agrobacterium*-mediated transformation of embryogenic suspension cultures. *Plant Cell Rep.* **17,** 827–831.
12. Zhang, P. and Puonti-Kaerlas, J. (2000) PIG-mediated cassava transformation using positive and negative selection. *Plant Cell Rep.* **19,** 1041–1048.
13. Raemakers, K. C. J. M., Schreuder, M., Pereira, I., Munyikwa, T. R. I., Jacobsen, E., and Visser, R. G. F. (2001) Progress made in FEC transformation of cassava. *Euphytica* **120,** 15–24.
14. Munyikwa, T. R. I., Raemakers, K. C. J. M., Schreuder, M., et al. (1998) Pinpointing towards improved transformation and regeneration of cassava (*Manihot esculenta* Crantz). *Plant Sci. 135,* 87–101.
15. Murashige, T. and Skoog, F. (1962) A revised medium for rapid growth and bioassays with tobacco tissue cultures. *Physiol. Plant.* **15,** 473–497.
16. Schenk, R. U. and Hildebrandt, A. C. (1972) Medium and techniques for induction and growth of monocotyledonous and dicotyledonous plant cell cultures. *Can. J. Bot.* **50,** 199–204.
17. Gresshoff, P. and Doy, C. (1974) Derivation of a haploid cell line from *Vitis vinifera* and the importance of the stage of meiotic development of the anthers for haploid culture of this and other genera. *Z. Pflanzenphys.* **73,** 132–141.
18. Hoekema, A., Hirsch, P. R., Hooykaas, P. J. J., and Schilperoort, R. A. (1983) A binary plant vector strategy based on separation of vir- and T-region of the *Agrobacterium tumefaciens* Ti-plasmid. *Nature* **303,** 179–180.
19. Jefferson R. A. (1987) Assaying chimeric genes in plants: the GUS gene fusion system. *Plant Mol. Biol. Rep.* **5,** 387–405.

12

Genetic Transformation of Mature Citrus Plants

Magdalena Cervera, José Juárez, Luis Navarro, and Leandro Peña

Summary

Most woody fruit species have long juvenile periods that drastically prolong the time required to analyze mature traits. Evaluation of characteristics related to fruits is a requisite to release any new variety into the market. Because of a decline in regenerative and transformation potential, genetic transformation procedures usually employ juvenile material as the source of plant tissue, therefore resulting in the production of juvenile plants. Direct transformation of mature material could ensure the production of adult transgenic plants, bypassing in this way the juvenile phase. Invigoration of the source adult material, establishment of adequate transformation and regeneration conditions, and acceleration of plant development through grafting allowed us to produce transgenic mature sweet orange trees flowering and bearing fruits in a short time period.

Key Words: Adult plants; cell competence; genetic engineering; grafting; invigoration; organogenesis; woody fruit plants.

1. Introduction

Plant development involves a juvenile phase characterized by the production of vegetative organs with a different size and shape than adult organs and by the inability to produce flowers and fruits, and a mature phase characterized for the acquisition of meristematic competence to initiate flowering and fruiting *(1)*. In addition, in most tree species there is a transition phase between the juvenile and the adult periods that can last several years and that is characterized by scarce flowering and fruit production. In woody fruit species, the juvenile phase and the transition period can last as long as decades, drastically prolonging the time required to analyze mature traits. In many woody fruit genus, most commercial varieties are vegetatively propagated and are hybrids of unknown origin or budsports that have been selected by growers based on a

From: *Methods in Molecular Biology, vol. 286: Transgenic Plants: Methods and Protocols*
Edited by: L. Peña © Humana Press Inc., Totowa, NJ

good agronomic behavior. High heterozygosity, large size, genetic incompatibility limitations, and lack of knowledge on how the most important horticultural traits are inherited make conventional breeding a long-term effort. Even when traditional breeding may be effective, another major obstacle is the long period of time between generations.

Genetic engineering would allow the insertion of specific genes into the unknown genetic background of elite varieties, theoretically adding desirable traits without affecting existing horticultural characteristics. However, one limitation to improvement of woody fruit species through genetic engineering might be the necessity of working with juvenile tissues as source of material for transformation. Maturation and aging seem to be responsible for the explant regenerative potential decline found in plant tissue culture of most woody species *(2)*. Furthermore, juvenile tissues show higher sensitivity to *Agrobacterium*-mediated transformation than mature ones. For these reasons, juvenile material has been extensively used for genetic transformation of woody species. There are two options: (a) Material derived from seeds or seedling organs, such as zygotic embryos, hypocotyls, or cotyledons *(3,4)* can be used; however, this implies a passage through a sexual stage and therefore a drastic reshuffling of the genome and a subsequent alteration of cultivar properties; and (b) embryogenic cells of somatic origin, somatic embryos, or even in vitro micropropagated tissues can be used, which would result in regeneration of transgenic plants with cultivar properties *(5–7)*; however, they would be juvenile, and therefore would require several years of cultivation before horticultural and commercial traits of the transgenic plants could be evaluated.

As a careful evaluation of the horticultural characteristics of mature plants bearing fruits is a standard procedure for releasing any new fruit variety to the market, genetic engineering would have limited applications unless tissue from mature plants can be readily transformed. Development of transformation procedures that allow bypassing the juvenile phase could greatly reduce the time and costs involved in improving and evaluating transgenic woody fruit trees species. This is the case of important citrus genotypes, as sweet orange trees, for which up to 20 yr may be needed to lose these juvenile characters.

We describe here a method to transform mature material of citrus plants based on: (a) grafting adult buds onto juvenile rootstocks for invigorating the mature tissues; (b) optimizing tissue culture conditions and media to shift citrus cells at the explants to a competent state for *Agrobacterium*-mediated transformation and regeneration; and (c) grafting in vitro of regenerating shoots onto decapitated seedling rootstocks as a high efficient alternative to shoot elongation and rooting. This method has been proved successful for the transformation of adult sweet orange *(8)*, sour orange *(9)*, lime, and some mandarin genotypes. Moreover, this general approach could be applicable to transformation of mature tissues from other woody fruit plants. In the case of sweet orange, species we will use as a reference in this chapter, it has led to the production of transgenic plants flowering and bearing fruits in 14 mo after transferring to the greenhouse. This is the standard period of time required for mature plants to achieve the adequate size and vigor to flower and set fruits.

2. Materials

2.1. Plant Material

1. Source of plant material: plants propagated in the greenhouse (18–27°C) by grafting of buds from adult sweet orange (*Citrus sinensis* L. Osbeck cv. Pineapple) trees on a vigorous rootstock, such as *C. volkameriana* Ten & Pasq. From these propagated plants, only the newly elongated first flushes will serve the objective (*see* **Note 1**).
2. Material for in vitro graft: seedlings of Troyer citrange (*C. sinensis* L. Osbeck × *Poncirus trifoliata* L. Raf.) germinated in vitro on seed germination medium (SGM) and grown in the dark for 2 wk.
3. Material for greenhouse graft: seedlings of rough lemon (*C. jambhiri* Lush) germinated in nursery and grown under greenhouse conditions (18–27°C) for approx 5 mo.

2.2. Tissue Culture Media and Components

1. Seed germination medium (SGM): 4.3 g/L of Murashige and Skoog (MS) salts ***(10)*** (*see* **Note 2**), 10 g/L of agar (Bacto-agar, Difco, Detroit, MI), pH 5.7.
2. Inoculation medium (IM): 4.3 g/L of MS salts, 10 mL/L of vitamin stock solution, 30 g/L of sucrose, pH 5.7.
3. Cocultivation medium (CM): IM plus 2 mg/L of 2,4-dichlorophenoxyacetic acid (2,4-D), 2 mg/L of indole-3-acetic acid (IAA), 1 mg/L of 2-isopentenyl-adenine (2,i-P), 8 g/L of agar, pH 5.7.
4. Shoot regeneration medium (SRM): IM plus 3 mg/L of 6-benzylaminopurine (BAP). Medium is semisolidified using 10 g/L of agar at pH 5.7, supplemented with 100 mg/L of kanamycin sulfate, 500 mg/L of cefotaxime, and 250 mg/L of vancomycin.
5. Shoot-tip grafting medium: 4.3 g/L of MS salts, 10 mL/L of vitamin stock, 75 g/L of sucrose, pH 5.7.
6. Vitamin stock: 10 g/L of *myo*-inositol (Duchefa, Haarlem, The Netherlands), 20 μg/L of thiamine-HCl (Duchefa), 100 μg/L pyridoxine-HCl (Duchefa), 100 μg/L of nicotinic acid (Duchefa).
7. 2,4-D (Sigma, St. Louis, MO) stock solution: 5 mg/100 mL. Prepare by dissolving the powder in a few drops of dimethyl sulfoxide (DMSO). Adjust volume with double-distilled water. Store at 4°C.
8. IAA (Sigma) stock solution: 5 mg/100 mL. Prepare as for 2,4-D and store at 4°C.
9. 2,i-P (Sigma) stock solution: 5 mg/100 mL. Prepare as for 2,4-D and store at 4°C.

10. BAP (Sigma) stock solution: 5 mg/100 mL. Prepare by dissolving the powder in a few drops of 1 *N* NaOH. Complete final volume with double-distilled water. Store at 4°C (*see* **Note 3**).
11. Kanamycin sulfate (Duchefa) stock solution: 100 mg/mL. Prepare by dissolving 1 g of powder in 10 mL of double-distilled water. Sterilize by filtration through a 0.2-μm membrane (Minisart, Sartorius, Göttingen, Germany), make 1-mL aliquots in sterile Eppendorf tubes and store at –20°C.
12. Cefotaxime stock solution: 250 mg/mL. Prepare by dissolving 1 g of powder in 4 mL of double distilled water. Sterilize by filtration through a 0.2-μm membrane, make 1-mL aliquots in sterile Eppendorf tubes and store at –20°C.
13. Vancomycin stock solution: 250 mg/mL. Prepare as for cefotaxime, aliquote, and store at –20°C (*see* **Note 4**).

All media are sterilized by autoclaving at 121°C for 20 min. Antibiotics are added to the medium after autoclaving.

2.3. Bacterial Strain and Vector

1. Bacterial strain: *Agrobacterium tumefaciens* EHA105, which is a disarmed derivative of *A. tumefaciens* A281 ***(11)*** (*see* **Note 5**). This strain holds chromosomic resistance to nalidixic acid.
2. Binary vector: The T-DNA of the binary plasmid usually contains, apart from the expression cassette/s of interest, a selectable marker gene, such as neomycin phosphotransferose II (*nptII*), which confers resistance to kanamycin, and a reporter marker gene, such as β-D-glucuronidase (*uidA*) or green fluorescent protein (*gfp*), under the control of constitutive promoter and terminator sequences. The binary plasmid is introduced into *Agrobacterium* by electroporation (*see* **Note 6**).

2.4. Culture Media for A. Tumefaciens

1. Luria broth (LB) medium: 10 g/L of tryptone, 5 g/L of yeast extract, 10 g/L of NaCl, pH 7.5.
2. Kanamycin sulfate stock solution: 100 mg/mL.
3. Nalidixic acid (Sigma) stock solution: 25 mg/mL. Prepare by dissolving 250 mg of powder in a few drops of 1 *N* NaOH and then add water to complete 10 mL. Sterilize by filtration, make 1-mL aliquots in sterile Eppendorf tubes and store at –20°C.
4. Liquid culture medium: LB medium containing 25 mg/L of kanamycin sulfate, and 25 mg/L of nalidixic acid.
5. Agar culture medium: LB medium, plus 10 g/L of agar (Difco), pH 7.5, with 25 mg/L of kanamycin sulfate, and 25 mg/L of nalidixic acid.

All media are sterilized by autoclaving at 121°C for 20 min. Antibiotics are added to the medium after autoclaving.

2.5. Other Solutions

Surface sterilant: 2% (v/v; stems) or 0.5% (v/v; seeds) sodium hypochlorite solution containing 0.1% (v/v) Tween-20 (Merck, Darmstadt, Germany).

2.6. Special Equipment

1. Culture chamber allowing temperature, humidity, and illumination control. Standard conditions are fixed at 26°C, 60% relative humidity, and a 16-h photoperiod at 45 μE/m^2/s illumination.
2. Incubators allowing temperature control at 26–28°C.
3. Orbital shaker allowing temperature and speed control.
4. Spectrophotometer.

3. Methods

3.1. Agrobacterium Preparation

1. Grow *A. tumefaciens* on LB agar culture medium (with antibiotics) at 28°C for 2 d. Take one loopful of bacteria and transfer to 100 mL of LB liquid culture medium (with antibiotics) and grow overnight at 28°C on an orbital shaker at 200 rpm. Measure absorbance at 600 nm of a 1-mL aliquot of the bacterial overnight culture in a spectrophotometer to calculate bacterial concentration (*see* **Note 7**).
2. Centrifuge the bacterial culture at 2000*g* for 10 min in 40-mL sterile centrifuge tubes with cap (Beckman Instruments, Palo Alto, CA), discard the supernatant, resuspend and dilute the pellet with IM to a concentration of approx 4×10^7 cells/mL (*see* **Note 7**). Pour 25-mL aliquots of the diluted culture into sterile glass 10-cm diameter Petri plates.

3.2. Explant Preparation

1. Select stem pieces from first flushes of propagated adult sweet orange plants (*see* **Note 8**). Strip stem pieces of their leaves and small thorns, brush carefully with soap and water, disinfect for 10 min in surface sterilant, and rinse three times with sterile distilled water.
2. Cut transversely 1-cm-long internodal stem segments with forceps and sterile scalpel (or small garden scissors) and keep in sterile humid plates until all stem pieces have been prepared.

3.3. Inoculation, Cocultivation and Selection

1. Immerse explants (approx 40 explants per plate) into the *A. tumefaciens* culture and incubate for 15 min with gentle shaking.
2. Blot dry on sterile filter paper (*see* **Note 9**), and place horizontally on plates containing CM (approx 20 explants per plate) for a 3-d cocultivation period at 26°C at a low light intensity (10 μE/m^2/s, 16-h photoperiod) (*see* **Note 10**).
3. After cocultivation, transfer the explants to SRM (10 explants per plate) (*see* **Note 11**). Maintain cultures in the dark for 2–4 wk at 26°C and then transfer to a 16-h photoperiod, 45 μE/m^2/s illumination at 26°C (*see* **Note 12**). Explants will be subcultured every 3–4 wk and any fungus- or bacteria-contaminated explant should be discarded.

3.4. Recovery of Whole Transgenic Plants

1. Shoots should develop from the cut ends of explants 3–5 wk after cocultivation. Check the transgenic nature of the regenerated shoots by performing a histochemical GUS assay (*see* Chapter 14) or by testing GFP expression. GUS- or green fluorescent protein (GFP)-negative shoots are considered as nontransformed, and commonly called escape shoots (*see* **Note 13**).
2. Graft in vitro apical portions of the GUS- or GFP-positive shoots onto decapitated seedlings of Troyer citrange. Rootstock preparation is as follows: peel seeds, remove both seed coats, disinfect for 10 min in surface sterilant, and rinse three times with sterile distilled water. Sow individual seeds onto 25-mL aliquots of SGM contained in 25 × 150 mm glass tubes and incubate at 27°C in the dark for 2 wk (*see* **Note 14**). Decapitate seedlings leaving 1–1.5 cm of the epicotyls. Shorten the roots to 4-6 cm and remove the cotyledons and their axillary buds. Place the regenerated shoot onto the apical end of the cut surface of the decapitated epicotyl, so that contact is with the vascular ring (*see* **Note 15**) (**Fig. 1A**).
3. Culture grafted plants in shoot-tip grafting medium and maintain at 25°C, 16 h of photoperiod and 45 μE/m^2/s of illumination (*see* **Note 16**). Scions develop two to four expanded leaves 3–4 wk after grafting.
4. Grafting of in vitro-grown plants onto vigorous rough lemon rootstocks in the glasshouse allows the rapid acclimatization and development of the plants (*see* **Note 17**).
5. Monitor plant growth and development. Putative mature transgenic sweet orange plants show morphology and growth habits of an adult plant, as compared to control mature plants. In fact, whereas juvenile plants show a pronounced thorniness, transgenic mature plants are almost thornless, similar to the mature plants from which the explants are taken for transformation (**Fig. 1B**). After fourteen months in the greenhouse, the transgenic and control plants usually start to flower and set fruit, confirming their mature nature (**Fig. 1C,D**) (*see* **Note 18**).
6. Putative transgenic plants should be assayed by polymerase chain reaction (PCR) to detect the presence of the transgene(s). Southern blot analyses must be performed to confirm the stable integration of the transgene(s), and Northern blot and Western blot analyses would confirm their expression in the transgenic plants.

4. Notes

1. In a preliminary study *(8)*, the regenerative potentials of juvenile explants and explants coming from the first, second, and third flushes of mature sweet orange plants were compared. Explants from first and second flushes of mature plants showed a reduction of regeneration frequency of 50–70% compared to juvenile explants. The third flush showed a more pronounced regenerative decline that made it useless for our purpose.
2. We use to keep separate stock solutions of MS macroelements and microelements, but a good laboratory practice is required to prepare and keep clean and uncontaminated solutions. A comfortable method is using commercial MS salts, weighed and presented in individual bags for 1 L of medium.

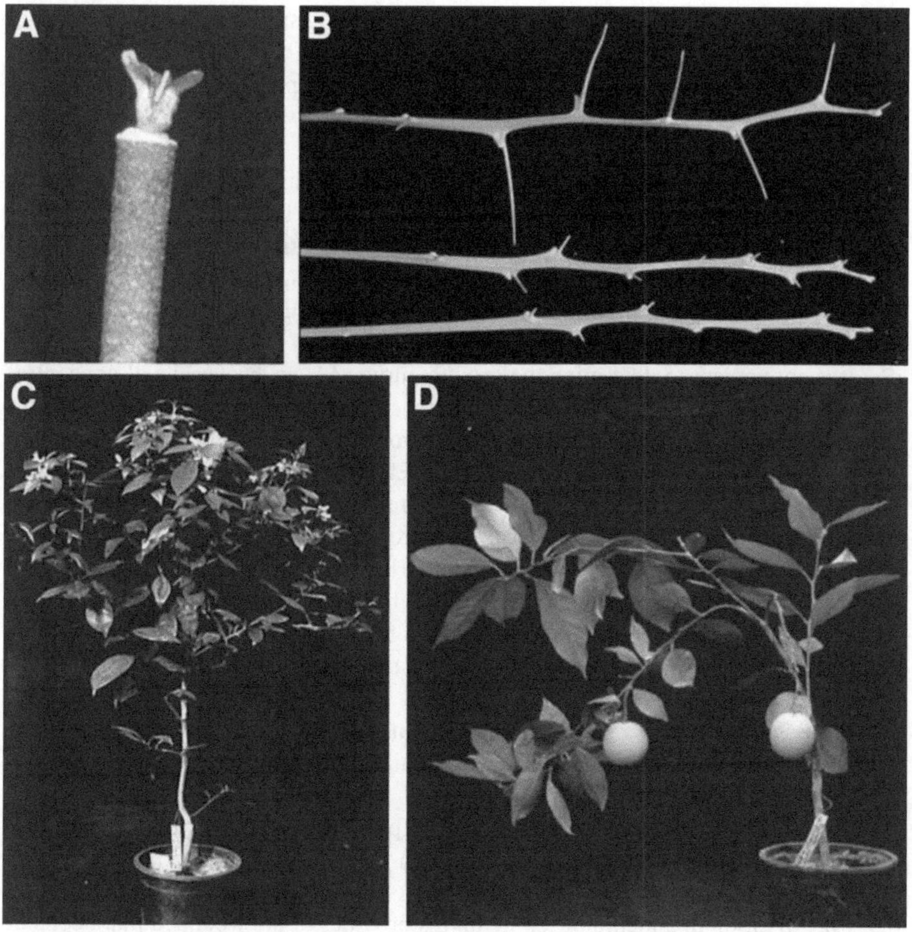

Fig. 1. **(A)** Transgenic shoot grafted in vitro on a decapitated Troyer rootstock. **(B)** Comparison of thorniness among flushes of: a juvenile seedling (*top*), a transgenic mature plant (*middle*), and a source mature plant from which the shoots were taken for transformation experiments (*bottom*). **(C)** Transgenic mature sweet orange plant flowering after 15 mo in the greenhouse. **(D)** Transgenic mature plant with ripe fruits.

3. Hormone solutions are not kept sterile in the refrigerator, so the possibility of contamination exists. If the stock solution is contaminated, discard and prepare a new one before mixing with the other components of the media.
4. We prefer to make 1-mL aliquots of antibiotic solutions to avoid possible contaminations. In the case of tissue culture media, one or two full aliquots will serve to reach the final desired concentration for 1 L of medium.

5. In a previous study performed in vivo by inoculating sweet orange seedlings with different nondisarmed *Agrobacterium* strains ***(8,12)***, strain A281 was shown to be the most virulent in the infection of this genotype and others of the genus *Citrus*. This is the reason of using a disarmed derivative of A281 for our transformation experiments.
6. Plasmids of reference, with a T-DNA only containing marker genes, could be p35SGUSINT ***(13)*** or pBIN19-sGFP ***(14)***. Both have been used in our laboratory and work very efficiently for the genetic transformation of many plants. Bacterial resistance to kanamycin, present in both plasmids, together with the chromosomic resistance of EHA105 to nalidixic acid, is used to select the bacteria (described in **Subheading 2.4.**).
7. It is convenient to determine the growth curve (A_{600} vs bacterial cell concentration) for the bacterial strain used in the transformation experiment. Bacterial culture should grow to the exponential phase to play all its infectious potential (A_{600} between 0.1 and 1.0, in the case of strain EHA105). For sweet orange, calculate the volume of bacterial culture necessary to prepare 40 mL of 4×10^7 cells/mL suspension in the centrifuge tubes. If it is too small, prepare an intermediate 4×10^8 cells/mL suspension in IM. A higher bacterial concentration (approx 10^8) results in lower transformation frequency because plant cells become stressed; lower bacterial concentration (approx 10^6) results in a lower transformation frequency and in this case fewer cells at the cut end of the explant become transformed ***(15)***.
8. Flushes should be in a good ontological state, neither too tender (they would not bear *Agrobacterium* infection) nor too lignified (as to keep an acceptable regenerative potential).
9. We use sterile soft paper towels to help explants to dry. It is important to eliminate any bacterial liquid residue, as it can be a source of bacterial overgrowth during cocultivation.
10. Cocultivation in a medium rich in auxins provides to the wounded plant cells of explants an appropriate treatment to shift them to a competent state for transformation, involving dedifferentiation, induction of cell division, and callus proliferation ***(16)***. Prolonging cocultivation period does not increase transformation frequency, but it frequently results in *Agrobacterium* overgrowth and subsequent decrease in regeneration frequency of transformed shoots. Therefore, a 3-d cocultivation is routinely used.
11. If explants were carefully dried after inoculation, they will not show an excess of bacterial growth at this point. But if this is not the case, immerse them in sterile water with cefotaxime at 250 mg/L for several minutes and blot them dry again before transferring to SRM.
12. Culture of explants in the dark improves callus formation and the progress of transformation events to regenerate transgenic shoots and avoids the regeneration of escape shoots that could be stimulated by the exposure of explants directly to light ***(16)***. Two weeks in the dark is normally the most appropriate period to favor callus formation in the case of sweet orange ***(8)***. This may be different for other citrus

genotypes. Indeed, the explants should be kept in darkness until they develop a prominent visible callus formed at the cambial ring.

13. Considering GUS- or GFP-negative shoots as escapes is convenient, but it should be noted that it can lead to errors in the actual number of transformants, because silencing or low expressing events or even partial T-DNA integrations are not accounted with this criterion.
14. After 2 wk in the dark at 27°C, Troyer citrange seedlings should be transferred to the refrigerator at 4–8°C to slow growth. They can be used within 15 d or 1 mo without appreciable loss in grafting efficiency.
15. For the in vitro grafting of long shoots (0.5–1 cm), cutting the basal end as a wedge and introducing it into a small longitudinal incision practiced on the upper part of the rootstock can also be helpful to facilitate vascular contact and success of the graft.
16. During development of the grafts, it is necessary to check them periodically and to remove, by using sterile small scissors, any shoot not coming from the grafted scion. The growth of other shoots could weaken the connection between rootstock and transgenic scion.
17. To ensure a rapid and successful acclimatization, it is important to follow good greenhouse practices. We recommend working with sterile potting substrate, vigorous seedlings, and keeping grafted plants in plastic bags that will be progressively opened over approx 1 mo. This will help to maintain an optimal degree of moisture and temperature and will facilitate a gradual process of acclimatization.
18. An alternative approach to shorten the juvenile period is the transformation of juvenile citrus plants with the *APETALA1* gene from *Arabidopsis thaliana.* Transgenic plants show a drastic reduction in the juvenile period, flowering and setting fruits within the first year after their transfer to the greenhouse ***(17)***. Genetic retransformation of plant material coming from these plants would allow the rapid evaluation of the expression of transgenes incorporated into the plant in a second transformation round.

Acknowledgments

We thank Carmen Ortega, Antonio Navarro and José Antonio Pina for their excellent technical assistance. Part of this work was supported by grants from the Generalitat Valenciana No. CTIDIA/2002/89, from the Instituto Nacional de Investigaciones Agrarias No. RTA-01-120, and from CICYT No. 2003-01644.

References

1. Poethig, R. S. (1990) Phase change and the regulation of shoot morphogenesis in plants. *Science* **250,** 923–930.
2. Durzan, D. (1990) Adult vs. juvenile explants: directed totipotency, in *Plant Aging. Basic and Applied Approaches* (Rodríguez, R., Sánchez-Tamés, R., and Durzan, D. J., eds.), Series A: Life Sciences, Vol. 186, NATO ASI Series, Plenum Press, New York, NY, pp. 19–25.

3. Mullins, M. G., Tang, F. C. A., and Facciotti, D. (1990) *Agrobacterium*-mediated genetic transformation of grapevines: transgenic plants of *Vitis rupestris* Scheele and buds of *Vitis vinifera* L. *Biotechnology* **8,** 1041–1045.
4. Mante, S., Morgens, P. H., Scorza, R., Cordts, J. M., and Callahan, A. M. (1991) *Agrobacterium*-mediated transformation of plum (*Prunus domestica* L.) hypocotyl slices and regeneration of transgenic plants. *Biotechnology* **9,** 853–857.
5. Brasileiro, A. C. M., Leple, J. C., Muzzin, J., Ounnoughi, D., Michel, M. F., and Jouanin, L. (1991) An alternative approach for gene transfer in trees using wild-type *Agrobacterium* strains. *Plant Mol. Biol.* **17,** 441–452.
6. Perl, A., Lotan, O., Abu-Abied, M., and Holland, D. (1996) Establishment of an *Agrobacterium*-mediated transformation system for grape (*Vitis vinifera* L.): the role of antioxidants during grape–*Agrobacterium* interactions. *Nat. Biotechnol.* **14,** 624–628.
7. James, D. J., Passey, A. J., Baker, S. A., and Wilson, F. M. (1996) Transgenes display stable patterns of expression in apple fruit and mendelian segregation in the progeny. *Biotechnology* **14,** 56–60.
8. Cervera, M., Juárez, J., Navarro, A., et al. (1998) Genetic transformation and regeneration of mature tissues of woody fruit plants bypassing the juvenile stage. *Transgen. Res.* **7,** 51–59.
9. Ghorbel, R., Domínguez, A., Navarro, L., and Peña, L. (2000) High efficiency genetic transformation of sour orange (*Citrus aurantium* L.) and production of transgenic trees containing the coat protein gene of citrus tristeza virus. *Tree Physiol.* **20,** 1183–1189.
10. Murashige, T. and Skoog, F. (1962) A revised medium for rapid growth and bioassay with tobacco tissue cultures. *Physiol. Plant.* **15,** 473–479.
11. Hood, E. E., Gelvin, S. B., Melchers, L. S., and Hoekema, A. (1993) New *Agrobacterium* helper plasmids for gene transfer to plants. *Transgen. Res.* **2,** 208–218.
12. Cervera, M., López, M. M., Navarro L., and Peña, L. (1998) Virulence and supervirulence of *Agrobacterium tumefaciens* in woody fruit plants. *Physiol. Mol. Plant Pathol.* **52,** 67–78.
13. Vancanneyt, G., Schmidt, R., O'Connor-Sánchez, A., Willmitzer, L., and Rocha-Sosa, M. (1990) Construction of an intron-containing marker gene: splicing of the intron in transgenic plants and its use in monitoring early events in *Agrobacterium*-mediated plant transformation. *Mol. Gen. Genet.* **220,** 245–250.
14. Chiu, W., Niwa, Y., Zeng, W., and Hirano, T. (1996) Engineered GFP as a vital reporter in plants. *Curr. Biol.* **6,** 325–330.
15. Peña, L., Cervera, M., Juárez, J., Navarro, A., Pina, J. A., Durán-Vila, N., and Navarro, L. (1995) *Agrobacterium*-mediated transformation of sweet orange and regeneration of transgenic plants. *Plant Cell Rep.* **14,** 616-619.
16. Cervera, M., Pina, J. A., Juárez, J., Navarro L., and Peña, L. (1998) *Agrobacterium*-mediated transformation of citrange: factors affecting transformation and regeneration. *Plant Cell Rep.* **18,** 271–278.

17. Peña, L., Martín-Trillo, M., Juárez, J., Pina, J. A., Navarro, L., and Martínez-Zapater, J. M. (2001) Constitutive expression of *Arabidopsis LEAFY* and *APETALA1* genes in citrus reduces their generation time. *Nat. Biotechol.* **19,** 263–267.

IV

Selection

13

Selectable Markers

Antibiotic and Herbicide Resistance

Julia L. Goodwin, Gabriela M. Pastori, Michael R. Davey, and Huw D. Jones

Summary

The low efficiencies of most plant transformation methods necessitate the use of selectable marker genes to identify those cells that successfully integrate and express transferred DNA. Genes conferring resistance to various antibiotics or herbicides are commonly used in laboratory transformation research. They encode proteins that detoxify corresponding selection agents and allow the preferential growth of transformed cells. This chapter describes the application of two selection systems on the transformation of wheat. One is based on the *npt*II gene and corresponding aminoglycoside antibiotics, the other is based on the *bar* gene and corresponding glufosinate ammonium herbicides.

Key Words: Antibiotic resistance; *bar*; G418; herbicide resistance; phosphinothricin acetyltransferase; neomycin phosphotransferase; *npt*II, selection; wheat transformation.

1. Introduction

In the generation of transgenic plants, selection systems permit the preferential growth of transformed cells; selectable marker genes based on negative selection are commonly delivered alongside genes of interest. These marker genes encode proteins that confer resistance to a selection agent that inhibits growth or kills nontransgenic cells. Genes encoding resistance to specific antibiotics or herbicides have proved particularly effective for selection and provide a means of rapidly identifying transformed cells, tissues, and regenerated shoots that have integrated foreign DNA and that express the selectable gene product and, by inference, the gene(s) of interest. For example, the aminoglyco-

From: *Methods in Molecular Biology, vol. 286: Transgenic Plants: Methods and Protocols*
Edited by: L. Peña © Humana Press Inc., Totowa, NJ

side antibiotics, such as kanamycin, neomycin, and G418 kill cells by inhibiting protein translation. The *Eschericia coli npt*II gene, encoding neomycin phosphotransferase, inactivates these antibiotics by phosphorylation, thus allowing preferential growth of plant cells transformed with this gene on media containing these selection agents. The herbicide phosphinothricin (PPT) is an analog of glutamine and acts by irreversibly inhibiting glutamine synthetase, a key enzyme for ammonium assimilation and the regulation of nitrogen assimilation in plants. The *bar* gene, cloned from the bacterium *Streptomyces hygroscopicus*, encodes phosphinothricin acetyltransferase (PAT), which converts PPT into the nontoxic acetylated form and allows growth of transformed plant cells in the presence of PPT, or commercial glufosinate ammonium-based herbicides (*see* **Note 1**). Although recent concerns relating to potential spread of selectable marker genes from genetically modified organisms released to the environment are driving techniques that avoid the use of such genes or remove them (*see* Chapters 17 and 18), these selection regimes remain an important tool for laboratory-based transformation research.

This chapter uses the transformation of bread wheat (*Triticum aestivum* L.) and pasta wheat (*T. turgidum* ssp. durum Desf.) with the *npt*II (*neo*) and *bar* genes to illustrate the application of antibiotic and herbicide selection systems, aspects of which have been reviewed in several publications ***(1–10)*** .

2. Materials

1. Plasmid constructs (pAHC20 ***[9]***, pCalNeo ***[11]***).
2. Donor plants of *T. aestivum* (*see* **Note 2**).
3. Petri dishes (9 cm in diameter, triple-vented).
4. Biolistic transformation equipment.
5. Induction medium: Murashige and Skoog (MS) basal salts ***(12)***, 0.5 mg/L of nicotinic acid, 0.1 mg/L of thiamine-HCl, 0.5 mg/L of pyridoxine-HCl, 100 mg/L of *myo*-inositol, 375 mg/L of L-glutamine, 75 mg/L of L-proline, 50 mg/L of L-asparagine, 90 g/L of sucrose, 0.5 mg/L of 2,4-dichlorophenoxyacetic acid (2,4-D), and 10 mg/L of $AgNO_3$ (*see* **Note 3**).
6. Regeneration medium: MS basal salts ***(12)***, 200 mg/L of *myo*-inositol, 10 mg/L of thiamine-HCl, 1 mg/L of pyridoxine-HCl, 1 mg/L of nicotinic acid, 1 mg/L of Ca-pantothenate, 1 mg/L of L-ascorbic acid and 30 g/L of maltose (*see* **Note 4**).
7. Agargel (Sigma Aldrich, Poole, UK) (*see* **Note 5**).
8. 2,4-D (*see* **Note 6**).
9. $AgNO_3$ (*see* **Note 7**).
10. Zeatin (*see* **Note 8**).
11. Glufosinate ammonium (*see* **Note 1**).
12. G418 (*see* **Note 9**).
13. DNA extraction reagents and equipment.
14. Polymerase chain reaction (PCR) and electrophoresis solutions and equipment.

15. 25-Multiwell plates.
16. 1.5-mL plastic cuvets.
17. Ammonium assay incubation medium: 50 m*M* potassium phosphate buffer, pH 5.8, 2% sucrose, 1.0 mg/L of 2,4-D, 25 mg/L of gluphosinate ammonium, and 0.1% Tween-20.
18. Ammonium assay reagent 1: 34 g/L of sodium salicylate, 25 g/L of trisodium citrate, 25 g/L of sodium tartrate, and 0.12 g/L of sodium nitroprusside (*see* **Note 10**).
19. Ammonium assay reagent 2: 30 g/L of sodium hydroxide and 0.52 g/L of sodium dichloroisocyanurate (*see* **Note 11**) .
20. Ammonium chloride.
21. Glufosinate ammonium-based herbicide (*see* **Note 12**).

3. Methods

The methods described in the following subheadings are applicable to both spring- and winter-sown commercial UK bread wheat (*T. aestivum*) varieties (e.g., Canon, Cadenza, Imp, and Buster). Variations in the protocol for tetraploid wheat (*T. turgidum* ssp. durum) varieties (e.g., Ofanto and Venusia) are included in the **Notes** subheading.

The wheat tissue culture and transformation procedures are based on the method described by Barcelo and Lazzeri (1995) ***(3)*** as modified subsequently by Rasco-Gaunt et al. (1999) ***(13)***, Pastori et al. (2001) ***(6)***, and Rasco-Gaunt et al. (2001) ***(7)***. This method allows the production of apparently phenotypically normal transgenic wheat plants in 15–18 wk, from the isolation of scutella through all tissue culture stages until plants are ready to be transferred to compost and evaluated for the presence of the transgene(s).

Immature scutella are isolated and cotransformed with the marker gene and gene of interest using biolistics (*see* Chapter 4). Two extra Petri dishes, each containing 10 explants, are prepared per experiment, one for nonbombarded controls and the other for scutella bombarded only with gold (*see* **Note 13**). After bombardment, explants are spread to 10/plate and callus induction from bombarded and nonbombarded explants is carried out on induction medium for 4–5 wk in the dark.

3.1. Plasmid Constructs

Typical configurations for the *bar* and *npt*II gene expression cassettes for wheat transformation are shown in **Figs. 1** and **2** (*see* **Note 14**).

3.1.1. Herbicide Resistance

The construct pAHC20 ***(9)*** contains the *bar* gene ***(1)*** (*see* **Note 15**) as the selectable marker under the control of the maize ubiquitin I promoter (**Fig. 1**). The *bar* gene encodes the enzyme PAT which confers resistance to PPT and glufosinate ammonium herbicides.

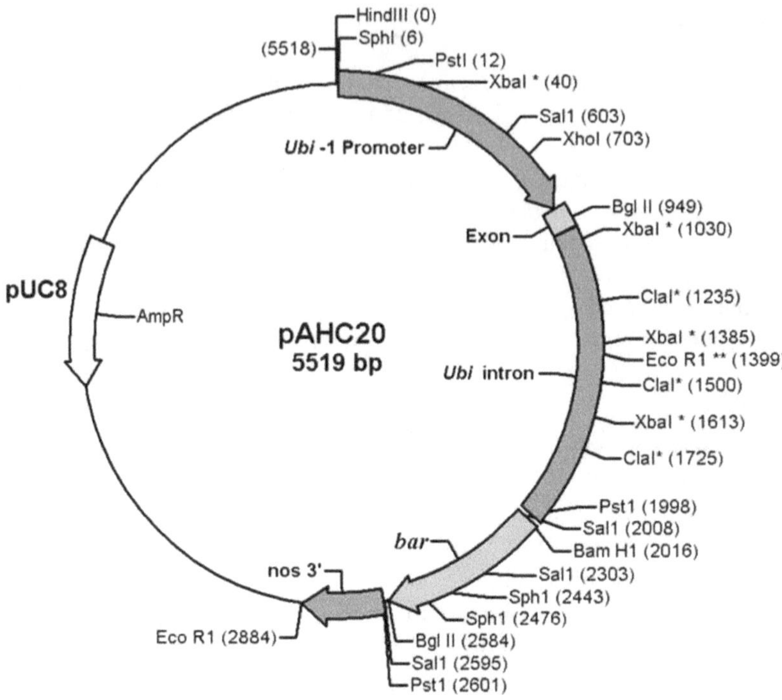

Fig. 1. Schematic diagram of pAHC20.

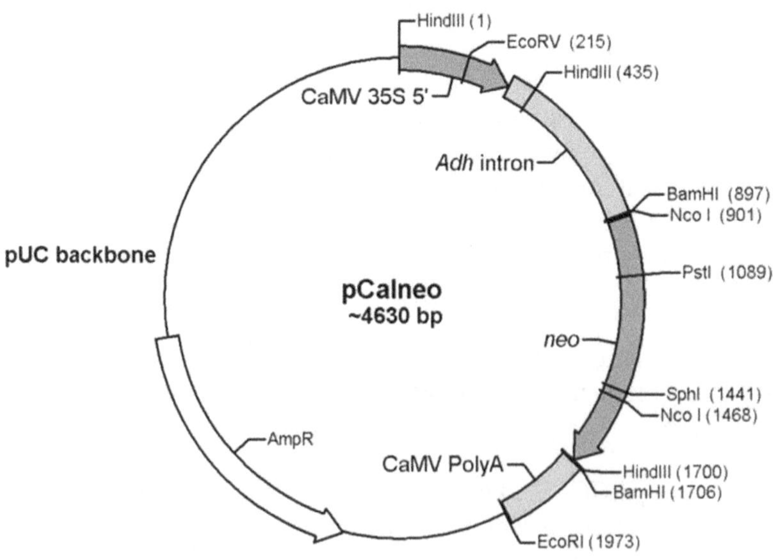

Fig. 2. Schematic diagram of pCaIneo.

3.1.2. Antibiotic Resistance

The construct pCaINeo *(2)* encodes neomycin phosphotransferase and confers resistance to aminoglycoside antibiotics such as kanamycin (**Fig. 2**).

3.2. Regeneration and Selection

The regeneration of shoots from embryogenic calli is performed at 26°C under a 12-h photoperiod with 10 calli/Petri dish. The regeneration response is evaluated in each culture cycle by scoring regenerants with only shoots, only roots, or shoots and roots. When choosing a selection agent, it is important to test for natural resistance of the target plant material by first performing a kill curve experiment and establishing the optimum concentration.

3.2.1. Selection Using bar

1. Culture calli in Petri dishes of regeneration medium supplemented with 0.1 mg/L of 2,4-D, 10 mg/L of $AgNO_3$ and 5 mg/L of zeatin (*see* **Note 16**) for 4 wk.
2. Evaluate the regeneration response (*see* **Note 17**).
3. Transfer regenerating calli to regeneration medium supplemented with 4 mg/L of the selection agent glufosinate ammonium and culture for 4 wk. Half of the tissues from each control plate should be cultured on regeneration medium with selection and the other half without selection to provide selected and nonselected controls (*see* **Note 18**).
4. Evaluate the regeneration and selection responses (*see* **Note 19**).
5. To ensure a low percentage of escapes, transfer only the healthy regenerating calli to regeneration medium in Magenta™ vessels (Sigma Aldrich) (*see* **Note 20**) supplemented with 4 mg/L of glufosinate ammonium and culture for a further 4 wk (*see* **Note 21**).
6. Repeat **steps 4** and **5** until the plants are of sufficient size to transfer to compost (*see* **Note 22**).
7. Transfer putative transgenic plants to compost and grow to maturity in a glasshouse (*see* **Note 23**).

3.2.2. Selection Using npt*II*

1. Culture calli in Petri dishes of regeneration medium supplemented with 0.1 mg/L of 2,4-D, 10 mg/L of $AgNO_3$, and 5 mg/L of zeatin (*see* **Note 16**) for 4 wk.
2. Evaluate the regeneration response (*see* **Note 17**).
3. Transfer regenerating calli to regeneration medium supplemented with 50 mg/L of G418 (*see* **Note 24**) and culture for 4 wk. One control plate should be cultured on regeneration medium with selection and the other without selection to provide selected and nonselected controls (*see* **Note 18**).
4. Evaluate the regeneration and selection responses (*see* **Note 19**).
5. To ensure a low percentage of escapes, transfer only the healthy regenerating calli to regeneration medium in Magenta™ vessels (*see* **Note 20**) supplemented with 50 mg/L of G418 and culture for a further 4 wk (*see* **Note 21**).

6. Repeat **steps 4** and **5** until the plants are of sufficient size to transfer to compost (*see* **Note 22**).
7. Transfer putative transgenic plants to compost and grow to maturity in a glass house (*see* **Note 23**).

3.3. Confirmation of Transformation by PCR

When plants are large enough for leaf material to be collected, DNA may extracted for PCR using the hexadecyltrimethylammonium bromide (CTAB) method ***(14)*** , or by using a commercial extraction kit (*see* **Note 25**). Both the *bar* and the *npt*II genes can be detected by PCR using the primers and conditions detailed below.

3.3.1. bar

Forward primer: 5' GTCTGCACCATCGTCAACC 3'
Reverse primer: 5' GAAGTCCAGCTGCCAGAAAC 3'
Annealing temperature: 57°C
Number of cycles: 30
Product size: 444 bp

3.3.2. npt*II*

Forward primer: 5' GAGGCTATTCGGCTATGACTG 3'
Reverse primer: 5' ATCGGGAGCGGCGATACCGTA 3'
Annealing temperature: 57°C
Number of cycles: 30
Product size: 679 bp

3.4. Marker Gene Expression Assays

The next step, after confirming the presence of the marker gene by PCR in DNA from selected antibiotic- or herbicide-resistant plants, is to analyze its expression. Resistance to antibiotics and herbicides can be evaluated in established primary transgenic plants (T_0 generation) and their progeny (T_1 and subsequent generations). Two methods commonly used for testing resistance to PPT are described in the following subheadings. In addition, enzyme-linked immunosorbent assay (ELISA) based kits for assessing expression of the *npt*II gene are commercially available.

3.4.1. Ammonium Assay

Stable expression of the *bar* gene can be analyzed by the ammonium assay *(8)*, which allows the qualitative and quantitative detection of PAT activity in leaf tissue.

3.4.1.1. Qualitative Assay

1. Place 4 × 8 mm leaf pieces in 25-multiwell plates containing 1 mL of incubation medium per well, leaving one or two wells without tissue as negative controls (*see* **Note 26**).
2. Incubate for 5 h at 24°C under a photosynthetic photon flux density of 250 μmol/m^2/s (*see* **Note 27**).
3. For each sample, take 200 μL of the incubation medium and add to 1 mL of reagent 1.
4. Add 1 mL of reagent 2 and mix by vortexing.
5. Incubate for 15 min at 37°C, followed by 15 min at room temperature.

The incubation medium with plant samples not expressing the *bar* gene and the negative controls will develop an emerald green to dark-blue color because of the presence of ammonium ions (**Fig. 3**). The medium of explants from transgenic plants expressing the *bar* gene will develop a light green or yellow color that occurs in the absence of ammonium ions.

3.4.1.2. Quantitative Assay

1. Proceed as for the qualitative assay.
2. Prepare a standard curve with increasing concentrations of ammonium chloride by adding 200 μL of standard solutions of ammonium (0–10 mg/L) to 1 mL of reagent 1 and 1 mL of reagent 2 and incubating as in **step 5**, **Subheading 3.4.1.1.**
3. Measure the absorbance of each solution at 655 nm.
4. Calculate the concentration of ammonium in each sample using the standard curve.

3.4.2. Leaf Painting Bioassay

Qualitative expression of the *bar* gene can also be scored using a herbicide leaf painting bioassay. This approach is simple and inexpensive and can be carried out on plants *in situ*. Application of herbicides based on glufosinate ammonium, such as Basta™, cause desiccation and browning in control plants, whereas plants expressing the *bar* gene are resistant and remain green.

1. Prepare dilutions of glufosinate ammonium-based herbicide using 0.1% Tween to give final concentrations of PPT of 0.2 g/L and 2 g/L (*see* **Note 28**).
2. Select healthy wheat plants at the tillering phase of growth. For each plant to be tested, select three approximately equal sized, healthy leaves, avoiding the flag leaf. Water the plants before herbicide application.
3. Label the chosen leaves on the stem immediately below the leaf to be painted (Tween only, 0.2 g/L of PPT or 2 g/L of PPT).
4. Mark each leaf with a ballpoint pen halfway along its length and paint the upper surface of the distal half of the leaf with the appropriate solution using a cotton bud. The application should be quite firm to ensure coating and some penetration of the solution into the leaf. All the control leaves should be painted first, fol-

Fig. 3. An example of the colorimetric ammonium test in cuvets as viewed from above. Control plants lacking the *bar* gene (*top row*) have a high concentration of free ammonium ions, giving a dark blue color. Three independent transgenic plants expressing the *bar* gene (*remaining rows*) assimilate ammonium ions and give a light green to yellow color.

lowed by the lower concentration, then the higher concentration of herbicide to minimize carryover of the herbicide to other leaves.

5. The herbicide resistance of each plant is assessed 7 d after application by scoring each treated leaf according to the percentage desiccation/browning over the painted area and the percentage of the proximal region of the leaf that has been affected by the spread of the herbicide.

4. Notes

1. Glufosinate ammonium is synthetically produced PPT bound to ammonium and is the active component in herbicides such as Basta™ (Bayer Cropscience AG, Monheim am Rhein, Germany). Glufosinate ammonium can be dissolved in distilled water (10 mg/mL), filter sterilized, and stored in aliquots at –20°C.
2. Donor material is grown under controlled conditions. The age and condition of donor material for transformation is crucial; full details can be found in Pastori et al. (2001) ***(6)***.
3. Induction medium should be made up at a 2X concentration, filter sterilized, and stored at 4°C. Induction medium for *T. turgidum* ssp. *durum* is the same as that for *T. aestivum* except with 30 g/L of sucrose (instead of 90 g/L) and 1.0 mg/L of picloram instead of 0.5 mg/L of 2,4-D.
4. The regeneration medium should be made up at a 2X concentration, filter sterilized and stored at 4°C. The regeneration medium for *T. turgidum* ssp. durum is the same as that for *T. aestivum*.
5. Agargel™ is made up at a 2X concentration of 10 g/L and autoclaved before use.
6. A working solution of 2,4-D (1.0 mg/mL) can be prepared by dissolving 100 mg in 70 mL of ethanol, adding 30 mL of distilled water, vortex-mixing, filter sterilizing, and storing 1 mL aliquots at –20°C.
7. $AgNO_3$ can be dissolved in distilled water (20 mg/mL), filter sterilized, and stored in aliquots at –20°C.
8. A working solution (10 mg/mL) of zeatin can be prepared by dissolving 100 mg in 1 mL of 1 *M* HCl, adding 9 mL of distilled water, vortexing to mix, filter sterilizing, and storing in 1-mL aliquots at –20°C.
9. G418 can be dissolved in sterile distilled water (50 mg/mL) and stored in aliquots at –20°C.
10. Reagent 1 can be stored at 4°C in the dark.
11. Reagent 2 can be stored at 4°C for 2 mo.
12. Several glufosinate ammonium based herbicides are commercially available, any of which are suitable (*see also* **Note 1**).
13. Bombarded and nonbombarded controls are included to determine whether the process of bombardment has a detrimental effect on callus induction and regeneration.
14. The methods described are for use with the specific constructs shown. Protocols using the *npt*II or *bar* genes with different promoter/intron/terminator configurations, or different marker genes, may require optimization for specific species/varieties.
15. The *bar* gene from *S. hygroscopicus* and the *pat* gene from *S. viridiochromogenes* both code for PAT; either of these genes can be used as selectable markers.
16. Agargel (*see* **Note 4**) is melted in a microwave oven and added to an equal volume of medium. The supplementary components are added before the mixture is poured into Petri dishes. The tissue culture response is improved if the medium in the dishes is relatively deep (approx 30 mL/dish).
17. If the regeneration response is poor, repeat one round of regeneration (up to 4 wk) on Petri dishes of fresh medium without the selection agent.

18. The selected control is used to confirm that selection is effective (i.e., the controls should all die in the presence of the selection agent); the nonselected control is used to monitor the response of the donor material (i.e., the nonselected controls should survive).
19. Nontransgenic material should show typical symptoms of selection pressure, including slow growth, yellow-brown shoots, and poor root development.
20. Prepare the medium for Magenta™ vessels in the same way as for Petri dishes (*see* **Note 16**) with 50 mL of medium per Magenta™ vessel. Magenta™ vessels should each contain a maximum of five plants.
21. Escapes are defined as plants that survive exposure to the selection agent, but which lack the transgene. The percentage of escapes depends on the selection pressure. If the percentage of escapes is high, greater selection pressure can be applied by increasing the concentration of the selection agent. However, if the selection pressure is increased too much, transgenic plants may also be lost. This may not be a problem if transformation frequency is high. However, if the transformation frequency is low, it is preferable to accept a higher percentage of escapes to avoid loss of transgenic plants.
22. Usually three rounds of selection are adequate.
23. Plants that survive the selection pressure with a good root system are considered putative transgenic individuals.
24. The *npt*II gene confers resistance to several aminoglycoside antibiotics including neomycin, kanamycin, G418, and paromomycin. In the majority of plant species, kanamycin is used as the selection agent, but it is not effective for use with wheat (and other cereals) as nontransformed cells exhibit some natural resistance. Kanamycin can also inhibit shoot regeneration, although G418 and paramomycin have been used successfully ***(5)***.
25. To extract DNA for PCR, kits such as Wizard™ Genomic DNA Purification Kit (Promega, Madison, WI) or Extract'n'Amp (Sigma Aldrich) may be used. If the DNA is also to be used for Southern blot analysis, it is preferable to use the CTAB method.
26. Young, green leaves are preferable for this assay. Set incubations up in a laminar flow hood to avoid contamination. Ensure leaf pieces are fully immersed in the incubation medium; this can be achieved by briefly shaking the plates by hand.
27. Tissue culture rooms are often adequate for this purpose.
28. The various commercially available glufosinate-ammonium herbicide formulations contain different concentrations of the active ingredient PPT. Therefore, the volume of herbicide must be adjusted accordingly to give 0.2 g/L and 2 g/L concentrations of the PPT.

References

1. Wohlleben, W., Arnold, W., Broer, I., Hillemann, D., Strauch, E., and Puhler, A. (1988) Nucleotide-sequence of the phosphinothricin *N*-acetyltransferase gene from *Streptomyces viridochromogenes*-Tu494 and its expression in *Nicotiana tabacum*. *Gene* **70,** 25–37.

2. Bevan, M. W., Flavell, R. B., and Chilton, M. D. (1983) A chimaeric antibiotic-resistance gene as a selectable marker for plant-cell transformation. *Nature* **304,** 184–187.
3. Barcelo, P. and Lazzeri, P. (1995) Transformation of cereals by microprojectile bombardment of immature inflorescence and scutellum tissues, in *Methods in Molecular Biology: Plant Gene Transfer and Expression Protocols* (Jones, H., ed.), Humana Press, Totowa, NJ, pp. 113–123.
4. Barcelo, P., Rasco-Gaunt, S., Thorpe, C., and Lazzeri, P. A. (2001) Transformation and gene expression, in *Advances in Botanical Research Incorporating Advances in Plant Pathology*, Vol. 34, Academic Press, London, UK, pp. 59–126.
5. Nehra, N. S., Chibbar, R. N., Leung, N., et al. (1994) Self-fertile transgenic wheat plants regenerated from isolated scutellar tissues following microprojectile bombardment with 2 distinct gene constructs. *Plant J.* **5,** 285–297.
6. Pastori, G. M., Wilkinson, M. D., Steele, S. H., Sparks, C. A., Jones, H. D., and Parry, M. A. J. (2001) Age-dependent transformation frequency in elite wheat varieties. *J. Exp. Bot.* **52,** 857–863.
7. Rasco-Gaunt, S., Riley, A., Cannell, M., Barcelo, P., and Lazzeri, P. A. (2001) Procedures allowing the transformation of a range of European elite wheat (*Triticum aestivum* L.) varieties via particle bombardment. *J. Exp. Bot.* **52,** 865–874.
8. Rasco-Gaunt, S., Riley, A., Lazzeri, P., and Barcelo, P. (1999). A facile method for screening for phosphinothricin (PPT)-resistant transgenic wheats. *Mol. Breed.* **5,** 255–262.
9. Christensen, A. H. and Quail, P. H. (1996) Ubiquitin promoter-based vectors for high-level expression of selectable and/or screenable marker genes in monocotyledonous plants. *Transgen. Res.* **5,** 213–218.
10. Vasil, V., Castillo, A. M., Fromm, M. E., and Vasil, I. K. (1992) Herbicide resistant fertile transgenic wheat plants obtained by microprojectile bombardment of regenerable embryogenic callus. *Biotechnology* **10,** 667–674.
11. Müller, E., Lörz, H., and Lütticke, S. (1996) Variability of transgene expression in clonal cell lines of wheat. *Plant Sci.* **114,** 71–82.
12. Murashige, T. and Skoog, F. (1962) A revised medium for rapid growth and bioassays with tobacco tissue cultures. *Physiol. Plant.* **15,** 473–497.
13. Rasco-Gaunt, S., Riley, A., Barcelo, P., and Lazzeri, P. A. (1999) Analysis of particle bombardment parameters to optimise DNA delivery into wheat tissues. *Plant Cell Rep.* **19,** 118–127.
14. Stacey, J. and Isaac, P. (1994) Isolation of DNA from plants, in *Methods in Molecular Biology: Protocols for Nucleic Acid Analysis by Nonradioactive Probes*, Vol. 28, (Isaac, P., ed.), Humana Press, Totowa, NJ, pp. 9–15.

14

Histochemical and Fluorometric Assays for *uidA* (GUS) Gene Detection

Magdalena Cervera

Summary

Transgenic plant production has been intimately connected to the β-glucuronidase (*uidA* or GUS) gene used as a reporter marker gene. The enzyme stability and the high sensitivity and amenability of the GUS assay to qualitative (histochemical assay) and to quantitative (fluorometric or spectrophotometric assay) detection are some of the reasons that explain the extensive use of *uidA* gene in plant genetic transformation. Methods for *uidA* gene detection have been thoroughly described in the literature. The aim of this chapter is to describe the basic protocols needed for GUS detection in a plant genetic transformation laboratory.

Key Words: Fluorometric GUS detection; β-glucuronidase; GUS; histochemical GUS detection; reporter marker genes; *uidA* gene.

1. Introduction

A reporter gene codes for an enzyme or other protein that can be detected directly or indirectly using a biochemical assay. The establishment of genetic transformation procedures has relied on, among other factors, the use of efficient reporter marker genes, which easily allows the detection of transgenic events after a transformation experiment, in either a transient or stable expression assay. The production of transgenic plants in many cases depends also on the use of reporter genes, as it facilitates the identification of stably transformed individuals once they have undergone a selection process based on the selectable marker gene used in the same experiment. It should also be mentioned that gene reporter systems have played a key role in many gene expression and regulation studies, in which expression of a reporter gene under, for instance,

From: *Methods in Molecular Biology, vol. 286: Transgenic Plants: Methods and Protocols*
Edited by: L. Peña © Humana Press Inc., Totowa, NJ

the direction of different promoters or the presence of different transcription factors may be investigated.

Since the β-glucuronidase (GUS) gene (*gus*, *gusA*, or *uidA*) was first isolated from *Escherichia coli* ***(1)***, many efforts have been made to develop the *E. coli uidA* gene as a reporter system for plant transformation ***(2,3)***. Indeed, it has become the most widely used marker system, mainly because of the enzyme stability and the high sensitivity and amenability of the assay to detection by fluorometric, spectrophotometric, or histochemical techniques. In addition, there is little or no detectable GUS activity in almost any higher plant tissues ***(2)***, with some exceptions ***(4–7)***. These compile a great part of the attributes a reporter marker gene must account for.

E. coli GUS has a monomeric molecular weight of 68,200 and appears to function as a tetramer ***(8)***. This enzyme hydrolyzes β-glucuronides as substrates and the detection method will vary depending on the specific substrate and product formed after the reaction. In plants it works as a fusion gene, where a promoter coming from a different organism directs the transcription of the *uidA* coding sequence, specifically regulating gene expression in time, quantity, and cell or tissue location. Some potential limitations have been reported in the use and subsequent detection of GUS activity in transformed plant tissues: background activity ***(4,9)***, normally because of diffusion of the reaction product or to endogenous activity; autofluorescence ***(9)***; quenching or inhibitors ***(9,10)***; or microbial contamination ***(11,12)***. In an *Agrobacterium*-mediated transformation system, it is adequate to work with a reporter gene modified by the presence of an intron, which will impede gene expression in bacteria and thus interferences in the detection assay ***(12)***.

One disadvantage of the *uidA* gene as a reporter marker is that commonly used GUS assays involve destruction of plant material. However, it is possible to detect glucuronidase activity in a nondestructive manner. Exposure of plant material to 5-bromo-4-chloro-3-indolyl-β-D-glucuronide (X-Gluc) or 4-methyl-umbelliferyl-β-D-glucuronide (MUG) for short periods of time reduces toxicity of these substrates as to allow later rescue of plant material ***(13,14)***. Nevertheless, in the past few years, other markers have been developed, for example, that encoding the green fluorescent protein (GFP) of *Aequorea victoria* ***(15)***, which has proved to be more useful for some applications than the *uidA* gene, as it can be visualized without demanding the destruction of plant material.

GUS assays can be performed in a wide variety of tissues, even in protoplasts, taking into account that differences in the tissue structure might require slight changes in the detection protocols. Besides, it is important to know that reproductive tissues may exhibit endogenous GUS activity ***(4)*** and that plant development may also affect *gus* gene expression ***(10)***. Although there are other substrates available, two of them are the most currently used, X-Gluc for GUS histochemical localization and MUG for GUS fluorometric quantitation. This chapter focuses our attention on describing histochemical and fluorometric detection assays involving these substrates.

2. Materials

2.1. Histochemical GUS Detection Assay

2.1.1. Stock Solutions

1. 1 *M* Tris-HCl buffer, pH 7.0.
2. 0.1 *M* Tris-HCl buffer, pH 7.0.
3. 5 *M* NaCl.
4. 1% (v/v) Triton X-100.
5. 5 m*M* Potassium ferricyanide, pH 7.0.
6. 5 m*M* Potassium ferrocyanide, pH 7.0.
7. 1% (v/v) Glutaraldehyde in 0.1 *M* Tris-HCl buffer, pH 7.0.

2.1.2. Substrate Solution (X-Gluc) (see **Note 1**)

1. 100 m*M* Tris-HCl buffer, pH 7.0 (*see* **Note 2**), 50 m*M* NaCl, 0.01% Triton X-100, 0.5 m*M* potassium ferricyanide, pH 7.0 (*see* **Note 3**), 0.5 m*M* potassium ferrocyanide, pH 7.0, 2 m*M* X-Gluc (Duchefa Biochemie BV; Haarlem, The Netherlands) (powder; for 10 mL of reagent mix dissolve 10.41 mg of X-Gluc in 0.4 mL of *N,N*-dimethylformamide [DMF] before mixing with the other components) (*see* **Note 4**).
2. Although there are other ways, we recommend keeping separate stock solutions of each component, storing them in the refrigerator. We do not keep a substrate stock solution, but weigh the required amount for each assay. Once substrate is dissolved in DMF, reagent mix is prepared fresh from stocks and adjusted to the final volume. If GUS assays are not often performed in the laboratory, we also prefer not to store a potassium ferrocyanide stock, as it oxidizes quickly (it does not last more than 2 mo in the refrigerator). It is better to weigh the amount required for each assay.

2.2. Fluorometric GUS Detection Assay

2.2.1. Stock Solutions

1. 1 *M* Sodium phosphate buffer, pH 7.0.
2. 0.25 *M* Na_2EDTA, pH 8.0.
3. 0.2 *M* Na_2CO_3, pH 9.5.
4. 1% (v/v) Triton X-100.
5. 1 mg/mL of bovine serum albumin (BSA).
6. Bradford solution for protein determination (Bio-Rad, Hercules, CA).
7. Diluted Bradford solution (1:5) in sterile water (prepare just before use).

8. Extraction buffer: 50 m*M* Na phosphate buffer, pH 7.0, 10 m*M* Na_2 ethylenediaminetetraacetic acid (EDTA), pH 8.0, 0.1% Triton X-100, 10 m*M* β-mercaptoethanol.

2.2.2. Substrate and Product Solutions

1. Substrate solution: 2 m*M* 4-MUG (Sigma, St. Louis, MO) (3.5 mg of 4-MUG in 5 mL of extraction buffer).
2. Product solution: 1 m*M* 4-methylumbelliferone (MU) (Sigma) (1.98 mg of MU in 10 mL of 0.2 *M* Na_2CO_3, pH 9.5). This is the stock solution, from which working solution will be prepared.

2.2.3. Equipment

Fluorometer (DNA fluorometer model TKO 100, Bio-Rad).

3. Methods

The methods described here are based on the protocols by Jefferson et al. *(2)* and Gallagher ***(16)*,** and some slight modifications have been introduced on our experience working with transgenic plant material. They are indicated and explained in the **Notes**.

3.1. Histochemical GUS Detection Assay

In the histochemical assay, hydrolysis of X-Gluc by GUS gives an insoluble and highly colored indigo dye, visualized as a blue precipitate at the site of enzyme activity that is easily detectable (**Fig. 1**).

3.1.1. Staining Assay

1. Cut plant sections to be assayed and place them inside testing tubes, Eppendorf tubes, small beakers, or multiple-well plates.
2. Add a generous volume of substrate solution to plant sections in the wells to cover them completely.
3. If necessary, infiltrate tubes or plates under vacuum for about 1–2 min to enhance penetration of substrate.
4. Cover tubes or plates with Parafilm or similar to avoid evaporation during incubation.
5. Incubate at 37°C a minimum of 3–4 h to a maximum of 14–18 h (*see* **Note 5**).
6. Uncover plates and remove carefully substrate solution using a Pasteur pipet. Wash tissues three times with 0.1 *M* Tris-HCl, pH 7.0.
7. Fix tissues, if necessary, with 1% glutaraldehyde in 0.1 *M* Tris-HCl, pH 7.0. Incubate at 15°C for 2–3 h (*see* **Note 6**).
8. Wash three times with 0.1 *M* Tris-HCl, pH 7.0.
9. Proceed to destaining of tissues (*see* **Note 7**) by following a series of increasing ethanol mixtures (30, 50, 70, 90, and 100%; 5 min each) to achieve total dehydra-

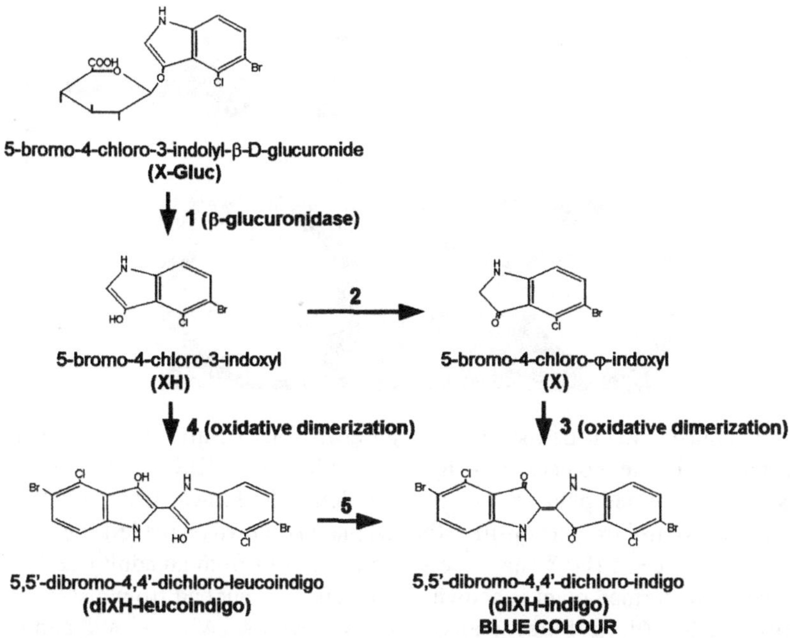

Fig. 1. Reaction taking place in the histochemical GUS assay (from Guivarc'h et al. *[20]*). 5-Bromo-4-chloro-3-indolyl-β-D-glucuronide (X-Gluc) is used as the substrate for GUS, and cleavage of X-Gluc leads to precipitation of a blue product (diXH-indigo or ClBr-indigo) at the site of enzyme activity.

tion of tissues. Keep tissues with pure ethanol for about 1 h. This will remove the chlorophyll and will allow an easier detection of GUS-positive events.

10. Proceed to rehydrate tissues by submerging them in decreasing ethanol mixtures (70, 50, and 30%; 5 min each) and diluted Tris-HCl buffer in the last step. Tissue samples are now ready to be observed under a stereomicroscope (*see* **Note 8**). If desired, samples can be mounted for microscopy. **Figure 2** shows the aspect of different transgenic citrus tissues after GUS staining assays; all were performed in our laboratory.

3.2. Fluorometric GUS Quantitation Assay

Unlike the histochemical detection, fluorometric analysis allows quantitation of GUS activity. In the presence of GUS, MUG is hydrolyzed to a fluorescent product, 4-methylumbelliferone (MU) (**Fig. 3**). After the reaction, total fluorescence is measured and product concentration is calculated based on a previous MU standardization curve. The fluorometric assay is highly reliable and simple to use. However, precautions must be taken to perform the analysis in steady conditions and achieve maximum assay repeatability.

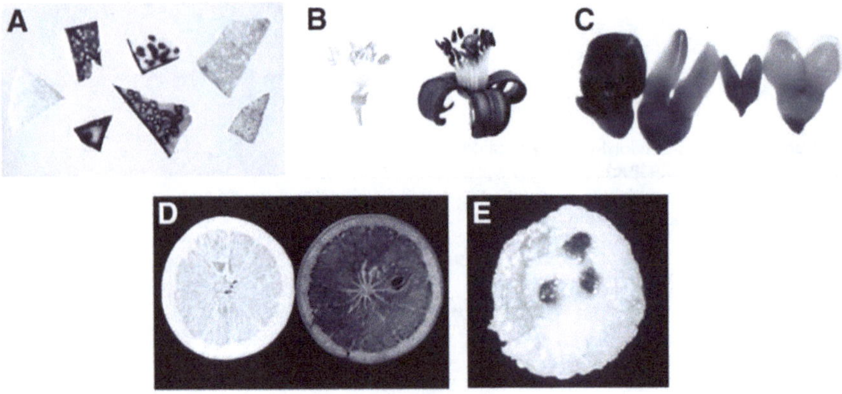

Fig. 2. Transgenic citrus tissues showing GUS staining after histochemical *uidA* gene detection: **(A)** leaf pieces showing different levels of GUS expression (a control from a nontransformed plant is shown on the left). **(B)** Flower from an adult sweet orange tree (a control from a non-transformed plant is shown on the left). **(C)** Embryos from transgenic seeds. **(D)** Transverse section of a fruit from an adult tree (a control from a non-transformed plant is shown on the left). **(E)** cut end from a stem segment showing GUS+ (dark blue) transformation events. All tissues came from *Agrobacterium*-mediated transformation and regeneration experiments, where *A. tumefaciens* EHA105 p35SGUSINT was used as the vector system.

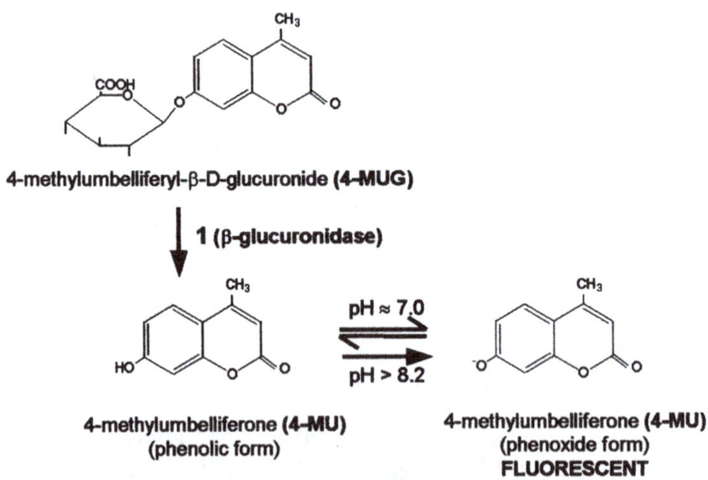

Fig. 3. Reaction taking place in the fluorometric GUS assay. 4-Methylumbelliferyl-β-D-glucuronide (MUG) is used as the substrate for GUS in a GUS fluorometric quantitation. Cleavage of MUG leads to the formation of fluorescent 4-methylumbelliferone (MU). At pH > 8.2 the phenoxide form is predominant and fluorescence at 460 nm is measured to quantitate GUS activity.

3.2.1. Fluorometric Assay

3.2.1.1. EXTRACTION METHOD

1. Use fresh material or material frozen in liquid nitrogen.
2. Extract 20–50 mg of tissue in an Eppendorf tube in 400 µL of extraction buffer and homogenize using a driller. Work on ice.
3. Centrifuge 10 min at 4°C at 13,000*g* to remove unlysed cells and debris (*see* **Note 9**).
4. Transfer the supernatant to a fresh tube and centrifuge again 10 min at 4°C at 13,000*g*.
5. Samples may be stored at –80°C, but it is not convenient to thaw and freeze them more than twice, as activity is partially lost.

3.2.1.2. BRADFORD ASSAY (PROTEIN CONCENTRATION)

1. Prepare diluted Bradford solution and distribute 1 mL in Eppendorf tubes.
2. Make dilutions of the BSA stock solution for calibration (in triplicate). Use this range: 0, 1.0, 2.0, 4.0, 6.0, 10.0, and 20.0 µg of BSA (*see* **Note 10**). Add appropriate volumes of BSA stock to the Bradford solution (for a more accurate calibration, previously remove the same volume of Bradford solution from the Eppendorf tubes) and mix well.
3. Prepare samples by adding 2–10 µL of protein extracts to 1 mL of diluted Bradford solution. Make sure that colors of solutions are in the range of BSA standards.
4. Determine the absorbance at 600 nm in a spectrophotometer (*see* **Note 11**).
5. Calculate the total protein concentration in the samples using the BSA standardization curve.

3.2.1.3. MUG ASSAY (ENZYME ACTIVITY)

1. Keep a bath at 37°C and prepare three Eppendorf tubes per sample with 900 µL of 0.2 *M* Na_2CO_3 to stop the reaction at different times.
2. Prepare the substrate solution to a final concentration of 2 m*M* 4-MUG (3.5 mg of 4-MUG in 5 mL of extraction buffer).
3. Mix protein extracts with substrate in a 1:1 proportion to a total volume of 300 µL. Incubate at 37°C.
4. Take 100 µL from mixtures at three different times (e.g., 10, 20, and 30 min) (*see* **Note 12**) and mix well with 900 µL of Na_2CO_3 (ready in Eppendorf tubes) (*see* **Note 13**). Keep in dark until measurement is performed.
5. Prepare stock product solution to a concentration of 1 m*M* MU (1.98 mg of MU in 10 mL of 0.2 *M* Na_2CO_3). Prepare a working dilution of 0.1 µ*M* MU making an intermediate 10 µ*M* dilution from the 1 m*M* stock. Keep both refrigerated in the dark.
6. Calibrate the fluorometer using the diluted MU solution as standard and measure MU fluorescence in the samples (*see* **Note 14**). Calculate GUS activities as pmol of MU/min/µg of total protein.

4. Notes

1. We usually use in the laboratory for GUS assays Tris-HCl buffer (+ NaCl) instead of phosphate buffer, as described originally by Jefferson et al. ***(2)***, but both have given good results in our case. At any rate, as substrate solution in phosphate buffer is obviously the most used reagent mixture, it is probably convenient to mention it here: 0.1 *M* Na phosphate buffer pH 7.0, 10 m*M* EDTA (*see* **Note 3**), 0.5 m*M* potassium ferricyanide, 0.5 m*M* potassium ferrocyanide, 1 m*M* X-Gluc, 0.1% Triton X-100 (final concentrations) ***(17)***.
2. Apparently some buffer substances might have effects on certain promoters ***(14)***, if fixation is not performed before staining, leading to equivocal results. This is not a substantial issue for the most usual promoters, and Tris-HCl, phosphate, or other buffers at pH 7.0 give good results. pH is another important factor in the case of the GUS staining assay, as most endogenous plant GUS activities exhibit at pH 5.0. However, it can occur at higher pH; for instance, we found endogenous GUS activity in citrus seeds at pH 7.0. This can be normally avoided by increasing working pH to pH 8.0 or even pH 9.0 ***(14)***, adding methanol ***(5,14,18)***, or polyvinylpolypyrrolidone (PVPP) ***(10)***.
3. X-Gluc hydrolysis is usually restricted to the site of GUS activity. However, at sites presenting local oxidative processes, such as high peroxidase activity, precipitation of diX-indigo may occur. A slow oxidation step (and following dimerization into insoluble indigo) allows the soluble intermediate to diffuse away from the site of reaction, making an accurate localization of GUS activity (background activity) difficult. Addition of a mixture of 0.5 *M* potassium ferrocyanide, and 0.5 *M* potassium ferricyanide ***(6)*** in the substrate solution accelerates oxidation of the reaction intermediates to diXH-indigo. It is convenient to note that a high concentration of potassium ferricyanide may inhibit GUS activity ***(6)***, so this should be taken into account when the *uidA* gene is driven by a weak promoter. EDTA is added to mitigate the partial inhibition of the enzyme by the oxidation catalyst.
4. Higher substrate concentrations may be used to increase signal in case of weak staining. Increasing potassium ferrocyanide/potassium ferricyanide concentrations or the incubation periods may also be useful in these cases.
5. Substrate penetration may be a problem in the case of some tissues, such as leaf pieces. Leaf cuticle may difficult the penetration of X-Gluc in this tissue, so normally it is better to punch the leaf with a punctilious object, add Triton X-100 to the substrate mixture and infiltrate tissue in the substrate solution by using a vacuum pump. Calluses, stem or young root pieces do not usually present this problem. Results can be visible after 1–2 h of incubation, but normally it is better to let the reaction finish. Infiltration problems or low expression could lead to errors in the final reading of results. Nevertheless, owing to the stability of the enzyme and extreme sensitivity of the assay, the product is accumulated during the entire period and relative quantification could lead as well to overinterpretation of the data.

6. Fixation can be performed before or after staining. If done before staining, fixation time must be short so as not to lose the activity of the glucuronidase ***(14).*** Fixation after staining can be performed as explained in the text or with other typical fixatives, such as ethanol.
7. The process of destaining is not normally necessary in the case of calluses or other tissues with low chlorophyll content, but it is highly recommended in leaf or stem tissues, where GUS-positive events may appear masked by chlorophyll.
8. Photographic records should be kept from these assays, so they can be useful for later comparisons. Stained samples can be stored in ethanol in well-sealed containers.
9. When working with protoplasts, collect them by centrifugation for 5 min at 85–90*g* (three or four times, 1.5 mL of protoplast suspension) in an Eppendorf tube. Discard supernatant each time. Add 0.4 mL of extraction buffer to the pellet and homogenize gently by ultrasound (10–20 s).
10. Values included in the BSA curve can be modified depending on the samples we are working with and their expression. The colors of the extract mixtures give an initial clue, they should fall among the colors of BSA standards.
11. If a reader of multiwell plates is available, the BSA standard curve and extracts can be prepared and read in the same plate, in volumes of 200 µL. This facilitates and even homogenizes readings. If it is not available in the laboratory, standards and samples must be measured one by one in the spectrophotometer.
12. Incubation times and extract:substrate proportion may vary depending on the samples. Longer incubation and higher extract volumes are probably needed when working with samples with low expression or in transient expression assays.
13. Fluorescent properties of phenolic and phenoxide forms of 4-MU are different: phenolic form (excitation 323, emission 386 nm) and phenoxide form (excitation 363, emission 447 nm). Owing to the equilibrium at physiological pH between phenolic and phenoxide forms of 4-MU (**Fig. 3**), the fluorescence value at the reaction pH is relatively low. Treatment of samples after incubation with 0.2 *M* Na_2CO_3 buffer, pH 9.5 stops the enzyme reaction and raises the final pH above the pK_a of 4-MU (pK_a 8.2), shifting the reaction to produce the maximum amount of fluorescence at 447 nm (for more details, *see* **ref. *19***).
14. At the end of the assay, there should be three fluorescence values per sample, corresponding to different reaction times. Final data are expressed as the slope of the amount of product formed in the reaction (starting from an extract with a determined concentration of total protein) vs time (pmol of MU/min/µg of total protein). In our laboratory, we use the Bio-Rad DNA fluorometer model TKO 100. It is a filter fluorescence photometer with a fixed excitation bandpass source (365 nm) and an emission (460 nm) bandpass filter. Calibration and measurements are easy to perform and sensitivity is acceptable.

Acknowledgments

The author thanks A. Navarro for his excellent help in many GUS assays. Part of this work was supported by grants from the Generalitat Valenciana No.

CTIDIA/2002/89, from the Instituto Nacional de Investigaciones Agrarias No. RTA-01-120 and from CICYT AGL2003-01644.

References

1. Novel, G. and Novel, M. (1973) Mutants d'*Escherichia coli* affectés pour leur croissance sur methyl β-glucuronide: localisation du gene de structure de la β-glucuronidase (*uidA*). *Mol. Gen. Genet.* **120,** 319–335.
2. Jefferson, R. A., Kavanagh, T. A., and Bevan, M. W. (1987) GUS fusions: beta-glucuronidase as a sensitive and versatile gene fusion marker in higher plants. *Embo J.* **6,** 3901–3907.
3. Jefferson, R. A., Bevan, M., and Kavanagh, T. (1987) The use of the *Escherichia coli* beta-glucuronidase as a gene fusion marker for studies of gene expression in higher plants. *Biochem. Soc. Trans.* **15,** 17–18.
4. Hu, C.-Y., Chee, P. P., Chesney, R. H., Zhou, J. H., and Miller, P. D. (1990) Intrinsic GUS-like activities in seed plants. *Plant Cell Rep.* **9,** 1–5.
5. Kosugi, S., Ohashi, Y., Nakajima, K., and Arai, Y. (1990) An improved assay for β-glucuronidase in transformed cells: methanol almost completely suppresses a putative endogenous β-glucuronidase activity. *Plant Sci.* **70,** 133–140.
6. Mascarenhas, J. P. and Hamilton, D. A. (1992) Artifacts in the localization of GUS activity in anthers of petunia transformed with a CaMV 35S-GUS construct. *Plant J.* **2,** 405–408.
7. Muhitch, M. J. (1998) Characterization of pedicel β-glucuronidase activity in developing maize (*Zea mays*) kernels. *Physiol. Plant.* **104,** 423–430.
8. Jefferson, R. A., Burgess, S. M., and Hirsh, D. (1986) β-Glucuronidase from *Escherichia coli* as a gene-fusion marker. *Proc. Natl. Acad. Sci. USA* **83,** 8447–8451
9. Thomasset, B., Ménard, M., Boetti, H., Denmat, L. A., Inzé, D., and Thomas, D. (1996) β-Glucuronidase activity in transgenic and non-transgenic tobacco cells: specific elimination of plant inhibitors and minimization of endogenous GUS background. *Plant Sci.* **113,** 209–219.
10. Serres, R., McCown, B., and Zeldin, E. (1997) Detectable β-glucuronidase activity in transgenic cranberry is affected by endogenous inhibitors and plant development. *Plant Cell Rep.* **16,** 641–646.
11. Tör, M., Mantell, S. H., and Ainsworth, C. (1992) Endophytic bacteria expressing β-glucuronidase cause false positives in transformation of *Dioscorea* species. *Plant Cell Rep.* **11,** 452–456.
12. Vancanneyt, G., Schmidt, R., O'Connor-Sanchez, A., Willmitzer, L., and Rocha-Sosa, M. (1990) Construction of an intron-containing marker gene: splicing of the intron in transgenic plants and its use in monitoring early events in *Agrobacterium*-mediated plant transformation. *Mol. Gen. Genet.* **220,** 245–250.
13. Kirchner, G., Kinslow, C. J., Bloom, G. C., and Taylor, D. W. (1993) Non-lethal assay system of β-glucuronidase activity in transgenic tobacco roots. *Plant Mol. Biol. Rep.* **11,** 320–325.
14. Martin, T., Wöhner, R.-V., Hummel, S., Willmitzer, L., and Frommer, W. B. (1992) The GUS reporter system as a tool to study plant gene expression, in *GUS*

Protocols: Using the GUS *Gene as a Reporter of Gene Expression* (Gallagher, S. R., ed.), Academic Press, San Diego, CA, pp. 23–43.

15. Stewart, C. N., Jr. (2001) The utility of green fluorescent protein in transgenic plants. *Plant Cell Rep.* **20,** 376–382.
16. Gallagher, S. R., ed. (1992) *GUS Protocols: Using the* GUS *Gene as a Reporter of Gene Expression*, Academic Press, San Diego, CA.
17. Stomp, A.-M. (1992) Histochemical localization of β-glucuronidase, in *GUS Protocols: Using the* GUS *Gene as a Reporter of Gene Expression* (Gallagher, S. R., ed.), Academic Press, San Diego, CA, pp. 103–113.
18. Wilkinson, J. E., Twell, D., and Lindsey, K. (1994) Methanol does not specifically inhibit endogenous β-glucuronidase (GUS) activity. *Plant Sci.* **97,** 61–67.
19. Naleway, J. J. (1992) Histochemical, spectrophotometric, and fluorometric GUS substrates, in *GUS Protocols: Using the* GUS *Gene as a Reporter of Gene Expression* (Gallagher, S. R., ed.), Academic Press, San Diego, CA, pp. 61–76.
20. Guivarc'h, A., Caissard, J. C., Azmi, A., Elmayan, T., Chriqui, D., and Tepfer, M. (1996) *In situ* detection of expression of the *gus* reporter gene in transgenic plants: ten years of blue genes. *Transgen. Res.* **5,** 281–288.

15

Green Fluorescent Protein Quantification in Whole Plants

Matthew D. Halfhill, Reginald J. Millwood, and C. Neal Stewart, Jr.

Summary

As future biotechnology applications utilize recombinant proteins as commercial products, nondestructive assays will be necessary to determine protein concentrations accurately within plant tissues. Green fluorescent protein (GFP) has been proposed as a potential marker for the monitoring of transgenic plants and quantifying recombinant protein levels under field conditions. This chapter discusses the utility of using GFP fluorescence as an indicator of protein concentrations and the methods used to quantify GFP fluorescence in whole plant tissues. Furthermore, we discuss the accuracy and effectiveness of the portable General Fluorescence Plant Meter (GFP Meter, Opti-Sciences, Inc.) compared to a laboratory-based spectrofluorometer (Fluoro-Max2, Jobin Yvon & Glen Spectra). In whole plants, GFP fluorescence was shown to be variable at each leaf position over time and among different leaves on the same plant. A leaf had its highest GFP fluorescence after emergence, and subsequently, its fluorescence intensity decreased over time. Younger leaves were significantly more fluorescent than older leaves on the same plant. GFP fluorescence intensity was directly correlated with the concentration of soluble protein per unit wet mass and with another genetically linked recombinant protein (*Bacillus thuringiensis* [Bt] *cry1Ac* endotoxin protein).

Key Words: *Bacillus thuringiensis* (Bt); green fluorescent protein (GFP); soluble protein concentration; spectrofluorometer; transgene monitoring.

1. Introduction

Monitoring transgenic plants under field conditions will become increasingly important as various new genetically modified (GM) crops are implemented in large-scale agriculture. At present, recombinant proteins produced within GM crops provide important production characteristics to plant cultivars, such as

From: *Methods in Molecular Biology, vol. 286: Transgenic Plants: Methods and Protocols*
Edited by: L. Peña © Humana Press Inc., Totowa, NJ

herbicide tolerance and insect resistance, but the proteins themselves are not commercial products. In many future applications, however, recombinant proteins produced within plant materials will be economically important products. Future transgenic plants will function as "biofactories," and will manufacture a wide array of products ranging from pharmaceuticals to plastics. Real-time, nondestructive assays to determine the concentrations of these economically important recombinant proteins will be essential technologies for the profitable use of future biotechnology products.

For a monitoring system to be effective, the genetic marker technology should be accurate with few false positives and negatives, detectable throughout the life cycle of the plant, and able to inform on the status of genetically linked or fused transgenes of interest. Green fluorescent protein (GFP) has been proposed as a whole-plant marker for field-level applications ***(1)***. The GFP gene was first cloned from jellyfish (*Aequorea victoria*) in 1992, and has since been modified for specific applications and transformed into many different organisms ***(2–5)***. GFP monitoring has the potential to track transgenes under large spatial scales utilizing visual or instrumental detection of the characteristic green fluorescence of transgenic material. The *mgfp5-er* variant gene has been shown to be a feasible transgene monitor in plants under field conditions ***(6,7)***. This gene was field-tested in tobacco (*Nicotiana tabacum*), and the plants synthesized the protein and remained fluorescent throughout the growing season ***(7)***. GFP has also been shown to be a feasible qualitative marker for the presence of a linked synthetic *Bacillus thuringiensis* (Bt) *cry1Ac* endotoxin transgene ***(7,8)***. With these beneficial characteristics, the next step in the development of a GFP monitoring system is to better describe the system and resolve weaknesses that could limit the utility of the monitoring system.

From our current research, GFP fluorescence in GFP transgenic plants followed two specific patterns ***(9)***. First, fluorescence intensity of 530-nm green light decreased at each leaf position over time. Second, fluorescence differed among leaf positions on the same plant with the highest fluorescence observed in young leaves. GFP fluorescence intensity was highest in young leaves up to 2 wk after emergence, then the fluorescence intensity decreased over time to levels observed in nontransgenic controls as leaves aged. Therefore, when a plant has a large number of leaves at various ages, a wide spectrum of GFP fluorescence can be detected, ranging from the highest level observed in that transgenic event at and near the apical meristem to those similar to nontransgenic levels in old leaves. In leaf tissues, in which the cauliflower mosaic virus 35S promoter produces a consistent percentage of recombinant protein per unit total soluble protein ***(10,11)***, GFP fluorescence consistently varied with the concentration of soluble protein in a mass of fresh leaf tissue. This research has shown that soluble protein per unit leaf mass changes during leaf development, and the GFP phenotype is correlated with this phenomenon.

One of the key aspects of using GFP fluorescence in a monitoring system for transgene expression is that it accurately quantifies the concentration of genetically linked proteins of interest. The results from current experiments showed that GFP fluorescence accurately predicted the concentration of Bt, even as both parameters changed over time ***(9)***. However, the results also imply that false negatives could be generated if one solely assays older leaves having lower concentrations of soluble proteins. That problem can be minimized if GFP fluorescence measurements are made on younger leaves near the apical meristem. In addition, it has been known for some time that sequential senescence and the decline in soluble protein proceeds at a faster pace under water stress or nitrogen deficiency ***(12)***. Therefore, the observation that changes in GFP fluorescence were correlated with those in soluble protein concentration may prove to be a significant tool for monitoring particular plant stresses.

There are several types of systems presently in use for the macroscopic detection and quantification of fluorescent compounds including: high-intensity UV lamps, spectrofluorometers (e.g., FluoroMax-2, Jobin Yvon & Glen Spectra, Edison, NJ), and scanning laser systems (e.g., FluorImager, FluorImager SI, Molecular Dynamics, Sunnyvale, CA). A handheld 365-nm UV lamp, such as a UVP Model B 100 AP (UVP, Upland, CA), allows for expeditious scanning of GFP fluorescence in whole plants. However, the UV light must be used in darkness, it is only effective for UV excitable GFP variants, and visual observation cannot be used to quantify GFP fluorescence. Spectrofluorometers and fluorescence imaging systems are capable of detecting the presence of GFP and also allow for quantification of fluorescent tissues ***(7,8,13–15)***. However, both systems are laboratory-based and expensive. For several years, plant researchers have sought a portable instrument that measures GFP in field plants under ambient lighting conditions. Millwood et al. ***(16)*** described the methods used to quantify GFP fluorescence with the Opti-Sciences General Fluorescence Plant Meter (GFP Meter). Performance of the GFP Meter was compared to the Fluoromax-2, a validated GFP fluorescence measurement tool ***(7,8,13)***. A comparison of instantaneous measurements from the GFP Meter (530-nm wavelength) and the Fluoromax-2 (standardized 510-nm wavelength fluorescence values) was completed to gage accuracy. The regression analyses of growth chamber-grown GFP canola, greenhouse GFP tobacco, and field-grown GFP tobacco produced high R^2 values (0.87, 0.88, and 0.89) and indicated a positive functional relationship between instruments.

This chapter discusses the methodology used to quantify GFP fluorescence in whole plant tissues. The two GFP detection systems that we use are introduced, and we describe how to collect and analyze the in vivo plant fluorescence values produced from these spectrofluorometers. We also introduce the method used to correlate GFP fluorescence and another genetically linked recombinant protein of interest (Bt). In the **Notes** subheading, we discuss problems that might arise with measuring GFP fluorescence and how plant health and life cycle status play an important role in GFP fluorescence.

2. Materials

2.1. GFP Meter

The GFP Meter (Opti-Sciences, Tyngsboro, MA) is a self-contained, field-portable fluorescence detection and data logging instrument powered by an internal 1.2-ampere h 12-V gel lead acid battery (**Fig. 1**). A filtered light emitting diode (LED) generates excitation light-when powered on. A small portion of this light is monitored to compensate for temperature drift. The output of the LED is focused on one of three fiber ports. A driver, controlled by a microprocessing unit (MPU), regulates the LED power level and compensates for changes in battery voltage. This excitation light travels through a bandpass filter to a fiber-optic cable and is then delivered to the sample. Attached to the end of this cable is a leaf clip, which has been installed to keep the cable in place. The light emitted from the sample enters back into the fiber-optic cable and is directed through a bandpass filter into a low-noise preamplifier. This signal is then fed into an analog/digital signal processing and filtering unit slaved to the MPU. Fluorescence measurements appear in real time on a liquid crystal display in units of counts per second (cps). A 12-selection keypad provides user management of test functions and setup. A nonvolatile memory chip (capable of storing 1020 sample points) assures that data will not be lost when power is turned off or the battery removed and a RS-232 port enables downloading data to a computer. The GFP Meter uses a modulated detection system to minimize the effects of temperature drift and stray light. Virtually any bandpass filter combination can be used for excitation and emission. For this study, a 465-nm filter with a bandwidth of 35 nm was used for excitation. Channel 1 (GFP channel) used a 530-nm filter with a bandwidth of 35 nm for emission and channel 2 (chlorophyll channel) used a 680-nm filter with a bandwidth of 35 nm for emission. Chlorophyll data are not reported in this study.

2.2. FluoroMax-2

The FluoroMax-2 (Jobin Yvon & Glen Spectra) is a laboratory-based spectrofluorometer system that uses a computer to process data. All FluoroMax-2 functions are under control of Datamax spectroscopy software (Galactic Industries Corporation, Salem, NH). Light from a 150-W xenon lamp enters an excitation spectrometer, which delivers monochromatic light to a bifurcated fiber-optic cable. A rubber protector surrounds the external end of the cable and extends 4

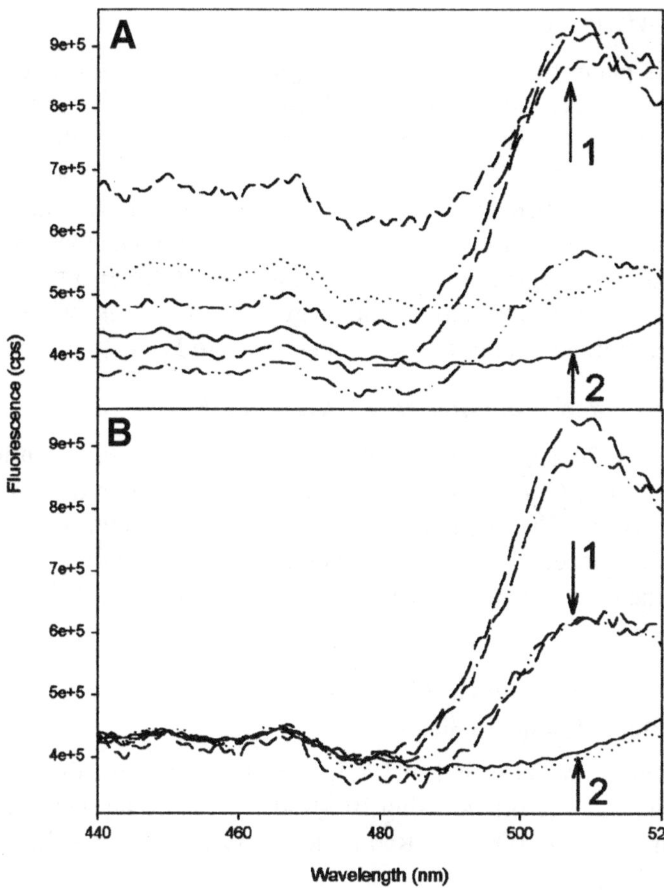

Fig. 1. Standardization of spectrofluorometer (FluoroMax-2) measurements of field-grown GFP tobacco excited with 385-nm UV light. Each line represents an individual tobacco plant. When looking at the GFP maxima (508 nm) of nonstandardized spectral data **(A)**, three plants appeared to be expressing high amounts of GFP with respect to the other plants analyzed. However, after standardization **(B)** to a nontransgenic control *(2)*, it is revealed that one of these plants *(1)* is actually expressing lower levels of GFP.

mm beyond to prevent signal disruption from dirt and debris and to provide a dark environment for measurements. Light flows through the cable only when a sample is being scanned. When scanning, the cable is then placed onto the sample and light emitted from the sample flows back through the fiber-optic cable to the emission spectrometer where it is dispersed and directed to a signal photomultiplier detector. This fluorescence signal is then amplified and displayed on a computer monitor in units of cps.

2.3. GFP Plant Material

In our research, plants transgenic for GFP (*mgfp5-er*) ***(6)*** and GFP/Bt (*mgfp5-er/Bacillus thuringiensis*) under the control of constitutive *CaMV 35S* promoter were used ***(7–9)***. This GFP variant is excited equally by both blue (465-nm) and UV (395-nm) wavelengths. We have experience with three species of GFP transgenic plants: canola (*Brassica napus* cv. Wester), tobacco (*Nicotiana tabacum* cv. Xanthi), and *Arabidopsis thaliana* ecotype Columbia. Nontransgenic control plants from the original plant cultivar or ecotype were also included in each experiment. The zygosity status (homozygous, hemizygous, or mixed) of each line was determined in previous research (*see* **Note 1**).

2.4. Bt Enzyme-Linked Immunosorbent Assay (ELISA)

The EnviroLogix Cry1Ab/Cry1Ac Plate Kit is designed for the quantitative laboratory detection of Cry1Ab and Cry1Ac residues in corn and cotton leaf tissue samples (Cry1Ab/Cry1Ac Plate Kit, cat. no. AP 003, Envirologix, Portland, ME). Previous research within our laboratory has also shown this kit to be accurate for the determination of Bt concentrations from canola leaf tissue samples.

2.5. Protein Extraction from Leaf Tissues and Soluble Protein Quantification

0.1 *N* NaOH and 1 *M* Tris-HCl, pH 4.5, were used as extraction buffers. Bradford analysis was performed using Bio-Rad Protein Assay Dye Reagent Concentrate (cat. no. 500-0006, Bio-Rad Laboratories, Hercules, CA).

3. Methods

3.1. GFP Fluorescence Quantification

1. Plant lines were germinated and grown under the same environmental conditions. The plant conditions varied based on the experimental design. Optimal plant health was important in GFP experiments (*see* **Note 2**), and therefore, conditions were selected based on the experimental plant species (*see* **Notes 3** and **4**).
2. After a specified time period, the plants were selected for GFP analysis. It was important to evaluate plants that were at the same developmental stage (*see* **Note 5**). The same leaf position was chosen for each individual plant. Time course studies were also an option, and in this case, the same leaf positions were tested at consistent time intervals (between 1 and 7 d) over the course of the experiment.
3. A spot (1-cm diameter, the size of the end of the fiber-optic cable) on the underside of sample leaves and adjacent to the leaf mid-vein was excited at 465 nm with the GFP Meter or 385 nm with the Fluoromax-2. Two fluorescence mea-

surements were recorded on opposite sides of the mid-vein on the underside of each leaf.

4. The GFP Meter recorded single 530-nm GFP fluorescence and 680-nm chlorophyll values. The Fluoromax-2 recorded a selectable emission range, and in our experiments included 420–600 nm.
5. Emissions spectra were recorded from five individual control plants and were averaged. The average control was used to standardize the Fluoromax-2 data.

3.2. Fluorescence Standardization

1. The FluoroMax-2 GFP fluorescence scans were standardized to control samples to account for baseline variation of each leaf measurement (*see* **Note 6**). The protocol for standardization involves selection of a wavelength outside the GFP fluorescence spectrum as a point of normalization for each FluoroMax-2 scan. For this study, the 450-nm wavelength was the anchor point.
2. Subsequently, each emissions scan was standardized to the average control for that species. Functionally, the 450-nm GFP value of the sample was subtracted from the 450-nm average control value. The resulting integer was then added to each wavelength value along the sample spectra (420–600 nm). This method eliminates differences outside the GFP emissions spectra, allowing for comparison of GFP magnitudes.
3. The GFP Meter did not have the option of standardization because the appropriate filter sets outside the GFP emissions range were not installed for this study.
4. The standardized GFP fluorescence values were used to compare treatments within each experiment.
5. In our experiments, GFP value for each leaf position was the unit of measure compared between treatment types. For example, the average GFP value at the fourth leaf position between plants could be compared for each experimental treatment.

3.3. GFP Fluorescence Correlations With Several Protein Concentrations

1. GFP fluorescence measurements were recorded as described in **Subheading 3.1., step 3**.
2. For the transgenic samples, two 1-cm diameter leaf punches were collected from fresh leaves at the same position as the GFP measurements from each leaf with a 1.5-mL microcentrifuge tube.
3. The mass of each punch was recorded, and was then flash frozen in liquid nitrogen (LN). The sample was homogenized with 0.5-mm glass beads in a mechanical amalgamator (Silamat S5, Ivoclar Vivadent Clinical, Austria).
4. After disruption of the leaf tissue, 400 mL of 0.1 *N* NaOH was added to each sample, and the sample was incubated on ice for 30 min. After incubation, 80 mL of 1 *M* Tris-HCl, pH 4.5 was added to each sample for neutralization. The sample was centrifuged for 7 min at 6200*g* (Denville 260D Microcentrifuge, 24 sample rotor, Denville Scientific, Metuchen, NJ), and the supernatant containing soluble protein was recovered.

5. Bradford Analysis (Bio-Rad Laboratories) was used to quantify soluble protein concentration in 96-well plates (EL 800 Universal Microplate Reader with the KC Junior software package, Bio-Tek Intruments, Winooski, VT).
6. Bt protein concentration was determined by the use of a Bt ELISA.
7. Regression analysis was performed for standardized GFP fluorescence per leaf sample by the concentration of soluble protein and Bt per unit wet mass (StatView 5.0 for Windows). The protein concentrations determined by these methods may also be compared with standard curves produced for each specific protein (*see* **Note 7**).

4. Notes

1. The zygosity status (either homozygous or hemizygous), for the experimental plants must be understood prior to the start of the experiment. In previous research, the GFP gene demonstrated additive transgene expression in 10 independent transformation events of canola ***(9)***. In all canola lines, homozygous individuals that contained two copies of the transgene locus fluoresced twice as much as hemizygous individuals above the background level of fluorescence. We caution against using segregating populations for any fluorescence-based experiments, because the plants may exhibit a wide range of fluorescence based on the number of transgene copies. Experimental treatments will rarely play a larger role than the zygosity status of the plant, and we suggest using strictly homozygous populations if GFP fluorescence is going to be measured between treatments.
2. Plant health is vitally important to utilize GFP fluorescence as a quantifiable tool. Sick or stressed plants will not have robust and reliable GFP fluorescence. Independent GFP plant lines may exhibit a wide range of fluorescence based on the plant health, and the introduction of any unintended plant stress during the course of the experiment will likely reduce the repeatability of the results. Plant stress caused by nutrient, water, or light deficiencies will dramatically alter the concentration of soluble protein in the leaf tissues. Several experiments ***(9–11)*** indicate the percentage of GFP within extractable soluble protein is expected to remain consistent, and therefore, the changes in GFP fluorescence in plant tissues is caused by changes in soluble protein concentration. If the plants undergo a period of stress and modulate soluble protein production, the GFP measurements will reflect this change and potentially invalidate the experiment.
3. The environmental conditions for all plants of the experiment must be highly consistent to compare the data between treatments. Variations between different environmental chambers, greenhouse rooms, or greenhouse locations can cause significant differences in GFP fluorescence, once again invalidating the data. Many experiments using large numbers of plants by necessity must use more than one location, and growing plants in different locations have caused significant location based in error in some of our preliminary work. Randomization of experimental treatments does not solve location-based variance, because the location effect often can be the largest cause of error in GFP fluorescence-based experiments. We suggest that emphasis on environmental and experimental standardiza-

tions during the development of the experimental design will lead to repeatable results.

4. Overall, the utility of GFP fluorescence as an analytical tool is limited to relative comparisons between plants of the same species grown under similar conditions. We have found that it is difficult to compare the actual GFP values between species and cases where the plants are grown in disparate conditions, that is field and in the laboratory. With this in mind, the experiments must have the appropriate controls to estimate relative changes in GFP fluorescence between treatments, and then the trends that are detected may be used to understand the differences in fluorescence between plant species and environmental conditions.
5. The plant life history stage alters the degree of GFP fluorescence for each plant tissue, and selecting tissues from the same stage is important in GFP experiments. The location of detectable fluorescence changes dramatically as a plant progresses through its life cycle. The 35S promoter produces high expression of GFP in young leaves and shoot meristems, and GFP yields similar patterns as GUS under the control of identical promoters ***(11)***. When plants have a large number of leaves, a full range of GFP fluorescence can be seen with the brightest possible young leaves to older leaves with wild-type levels of fluorescence. In mature plants, green fluorescence was detectable in meristems. GFP fluorescence has been shown to be visible in young leaves, stems, veins, and flowers, and specifically selecting tissues at the same stage of development will allow comparisons between treatments.
6. The spectrofluorometers used in this study may add variation to the GFP fluorescence measurements based on the techniques used by the experimenter. Both instruments use a cable to deliver the excitation light to the leaf surface, and different experimenters may add significant variation to the data by their interpretation of the methods. In our case using the Fluoromax-2, we have seen that the angle the cable is held in relation to the leaf surface can affect the magnitude of the GFP value. We have found that it may be beneficial to standardize the angle by clamping the cable in a fixed orientation. This potentially solves this problem and allows multiple users to produce similar GFP values. For the GFP Meter, the amount of time the cable is clipped to the leaf prior to excitation has been shown to change the GFP magnitude. We suggest that the GFP value should be measured rapidly after the leaf clip is placed on the leaf.
7. Producing standard curves for known amounts of GFP may be useful, because this procedure allows for in vivo estimations of protein amount based on GFP fluorescence values. Richards et al. ***(17)*** reported that the fluorescence intensity increases linearly as the amount of GFP increases. The resulting standard curves were then used to estimate the amount of GFP in unknown samples, in this case either protein extracts or direct leaf measurements. Data from ELISA supported the validity of the fluorescence-based estimates. In fact, it was possible to generate recombinant protein estimates *in planta* because the fluorescence properties of the intact leaf did not affect the GFP signal. This technique may facilitate future characterization of GFP and GFP-fusion transgenic plants by eliminating the need for laboratory-based protein quantification methods.

References

1. Stewart, C. N., Jr. (1996) Monitoring transgenic plants using *in vivo* markers. *Nat. Biotech.* **14,** 682.
2. Chalfie, M., Tu, Y., Euskirchen, G., Ward, W. W., and Prasher, D. C. (1994) Green fluorescent protein as a marker for gene expression. *Science* **263,** 802—805.
3. Prasher, D. C., Eckenrode, V. K., Ward, W. W., Pendergast, F. G., and Cormier, M. J. (1992) Primary structure of the *Aequorea victoria* green fluorescent protein. *Gene* **111,** 229–233.
4. Siemering, K. R., Golbik, R., Sever, R., and Haseloff, J. (1996) Mutations that suppress the thermosensitivity of green fluorescent protein. *Curr. Biol.* **6,** 1653–1663.
5. Chiu, W. L., Niwa, Y., Zeng, W., Hirano, T., Kobayashi, H., and Sheen, J. (1996) Engineered GFP as a vital reporter in plants. *Curr. Biol.* **6,** 325–330.
6. Haseloff, J., Siemering, K. R., Prasher, D., and Hodge, S. (1997) Removal of a cryptic intron and subcellular localization of green fluorescent protein are required to mark transgenic *Arabidopsis* plants brightly. *Proc. Natl. Acad. Sci. USA* **94,** 2122–2127.
7. Harper, B. K., Mabon, S. A., Leffel, S. M., et al. (1999) Green fluorescent protein as a marker for expression of a second gene in transgenic plants. *Nat. Biotech.* **17,** 1125–1129.
8. Halfhill, M. D., Richards, H. A., Mabon, S. A., and Stewart, C. N., Jr. (2001) Expression of GFP and Bt transgenes in *Brassica napus* and hybridization and introgression with *Brassica rapa. Theor. Appl. Genet.* **103,** 362–368.
9. Halfhill, M. D., Millwood, R. J., Weissinger, A. K., Warwick, S. I., and Stewart, C. N., Jr. (2003) Additive transgene expression in multiple GFP transgenic crop x weed hybrid generations. *Theor. Appl. Genet.* **107,** 1533–1540.
10. Blumenthal, A., Kuznetzova, L., Edelbaum, O., Raskin, V., Levy, M., and Sela, I. (1999) Measurement of green fluorescent protein in plants: quantification, correlation to expression, rapid screening and differential gene expression. *Plant Sci.* **142,** 93–99.
11. Harper, B.K. and Stewart, C.N., Jr. (2000) Patterns of green fluorescent protein expression in transgenic plants. *Plant Mol. Bio. Rep.* **18,** 1–9.
12. Woolhouse, H. W. (1967) The nature of senescence in plants. *Sym. Soc. Exp. Biol.* **21,** 179–213.
13. Leffel, S. M., Mabon, S. A., and Stewart, C. N., Jr. (1997) Applications of green fluorescent protein in plants. *BioTechniques* **23,** 912–918.
14. Niwa, Y., Hirano, T., Yoshimoto, K., Shimizu, M., and Kobayashi, H. (1999). Noninvasive quantitative detection and applications of non-toxic, S65T-type green fluorescent protein in living plants. *Plant J.* **18,** 455–463.
15. Stewart, C. N., Jr. (2001). The utility of green fluorescent protein in transgenic plants. *Plant Cell Rep.* **20,** 376–382.
16. Millwood, R. J., Halfhill, M. D., Harkins, D., Russotti, R., and Stewart, C. N., Jr. (2003) Instrumentation and methodology of GFP quantification in intact plant organs. *BioTechniques* **34,** 638–643.

17. Richards, H. A., Halfhill, M. D., Millwood, R. J., and Stewart, C. N., Jr. (2003) Quantitative GFP fluorescence as an indicator of recombinant protein synthesis in transgenic plants. *Plant Cell Rep.* **22,** 117–121.

16

Positive Selection

Allan Wenck and Geneviève Hansen

Summary

The use of a new mode of selection—positive selection—has been demonstrated to be successful in a large variety of monocot and dicot species. This selection differs from more traditional modes of selection in which compounds such as antibiotics or herbicides are used to kill nontransformed cells (negative selection). In the case of positive selection, a transformed cell acquires the ability to metabolize a substrate that it previously could not use (or not use efficiently) and thereby grows out of the mass of nontransformed tissue. Positive selection can be of many types from inactive forms of plant growth regulators that are then converted to active forms by the transferred enzyme to alternative carbohydrate sources that are not utilized efficiently by the nontransformed cells that become available upon transformation with an enzyme that allows them to be metabolized. Nontransformed cells either grow slowly in comparison to transformed cells or not at all. Using positive selection, nontransformed cells may die, but, typically, production of phenolic compounds observed with negative selection markers does not occur. In many cases, this effect contributes to higher transformation efficiencies, as these compounds can negatively influence the growth of transformed cells. The use of one form of positive selection—transformation with phosphomannose isomerase followed by selection on mannose containing media—is presented here as an example.

Key Words: Mannose; negative selection; phosphomannose isomerase; plant transformation; positive selection.

1. Introduction

Genetic transformation of a diverse range of plant species requires the use of tools for selection of transformed cells from a population of nontransformed cells. In most cases, "negative" selection markers are used to accomplish this goal. Within this report, the term negative selection will be used to differentiate the use of compounds that directly inhibit growth of nontransformed cells.

From: *Methods in Molecular Biology, vol. 286: Transgenic Plants: Methods and Protocols*
Edited by: L. Peña © Humana Press Inc., Totowa, NJ

Further, transformed cells obtain no benefit from the selection agent. Transformed cells are able to survive the actions of the selective agent by either deactivation of the selection compound as in the case of kanamycin and the neomycin phosphotransferase gene or by having an alternative form of an endogenous enzyme that no longer binds to the previous toxic compound as is the case for certain herbicide resistance genes. Negative selection agents include traditional antibiotics such as the aforementioned kanamycin and others such as hygromycin, gentamycin and spectinomycin. Other examples of negative selection systems include the use of herbicides such as glyphosate or phosphinothricin. An alternative selection system, termed "positive" selection relies on a novel concept in which the compound that is provided as the selection agent is not in itself directly toxic to plant cells and may even be completely devoid of biological activity. With positive selection systems, an enzyme or enzymes are provided that allow the transformed plant cell to utilize the selective compound for growth. Nontransformed cells cannot grow or grow slowly on media in which the compound is found. A recent review has been published in which both negative and positive selection agents are presented along with mode of action and use within plant transformation *(1)*.

The first example of positive selection was provided by Joersbo and Okkels *(2)*. These researchers demonstrated that transgenic tobacco plants could be obtained from leaf disks transformed with the β-glucuronidase gene (GUS) when a cytokinin glucuronide was provided as a substrate and cytokinin was absent from the media. Only cells expressing GUS could metabolize the cytokinin glucuronide. These cells could then proliferate and differentiate into shoots, whereas cells without GUS activity could not. This concept of positive selection was further expanded to include not only plant hormones, but also carbohydrate and nitrogen sources (**ref. *3***, and reviewed in **ref. *4***).

Carbohydrates represent one of the most readily used aspects of positive selection because plant cells in culture require the presence of a carbohydrate source. In addition, many carbohydrates are easily and cheaply obtained from commercial sources. Typically sucrose, glucose, or maltose is incorporated into plant culture media. However, if another carbohydrate such as mannose is introduced instead into the media, in the majority of cases studied, the plant cells will be unable to proliferate and may die. In the case of mannose and many other carbohydrates, the compound is metabolized but the product of that step cannot be further metabolized. For example, when mannose is used, it is quickly converted to mannose-6-phosphate (M6P) by the action of hexokinase (reviewed in **ref. *5***). A similar effect is observed when deoxyglucose is used in the media *(6)*. Further examples of the use of alternative carbohydrate sources as a means of selection for transgenic cells have utilized xylose *(7)* and ribitol *(8)*. The use of these sugars, although not toxic in themselves, may lead to cell death through the depletion of ATP. Events similar to apoptosis in animals have been documented in the presence of mannose ***(9)***.

The best documented of these carbohydrate sources is the use of mannose combined with the phosphomannose isomerase (PMI) gene of *E. coli* (*manA*). PMI catalyzes the conversion of M6P to fructose-6-phosphate, which can be utilized as a carbohydrate source. The PMI system has been shown to be effective for sugar beet ***(10)***, maize ***(11–13)***, rice ***(14)***, wheat ***(13)***, *Arabidopsis* ***(15)***, and many other dicotyledonous and monocotyledonous species (reviewed in **ref. *16***). This system not only has been shown to be useful for the selection of transgenic cells and plants, but higher efficiencies also have been reported over the more traditional negative selection techniques currently used. Higher efficiencies were reported for sugar beet compared to kanamycin selection ***(10)***, for rice compared to hygromycin selection ***(17)***, and for wheat compared to phosphinothricin selection ***(13)***. Todd and Tague ***(15)*** further documented that the use of mannose as a selection agent was significantly less expensive than other selection agents currently in use.

This chapter presents a procedure for the use of mannose selection with maize. The basic procedures will need to be modified for different crops, but the concepts presented should be applicable to other crop species. Although many crop species contain no detectable PMI activity ***(5)***, there are some exceptions. The presence of endogenous activity does not preclude the use of mannose for selection but may require the use of more stringent selection regimes than presented here. For example, barley does have a documented PMI activity, yet successful selection of transgenic barley has been obtained in both biolistic and *Agrobacterium*-mediated transformation methods ***(17)***.

2. Materials

1. Maize immature embryos Hi-II greenhouse or field grown ears, 8–12 d post-pollination (embryo size from 0.75–1.5 mm).
2. 10X Linsmaier and Skoog (LS) modified major salts (1 L [modified from **ref. *11*** as per ***18***]): 16.5 g of NH_4NO_3, 19.0 g of KNO_3, 4.4 g of $CaCl_2 \cdot 2H_2O$, 3.7 g $MgSO_4 \cdot 7H_2O$, 0.17 g KH_2PO_4.
3. 1000X LS minor salts (1 L ***[19]***): 0.83 g of KI, 6.2 g of H_3BO_3, 22.3 g of $MnSO_4 \cdot 4H_2O$, 8.6 g of $ZnSO_4 \cdot 7H_2O$, 0.25 g of $NaMoO_4 \cdot 2H_2O$, 0.025 g of $CuSO_4 \cdot 5H_2O$.
4. 200X Murashige and Skoog (MS)/LS iron stock (1 L ***[19]***): 8.26 g of Na_2 ethylenediamine tetraacetic acid (EDTA) $\cdot 2H_2O$, 5.56 g of $FeSO_4 \cdot 7H_2O$.
5. 100X Japan Tobacco (JT) additions (1 L ***[20]***): Filter sterilize: 0.1 g of thiamine-HCl, 0.05 g of pyridoxine-HCl, 0.05 g of nicotinic acid, 10 g of *myo*-inositol.
6. Modified MS (JMS) major salts 10X (1 L ***[21]***): 16.9 g of NH_4NO_3, 18.2 g of KNO_3, 2.1 g of $CaCl_2 \cdot 2H_2O$, 4.0 g of $MgSO_4 \cdot 7H_2O$, 3.5 g of KH_2PO_4.

7. 100X Schenk and Hildebrandt (SH) minor salts (1 L *[22]*): 0.1 g KI, 0.5 g of H_3BO_3, 1.0 g of $MnSO_4 \cdot 4H_2O$, 0.1 g of $ZnSO_4 \cdot 7H_2O$, 0.010 g of $NaMoO_4 \cdot 2H_2O$, 0.020 g of $CuSO_4 \cdot 5H_2O$, 0.010 g of $CoCl_2 \cdot 6H_2O$.
8. 100X G5 additions (an internally developed vitamin mixture (1 L *[19]*): Filter sterilize: 10 g of casein hydrolysate, 0.5 g of thiamine-HCl, 0.05 g of pyridoxine-HCl, 0.5 g of nicotinic acid, 10 g of *myo*-inositol, 10 g of proline.
9. 100X MS additions (1 L *[23]*): Filter sterilize: 0.01 g of thiamine-HCl, 0.05 g of pyridoxine-HCl, 0.05 g of nicotinic acid, 10 g of *myo*-inositol, 0.20 g of glycine.
10. Mannose: 1 g/mL of distilled water, heat to dissolve and filter sterilize.
11. Ticarcillin: 100 mg/mL of distilled water: filter sterilize (Ticar; Smith Kline Beecham, Philadelphia, PA).
12. Dicamba: 1 mg/mL in 0.1 *M* HCl.
13. Acetosyringone: 40 mg/mL DMSO: 500 μ*M*.
14. Ancimidol: Sepro A-rest Solution (0.0264% Ancimidol, Carmel, IN).
15. Plant Preservative Mixture (PPM): (Plant Cell Technology, Washington, DC).
16. 2-(*N*-morpholino)ethanesulfonic acid (MES): 100 mg/mL, pH 5.8.
17. Silver nitrate: 10 mg/mL of $AgNO_3$. Filter sterilize.
18. Spectinomycin: 50 mg/mL. filter sterilize.
19. Kinetin: 1 mg/mL in 0.1 *M* HCl.
20. LS Modified Dc5 – inf (1 L (modified from **ref. *11***): 100 mL of LS modified major salts, 1 mL of LS minor salts, 5 mL of LS Iron, 1 g of casamino acids, 5 mL of Dicamba, 68.5 g of sucrose, 36.0 g of glucose, 10 mL of JT additions. Adjust the pH to 5.2 with KOH and filter sterilize.
21. LS modified Dc5 As500 (1 L (modified from **ref. *11***): 100 mL of LS modified major salts, 1 mL of LS minor salts, 5 mL of LS Iron, 1.4 g of proline, 5 mL of Dicamba, 20g of sucrose, 10 g of glucose, 10 mL of MES, pH adjusted to 5.8: Add 8 g/L plant tissue culture agar. Autoclave. Additions after autoclave: 10 mL JT additions, 2.5 mL of acetosyringone. 100 × 15 Petri dishes.
22. 2JMS (1 L *[21]*): 100 mL JMS major salts, 10 mL of SH minor salts, 5 mL of LS iron, 5 mL of Dicamba, 20 g of sucrose, pH adjusted to 5.8. Add 2.4 g/L of Gelrite. Autoclave. Additions after autoclave: 10 mL of G5 additions, 2 mL of Ticarcillin, 0.5 mL of $AgNO_3$. 100 × 15 Petri dishes.
23. JMS 1M/0.5S (1 L *[21]*): 100 mL JMS major salts, 10 mL of SH minor salts, 5 mL of LS iron, 5 mL of Dicamba, 5 g of sucrose. pH adjusted to 5.8. Add 2.4 g/L of Gelrite. Autoclave. Additions after autoclave: 10 mL of G5 additions, 2 mL of Ticarcillin, 10 mL of mannose. 100 × 15 Petri dishes.
24. MSAK 0.5M/2S (1 L: reg. 1 *[11]*): 4.3 g of MS salts (Invitrogen, Carlsbad, CA), 10 mL of MS additions, 20 g of sucrose, 0.25 mg of Ancimidol, 0.5 mg of Kinetin. pH adjusted to 5.8. Add 2.4 g/L of Gelrite. Autoclave. Additions after autoclave: 2 mL of Ticarcillin, 5 mL of mannose. 100 × 25 Petri dishes.
25. MS 0.5M/2S (1 L: reg. 2 *[11]*): 4.3 g of MS salts, 10 mL of MS additions, 20 g of sucrose. pH adjusted to 5.8. Add 2.4 g/L of Gelrite. Autoclave. Additions after autoclave: 2 mL of Ticarcillin, 5 mL of mannose. 100 × 25 Petri dishes.
26. GA-7 Boxes (1 L: reg. 3 *[11]*): 3.2 g of MS salts, 10 mL of MS additions, 30 g of sucrose. pH adjusted to 5.8. Add 2.4 g/L of Gelrite. Autoclave. Additions after

autoclave: 2 mL of Ticarcillin, 5 mL of PPM. GA-7 plant cons or other appropriate container.

27. Yeast peptone (YP) (1 L *[11]*): 5 g of yeast extract, 5 g of NaCl, 10 g of peptone. pH adjusted to 6.8. Add 15 g/L of Bacto-agar. Autoclave. Additions after autoclave: 2 mL of spectinomycin. 100 × 15 Petri dishes.
28. *Agrobacterium*: *Agrobacterium* containing the superbinary plasmid pSB1 ***(11,20)*** with a cointegrated or separate binary plasmid containing the *pmi* cassette with or without an additional gene of interest ***(11)***. Grown on YP media.

3. Methods

1. Ears of greenhouse or field grown Hi-II are picked on the day of use. Ears are sterilized by shaking in a solution of 20% Clorox with several drops of Tween-20 for 20 min and rinsed with sterile water.
2. *Agrobacterium* is suspended in 5–10 mL of LS modified Dc5-inf containing 100 μ*M* acetosyringone. The optical density is adjusted between 0.5 and 0.8 (OD_{660}). This preinduction period is for a minimum of 30 min.
3. A long, sharp sterile forceps or other devise for holding the ear is run down the center from the top of the ear.
4. Caps are removed from the kernels with a scalpel. Care is taken to not cut so far down into the kernel as to damage the embryo (*see* **Note 1**) (**Fig. 1A**).
5. Embryos are gently removed from each kernel using a small, sterile, pointed spatula.
6. The embryo is "floated" off of the spatula into a centrifuge tube containing LS modified Dc5-inf solution containing 500 μ*M* acetosyringone.
7. Embryos are rinsed by briefly vortex-mixing and replacing with fresh infection media.
8. Embryos are heat shocked by immersing the tube in a water bath at 45°C for 5 min ***(24)*** (*see* **Note 2**).
9. Infection media is removed and 1 mL of preinduced *Agrobacterium* is added. The mixture is mixed by vortexing.
10. Embryos are allowed to settle in the solution for 5 min.
11. Liquid is removed and embryos are placed scutellum side up onto LS-modified Dc5 As500 for coculture (*see* **Note 3**) (**Fig. 1B**).
12. Coculture is at 22°C in the dark for 1–3 d.
13. Embryos are placed onto 2JMS media for callus induction. Cultures are kept in the dark at 28°C for 10–14 d (*see* **Note 4**).
14. Callusing embryos are placed onto selection media—JMS 1M/0.5S—for 2–3 wk (*see* **Note 5** on selection level) (**Fig. 1B**).
15. Proliferating callus is placed onto a second round of selection for an additional 2–3 wk.
16. Proliferating callus is placed on MSAK 0.5M/2S media in the dark for regeneration 10–14 d.
17. Regenerating sectors are placed on fresh MSAK 0.5M/2S and moved to the light for 7 d (**Fig. 1C**).

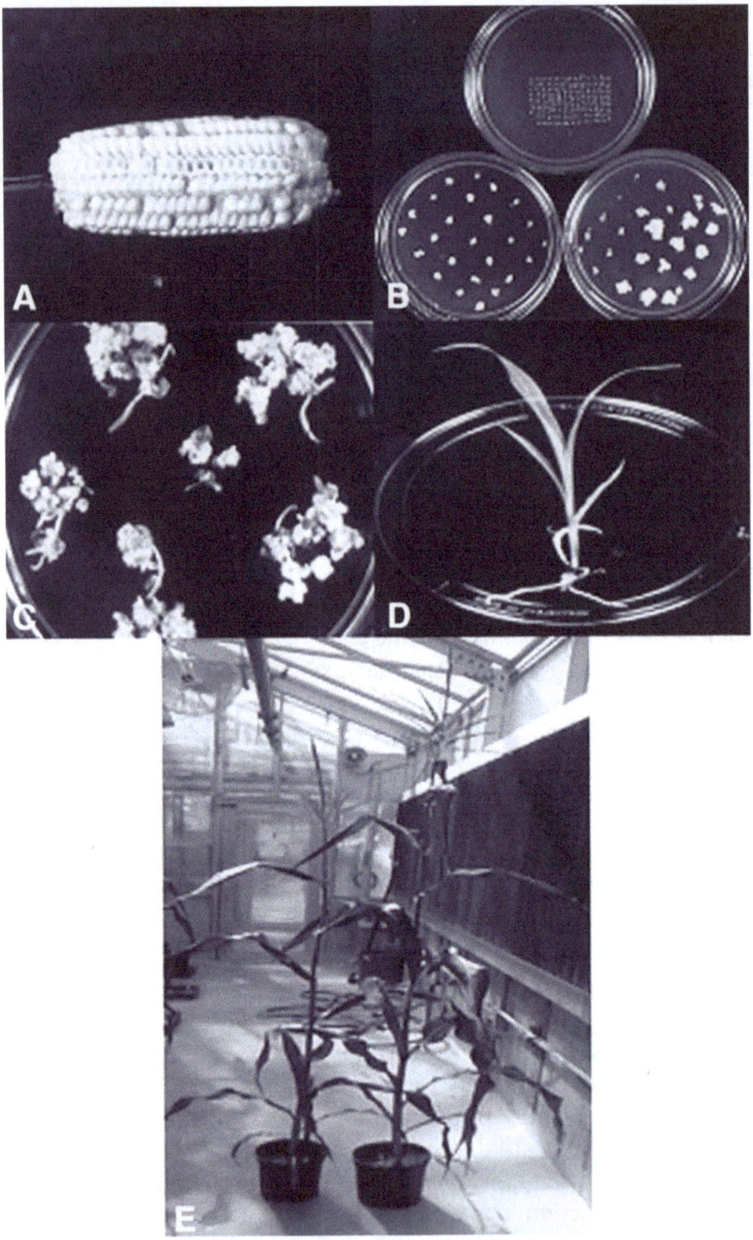

Fig. 1. Steps in the transformation, selection, and regeneration procedures. (**A**) Ear for embryo isolation. (**B**) Embryos cocultured with *Agrobacterium* (*top center*), callus induction (*bottom left*), and selection (*bottom right*). (**C**) Regeneration of selected transformants. (**D**) Regenerated plant. (**E**) Tissue culture produced (*left*) and seed grown (*right*) plants producing both tassels and ears.

Table 1
Experiments Using *Agrobacterium*-Mediated Transformation, Hi-II Immature Embryos, and Mannose Selection

Experiment no.	Number of embryos	Number of events [a]	Efficiency [b]
1	32	22	69
2	37	17	46
3	86	61	71
4	28	13	46
5	25	15	60
6	77	41	53
7	143	48	34

[a] Events confirmed as positive by polymerase chain reaction.
[b] Efficiency calculated as the percent of embryos that produced events (at least one plant produced/the number of embryos plated).

18. Shoots are placed onto MS 0.5M/2S for 7–14 d until rooted plants are produced (*see* **Note 6**) (**Fig. 1D** and **Table 1**).
19. Plants are transferred to GA-7 boxes or other appropriate containers for further development before being transferred to soil (*see* **Note 7**) (**Fig. 1E**).

4. Notes

1. The procedure is similar to the one described previously ***(11)***. Several of the steps are shown in **Fig. 1**. There were several modifications made from this basic procedure that assured higher and more consistent transformation efficiencies.
2. Heat shock contributes to a more vigorous response following coculture by limiting the amount of *Agrobacterium*-induced necrosis ***(24,25)***.
3. When placing the embryos for coculture, damaged embryos are not used because these typically will die during coculture.
4. We have found that further media modifications can help as was also noted by Zhao et al. ***(26)***. We have selected JMS media as it has given us superior embryogenic culture response in comparison with the previously utilized LS-based media ***(11)***.
5. The amount of mannose used during selection varies by the species and by the tissue culture stage. The stringency used in this example results in fewer than 1% escapes.
6. The procedure outlined above is extremely fast and reproducible in our hands. We have been able to average 40% efficiencies (based on the number of embryos initially plated divided by the number that give rise to at least one plant) over several hundred experiments. Examples of efficiencies obtained are given in **Table 1**.

7. The speed of recovery of plants from induction of embryos to regeneration of plants may contribute to the fact that few or no signs of somaclonal variation are observed (no observed albinos or sectoring and few infertile plants).

Acknowledgments

We would like to thank the maize *Agrobacterium* transformation group for development of the system described here. This research was carried out at Syngenta Biotechnology Inc., Research Triangle Park, NC.

References

1. Brasileiro, A. C. M. and Aragao, F. J. L. (2001) Marker genes for *in vitro* selection of transgenic plants. *J. Plant Biotech.* **3,** 113–121.
2. Joersbo, M. and Okkels, F. T. (1996) A novel principle for selection of transgenic plant cells: positive selection. *Plant Cell Rep.* **16,** 219–221.
3. Okkels, F. T. and Whenham, R. J. (1994) Method for the selection of genetically transformed cells and compounds for use in the method. European patent no. 0601092B1.
4. Joersbo, M. (2001) Advances in the selection of transgenic plants using non-antibiotic marker genes. *Physiol. Plant.* **111,** 269–272.
5. Harold, A. and Lewis, D.H. (1977) Mannose and green plants: occurrence, physiology and metabolism, and use as a tool to study the role of orthophosphate. *New Phytol.* **79,** 1–40.
6. Kunze, I., Ebneth, M., Heim, U., Geiger, M., Sonnewald, U., and Herbers, K. (2001) 2-Deoxyglucose resistance: a novel selection marker for plant transformation. *Mol. Breed.* **7,** 221–227.
7. Haldrup, A., Peterson, S. G., and Okkels, F. T. (1998) Positive selection: a plant selection principle based on xylose isomerase, an enzyme used in the food industry. *Plant Cell Rep.* **18,** 76–81.
8. LaFayette, P. R. and Parrott, W. A. (2001) A non-antibiotic marker for amplification of plant transformation vectors in *E. coli*. *Plant Cell Rep.* **20,** 338–342.
9. Stein, J. C. and Hansen, G. (1999) Mannose induces an endonuclease responsible for DNA laddering in plant cells. *Plant Physiol.* **121,** 1–9.
10. Joersbo, M., Donaldson, I., Kreiberg, J., Petersen, S. G., Brunstedt, J., and Okkels, F. T. (1998) Analysis of mannose selection for the production of transgenic sugar beet. *Mol. Breed.* **4,** 111–117.
11. Negrotto, D., Jolley, M., Beer, S., Wenck, A. R., and Hansen, G. (2000) The use of phosphomannose-isomerase as a selectable marker to recover transgenic maize plants (*Zea mays* L.) via *Agrobacterium* transformation. *Plant Cell Rep.* **19,** 798–803.
12. Wang, A. S., Evans, R. A., Altendorf, P. R., Hanten, J. A., Doyle, M. C., and Rosichan, J. L. (2000) A mannose selection system for production of fertile transgenic maize plants from protoplast. *Plant Cell Rep.* **19,** 654–660.
13. Wright, M., Dawson, J., Dunder, E., et al. (2001) Efficient biolistic transformation of maize (*Zea mays* L.) and wheat (*Triticum aestivum* L.) using the

phosphomannose isomerase gene, *pmi*, as the selectable marker. *Plant Cell Rep.* **20,** 429–436.
14. Lucca, P., Ye, X., and Potrykus, I. (2001) Effective selection and regeneration of transgenic rice plants with mannose as selective agent. *Mol. Breed.* **7,** 43–49.
15. Todd, R. and Tague, B. W. (2001) Phosphomannose isomerase: a versatile selectable marker for *Arabidopsis thaliana* germ-line transformation. *Plant Mol. Biol. Rep.* **19,** 307–319.
16. Melanson, D., Roussy, I., and Hansen, G. (1999) The use of phosphomannose isomerase as a selectable marker to recover transgenic *Arabidopsis* plants. *20th Annual Crown Gall Conference*, Houston, TX, Nov. 5–7, p. 59.
17. Reed, J., Privalle, L., Powell, M. L., et al. (2001) Phosphomannose isomerase: an efficient selectable marker for plant transformation. *In Vitro Cell. Dev. Biol. Plant.* **37,** 127–132.
18. Winans, S. C. (1990) Transcriptional induction of an *Agrobacterium* regulatory gene at tandem promoters by plant-released phenolic compounds, phosphate starvation, and acidic growth media. *J. Bacteriol.* **172,** 2433–2438.
19. Linsmaier, E. and Skoog, F. (1965) Organic growth factor requirements of tobacco tissue culture. *Physiol. Plant.* **18,** 100–127.
20. Ishida, Y., Saito, H., Ohta, S., Hiei, Y., Komari, T., and Humashiro, T. (1996) High efficiency transformation of maize (*Zea mays* L.) mediated by *Agrobacterium tumefaciens*. *Nat. Biotech.* **14,** 745–750.
21. Koziel, M. G., Desai, N. M., Lewis, K. S., et al. (2002) Method of producing transgenic maize using direct transformation of commercially important genotypes. US patent no. 6403865B1.
22. Schenk, R. V. and Hildebrandt, A. C. (1972) Medium and techniques for induction and growth of monocotyledonous and dicotyledonous plant cell cultures. *Can. J. Bot.* **50,** 199–204.
23. Murashige, T. and Skoog, F. (1962) A revised medium for rapid growth and bioassays with tobacco tissue cultures. *Physiol. Plant.* **15,** 473–497.
24. Hansen, G. (2002) Plant Transformation Methods. US patent application no. 2002/0088029A1.
25. Hansen, G (2000) Evidence for *Agrobacterium*-induced apoptosis in maize cells. *Mol. Plant Microbe Interact.* **13,** 649–657.
26. Zhao, Z-Y, Gu, W., Cai, T., et al. (2001) High throughput genetic transformation mediated by *Agrobacterium tumefaciens* in maize. *Mol. Breed.* **8,** 323–333.

17

Elimination of Marker Genes From Transgenic Plants Using MAT Vector Systems

Hiroyasu Ebinuma, Koichi Sugita, Saori Endo, Etsuko Matsunaga, and Keiko Yamada

Summary

We have developed an efficient system (Multi-Auto-Transformation [MAT] vectors) for the removal of marker genes and to increase the regeneration frequency of transgenic crops without using antibiotic selection, reducing their possible environmental impact. The MAT vector system is designed to use the oncogenes (*ipt*, *iaaM/H*, *rol*) of *Agrobacterium*, which control the endogenous levels of plant hormones and the cell responses to plant growth regulators, to differentiate transgenic cells, and to select marker-free transgenic plants. The oncogenes are combined with the site-specific recombination system (*R*/RS). At transformation, the oncogenes regenerate transgenic plants and then are removed by the *R*/RS system to generate marker-free transgenic plants. The choice of a promoter for the oncogenes and the recombinase (*R*) gene, the state of plant materials and the tissue culture conditions greatly affect efficiency of both the regeneration of transgenic plants and the generation of marker-free plants. We have evaluated these conditions in several plant species to increase their generation efficiency. This chapter describes our transformation protocols using MAT vectors.

Key Words: MAT vectors; marker-free; positive selection; site-specific recombination; transformation.

1. Introduction

Recently, there has been dramatic progress in the field of transgenic technology. These advances have been applied to crop improvement, and many transgenic crops with novel characteristics have been produced. Transgenic crops with novel traits have been widely used as breeding materials to produce

From: *Methods in Molecular Biology, vol. 286: Transgenic Plants: Methods and Protocols*
Edited by: L. Peña © Humana Press Inc., Totowa, NJ

commercial varieties. However, current transformation methods have four pitfalls regarding their incorporation into breeding programs: (a) The negative effects of selection agents decrease the ability of transgenic cells to proliferate and differentiate into transgenic plants. (b) Recent public concerns regarding the release of antibiotic-resistance genes limit their use for the commercialization of transgenic crops. (c) The presence of marker genes in transgenic plants precludes the use of the same marker genes for gene stacking through retransformation. (d) The stacking of highly expressed genes through sexual crossing enhances the possibility of homology-dependent gene silencing. Therefore, it would be desirable to develop a selection system that uses positive markers to reduce these negative effects and a system for removing selectable marker genes so that the same selectable marker gene can be reused for sequential transformation ***(1–3)***. We have developed removal systems combined with a positive marker, which are called Multi-Auto Transformation (MAT) vectors, to address these four pitfalls ***(4,5)***.

Agrobacteria can infect a wide range of plant species and induce crown galls or hairy roots. Their amazing capability for in vivo transformation depends on functions of the oncogenes on the T-DNA of Ti- or Ri-plasmids. The oncogenes manipulate the hormonal level and sensitivity of transgenic cells and induce their proliferation in vivo. The MAT vector system is designed to use oncogenes for cell proliferation and regeneration of transgenic plants. The MAT vectors combine these genes with the site-specific recombination system (*R*/RS) to remove them from transgenic plants after transformation that recovers the normal phenotype. The oncogenes of *Agrobacterium tumefaciens* include the *ipt* and iaa*M*/*H* genes, which catalyze cytokinin and auxin synthesis, respectively ***(6)***. Cytokinin and auxin are major plant growth regulators that control growth and development in plants. The *R*/RS system is derived from the plasmid pSR1 of *Zygosaccharomyces rouxii*, and consists of R recombinase and its recognition site (RS) ***(7,8)***. R recombinase mediates recombination between RS recognition sites and excises the DNA fragment flanked by the two RS sites in the same orientation. The oncogenes and the recombinase gene (*R*) are flanked by two directly oriented RS sites and placed on the T-DNA region of a binary vector plasmid. The gene of interest is inserted into the T-DNA region outside of the *R*/RS cassette. These genes on the T-DNA region are transferred to plant cells and integrated into the genome by infection with *Agrobacterium*. The oncogenes of the *R*/RS cassette differentiate transgenic cells. The *R* gene is expressed in transgenic cells and R recombinase excises the *R*/RS cassette from the plant genome. The resulting transgenic plants have only the gene of interest and one RS site in the genome. This chapter describes the application of MAT vector systems to tobacco and rice.

1.1. Cloning Method of the Gene of Interest into MAT Vectors

MAT vectors are derivatives of a disarmed binary vector plasmid pBI121 and have oncogenes and the recombinase gene (*R*) flanked by two directly oriented RS sites in the T-DNA region. The gene of interest is inserted into the T-DNA region outside of the *R*/RS. We have constructed different MAT vector plasmids and present here procedures for cloning the gene of interest.

1.2. Two-Step Transformation Method by ipt*-Type Vectors*

The *ipt*-type MAT vectors combine the *R*/RS system with the *ipt* gene, which catalyzes cytokinin synthesis (*see* **Note 1**). First, the *ipt* gene regenerates transgenic shoots, and then it is removed by the *R*/RS system to generate marker-free transgenic shoots.

1.3. Single-Step Transformation Method by ipt*-Type Vectors*

Most economically important crops, including rice, are regenerated through auxin-dependent embryogenesis. Single-step transformation methods were developed for the application of *ipt*-type MAT vectors to embryogenic plant species. First, the *ipt* gene induces proliferation of transgenic callus and embryogenesis, and then the *R*/RS system removes the *ipt* gene to regenerate marker-free transgenic plants directly, without the production of *ipt*-shooty intermediates.

1.4. Two-Step Transformation Method by rol*-Type Vectors*

The *rol*-type MAT vectors combine the *R*/Rs system with the *rol* genes, which increase auxin sensitivity (*see* **Note 2**). First, the *rol* genes induce transgenic roots, and the addition of cytokinin regenerates transgenic shoots (*see* **Note 3**). The *R*/RS system then removes *rol* genes to generate marker-free transgenic shoots.

2. Materials

2.1. MAT Vector Plasmids

Component plasmids to construct MAT binary vectors:

1. pTL7 binary plasmid: The vector pTL7, a derivative of pBI121 (Clontech, Franklin Lakes, NJ), has LacZ' multicloning sites and an *Sse8387I* site between the left and right border sequences (**Fig. 1A**). The *Pst*I, *Sse8387I* and *SphI* sites of LacZ' multicloning sites are deleted in the pTL7 plasmid.
2. pTSattp binary plasmid: The binary vector pTSattp, a derivative of pBI121, has a *ccdB* gene flanked by the *attP1* and *attP2* sites and an *Sse8387I* site between the left and right border sequences (**Fig. 1A**).
3. MAT cassette plasmids: The MAT cassette plasmids, derivatives of pHSG398 (Takara Shuzo, Kyoto, Japan), have the oncogenes (*ipt*, *rol*) and recombinase genes (*R*) flanked by two directly oriented RS sites at the *Sse8387I* site.

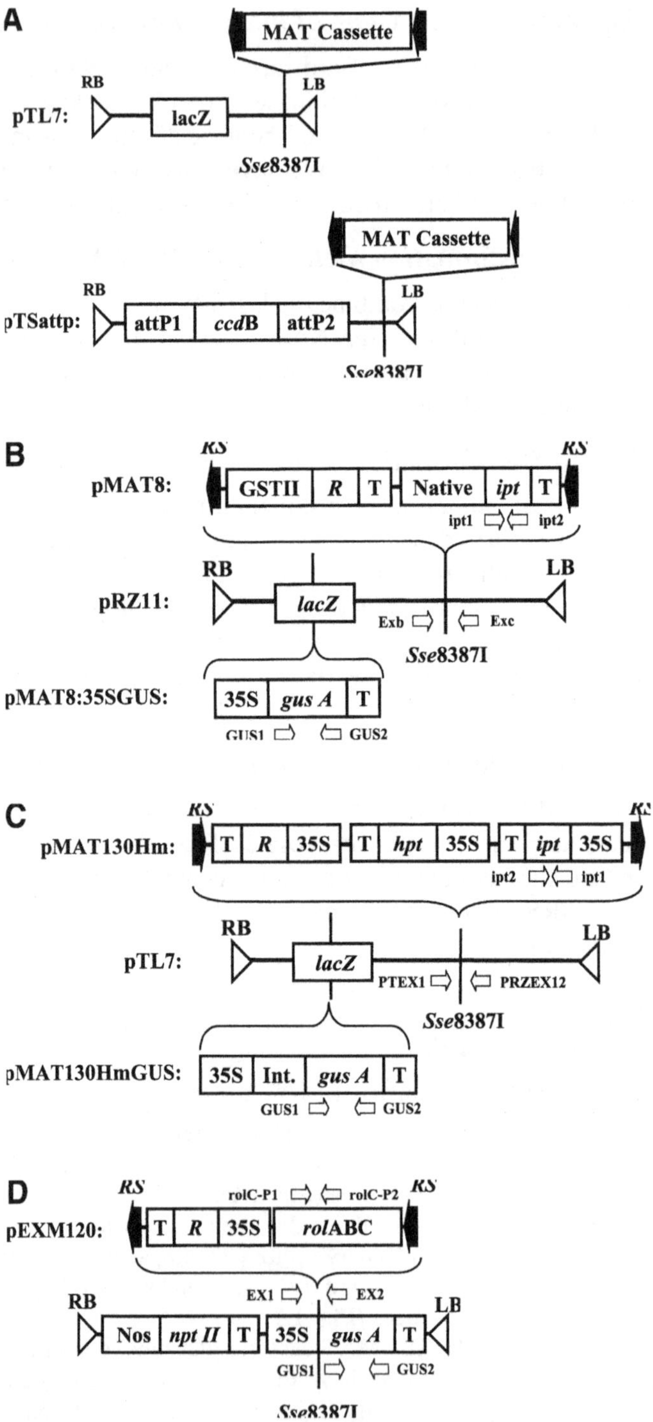
A
MAT Cassette
RB
LB
pTL7:
lacZ
Sse8387I
MAT Cassette
RB
LB
pTSattp:
attP1
ccdB
attP2
B
RS
RS
pMAT8:
GSTII
R
T
Native
ipt
T
ipt1
ipt2
RB
LB
pRZ11:
lacZ
Exb
Exc
Sse8387I
pMAT8:35SGUS:
35S
gus A
T
GUS2
C
RS
pMAT130Hm:
T
R
35S
T
hpt
35S
T
ipt
35S
ipt2
ipt1
RB
LB
pTL7:
lacZ
PTEX1
PRZEX12
Sse8387I
pMAT130HmGUS:
35S
Int.
gus A
T
GUS1
GUS2
D
RS
rolC-P1
rolC-P2
RS
pEXM120:
T
R
35S
rolABC
EX1
EX2
RB
LB
Nos
npt II
T
35S
gus A
T
GUS1
GUS2

4. MAT binary vector plasmids: The MAT binary plasmids, derivatives of pBI121, have LacZ' multicloning sites and a MAT cassette between the left and right border sequences. A list of MAT binary vector plasmids is shown in **Table 1**.
5. Plasmid containing the gene of interest.

2.2. Culture Media, Plasmids, and Reagents (Two-Step ipt*)*

1. Plant materials: *Nicotiana tabacum* cv. Petite Havana SR1 plants.
2. MS medium: Murashige and Skoog (MS) salts (Icn) and vitamins (Sigma Aldrich, St. Louis, MO), 20 g/L of sucrose, 0.8% agar (INA agar-BA30; Funakoshi), pH 5.6.
3. Acetosyringone (Sigma Aldrich).
4. α-Naphthalene acetamide (NAM) (Sigma Aldrich).
5. Kanamycin (Wako Chemical).
6. Carbenicillin (Wako Chemical).
7. Safener (R29148).
8. pMAT8:35SGUS: The pMAT8:35SGUS vector plasmid, a derivative of pBI121, has a 35S-*GUS* gene and a MAT cassette that contains a native *ipt* gene and a GST-II-*R* gene (**Fig. 1B**). A list of MAT binary vector plasmids is shown in **Table 1**.
9. YEB medium: 0.49 g/L $MgSO_4 \cdot 7H_2O$, 5 g/L of beef extract, 1 g/L of yeast extract, 5 g/L of peptone, 1.2% Bacto-agar, pH 7.2.
10. FastDNA™ Kit (Q-Bio gene).
11. Nylon membranes (Hybond-N, Amersham Pharmacia Biotech).
12. DIG-dUTP, DIG Easy Hyb (hybridization solution), DIG Wash and Block Buffer Set (Roche Diagnostics GmbH).
13. Cooled-CCD LAS 1000 system (Fuji Photo Film).

2.3. Culture Media, Plasmids and Reagents (Single-Step ipt*)*

1. Plant materials: mature rice seeds (*Oryza sativa* L. cv. Nipponbare).
2. N6 liquid medium: N6 salts (N6 basal salt mixture, Sigma) and vitamins (Sigma), 30 g/L of sucrose, pH 5.8.

Fig. 1. *(opposite page)* Constructions of MAT vectors. (**A**) The pTL7 plasmid has *LacZ* multicloning sites for the gene of interests and an *Sse8387I* site for the MAT cassette. The pTSattp plasmid has the *ccdB* gene flanked by *attB* sites for the gene of interests and an *Sse8387I* site for the MAT cassette. (**B**) The *pMAT8*:35SGUS has a 35S-*GUS* gene and a MAT cassette that contains a native *ipt* gene and a GST-II-*R* gene. PCR primers (IPT1-2, GUS1-2, Exb-c). (**C**) The pMAT130Hm plasmid has a 35S-*GUS* gene and a MAT cassette that contains a 35S-*ipt* gene, a 35S-*R* gene and a 35S-*HPT* gene. PCR primers (IPT1-2, GUS1-2, PTEX1-PRZEX12). (**D**) The pEXM120 plasmid has a Nos-*NPTII* gene and a 35S-*GUS* gene. A MAT cassette, that contains *rolABC* genes and a 35S-*R* gene, is placed between a 35S promoter and a start codon of GUS gene. PCR primers (rolC-P1-2, GUS1-2, EX1-2).

Table 1
MAT Binary Vector Plasmids

Plasmid	Gene of interest	MAT cassette	References
pIPT5	Nos-*nptII*, 35S-*gusA*, 35S-*ipt*		***14***
pIPT10	Nos-*nptII*, 35S-*gusA*, Native-*ipt*		***14***
pIPT20	Nos-*nptII*, 35S-*gusA*, rbcS-*ipt*		***14***
pIPTIMH	Nos-*nptII*, 35S-*gusA*, Native-*ipt*, *iaaM/H*		***15***
pTL7	*lacZ'*		***5***
PTSattp	*ccd*B		***5***
pNPI132	NOS-*nptII*, 35S-*gusA*	35S-*R*, 35S-*ipt*	***17***
pMAT8	*lacZ'*	GSTII-*R*, Native-*ipt*	***18***
pMAT8:GUS	35S-*gusA*	GSTII-*R*, Native-*ipt*	***18***
pMATIMH	Nos-*gusA*	GSTII-*R*, Native-*ipt*, *iaaM/H*	***15***
pRBI11	NOS-*nptII*, 35S-*gusA*	GSTII-*R*, rbcS-*ipt*	***14***
pNPIBOGFP	NOS-*nptII*, 35S-*gusA*, 35S-*hpt*	35S-*R*, 35S-*GFP*, 35S-*ipt*	***16***
pMAT130HmGUS	35S-*gusA*	35S-*R*, 35S-*hpt*, 35S-*ipt*	In preparation
pEXM120	35S-*gusA*	35S-*R*, *rol*ABC	In preparation

Oncogenes (*ipt*, *iaaM/H*) are inserted into the *Hin*dIII of the binary vector plasmid pBI121 (Clontech) to construct control vectors (pIPT5, pIPT10, pIPT20, pIPTIMH). A MAT cassette is inserted into the *Sse*8387I site of the binary vector plasid pTL7, a derivative of pBI121 (Clontech), and a gene of interest outside of the cassette to construct MAT vectors. MAT cassette: oncogenes (*ipt*, *iaaM/H*, *rol*), a recombinase (*R*) gene and a selectable marker gene are flanked by two directly oriented Rs sites.

3. Co-N6CL2 medium: N6 salts and vitamins, 30 g/L of sucrose, 10 g/L of glucose, 0.4% Gelrite (Gellan Gum; Wako Chemical), 10 mg/L of acetosyringone, pH 5.2.
4. N6CL2 medium: N6 salts and vitamins, 30 g/L of sucrose, 10 g/L of glucose, 0.4% Gelrite, pH 5.8.
5. MSR medium: MS salts and vitamins, 30 g/L of sucrose, 30 g/L of sorbitol, 2 g/L of casamino acids, 0.4% Gelrite, pH 5.8.
6. Hygromycin solution (Roche Diagnostics).
7. 2,4-Dichlorophenoxyacetic acid (2,4-D) (Sigma).
8. pMAT130HmGUS: The pMAT130HmGUS vector plasmid, a derivative of pBI121, has a 35S-*GUS* gene and a MAT cassette that contains a 35S-*ipt* gene, a 35S-*R* gene, and a 35S-*HPT* gene (**Fig. 1C**). A list of MAT binary vector plasmids is shown in **Table 1**.
9. YEB medium.
10. Kanamycin.
11. Acetosyringone.
12. Carbenicillin.
13. DNAeasy Plant System (Qiagen, Valencia, CA).
14. MS medium.

2.4. Culture Media, Plasmids, and Reagents (Two-Step rol*)*

1. Plant materials: *Nicotiana tabacum* cv. Petite Havana SR1 plants.
2. Culture media, plasmids, and reagents are the same as those for the *ipt* type MAT vector, except that culture is done under continuous dark.
3. α-Naphthalene acetic acid (NAA) solution (Sigma).
4. 6-Benzylaminopurine (BA) solution (Sigma).
5. 3-Indolebutyric acid (IBA) (Sigma).
6. NAM.
7. pEXM120: The pEXM120 vector plasmid, a derivative of pBI121, has a Nos-*NPTII* gene and a 35S-*GUS* gene. A MAT cassette, that contains *rolABC* genes and a 35S-*R* gene, is placed between a 35S promoter and a start codon of the *GUS* gene (**Fig. 1D**). Removal of the MAT cassette causes expression of the *GUS* gene. A list of MAT binary vector plasmids is shown in **Table 1**.
8. FastDNA™ Kit (Q-Bio gene).

3. Methods

3.1. Cloning Procedures (MAT Vectors)

We present three procedures for cloning the gene of interest.

3.1.1. Two-Component Type (Standard)

1. When the gene of interest has no *Sse8387I* site, the pTL7 plasmid is used for cloning.
2. Both the plasmid containing the gene of interest and pTL7 plasmid are digested with appropriate restriction enzymes.

3. The gene of interest is ligated into the multicloning sites of pTL7 plasmid.
4. Recombinants are identified by blue/white colony selection.
5. Both the MAT cassette plasmid and pTL7 plasmid containing the gene of interest are digested with *Sse8387I*.
6. The MAT cassette is ligated into the *Sse8387I* site of recombinant pTL7 plasmid.
7. The recombinant pTL7 plasmid containing both the gene of interest and the MAT cassette is identified by digestion with appropriate restriction enzymes.

3.1.2. Two-Component Type (GATEWAY)

1. When the gene of interest has *Sse8387I* sites, the pTSattp plasmid is used for cloning.
2. Both the MAT cassette vector plasmid and pTSattp plasmid are digested with *Sse8387I*.
3. The MAT cassette is ligated into the *Sse8387I* site of the pTSattp plasmid.
4. The gene of interest flanked by *attB* sites is amplified by polymerase chain reaction (PCR) (*see* **Note 4**).
5. The amplified gene of interest is incubated together with the recombinant pTS3 plasmid containing the MAT cassette and the recombinase BP clonase (*see* **Note 5**).
6. Recombinants are identified by negative selection (*see* **Note 6**).
7. The recombinant pTSattp plasmid containing both the gene of interest and MAT cassette is identified by digestion with appropriate restriction enzymes.

3.1.3. Binary Type

1. The MAT binary vector plasmids are digested at available multicloning sites.
2. The plasmid containing the gene of interest is digested with appropriate restriction enzymes.
3. The gene of interest is ligated into the multi-cloning sites of the MAT binary vector plasmid.
4. Recombinants are identified by blue/white colony selection.
5. The recombinant MAT binary vector plasmid containing the gene of interest is identified by digestion with appropriate restriction enzymes.

3.2. Two-Step Transformation Procedures (ipt-Type)

3.2.1. Agrobacterium Suspension Culture

1. The pMAT8:35SGUS vector plasmid is introduced into the disarmed *A. tumefaciens* strain LBA4404 by electroporation *(9)*.
2. Transformants are selected and maintained on YEB agar medium containing 50 mg/L of kanamycin.
3. A fresh colony is inoculated into 10 mL of YEB liquid medium without any antibiotics and cultured overnight at 27°C (*see* **Note 7**).
4. The suspension culture is diluted to $A_{630} = 0.25$ with sterilized water.

3.2.2. Explant Preparation

1. Leaves from greenhouse-grown tobacco plants are surface-sterilized by 1% (v/v) sodium hypochlorite solution for 5 min, followed by three rinses with sterile distilled water.
2. Leaves are cut into approx 8-mm square pieces and immersed in sterilized water until infection is performed.

3.2.3. Infection

1. Leaf segments are immersed in diluted *Agrobacterium* suspension culture for approx 1 min.
2. They are blotted dry on sterilized filter paper to remove excess of suspension culture.
3. They are placed on hormone-free MS agar medium containing 40 mg/L of acetosyringone for 3 d of cocultivation at 25°C.
4. The inoculated leaf segments are transferred to hormone-free MS agar medium containing 500 mg/L of carbenicillin and cultured in a culture room or growth chamber at 25–28°C under continuous light (*see* **Note 8**).
5. They are transferred to fresh medium every 2 wk.

3.2.4. Selection of Transgenic Shoots

1. One month after *Agrobacterium* infection, the regenerated adventitious buds are separated from the leaf segments and transferred to the same medium (*see* **Note 9**).
2. After 1 mo of cultivation, developed shoots are classified into two groups according to their phenotype: (a) normal shoots and (b) abnormal shoots (*see* **Note 10**).
3. Genomic DNA is extracted from the leaves of both normal and abnormal shoots using a FastDNA™ Kit, and used for PCR analysis.
4. PCR is performed under standard conditions with 1 min of denaturation, 1 min of annealing, and 2 min of extension at 94°C, 60°C, and 72°C, respectively, for 30 cycles (*see* **Note 11**).
5. Reaction products are resolved by electrophoresis in a 1.8% (w/v) agarose gel (*see* **Note 12**).
6. About 10–40% of normal shoots are marker-free transgenic plants. These transgenic plants are maintained in a flask on hormone-free MS agar medium containing 500 mg/L of carbenicillin (*see* **Note 13**).

3.2.5. Induction of Marker-Free Transgenic Plants

1. About half of *ipt*-shooty lines are excision- and β-glucuronidase (GUS)-positive ones by PCR (*see* **Note 14**). These lines are subcultured monthly to hormone-free MS agar medium containing 500 mg/L of carbenicillin and 30 mg/L of Safener.
2. Normal shoots develop from these *ipt*-shooty lines within 3 mo of induction with Safener. These shoots are transferred to hormone-free MS agar medium containing 500 mg/L of carbenicillin, grown normally and rooted.

3. Genomic DNA is extracted from leaves of these normal shoots using a FastDNA™ Kit, and used for PCR analysis.
4. About 80% of normal shoots that develop from *ipt*-shooty lines are marker-free transgenic plants.

3.2.6. Southern Analysis of Marker-Free Transgenic Plants

1. Genomic DNA is isolated from in vitro grown transgenic plants by a modified cetyltrimethyl-ammonium bromide (CTAB) method ***(10)***.
2. Ten to twenty microgram of DNA samples are digested by appropriate restriction enzymes, separated on 0.8% (w/v) agarose gel and blotted to nylon membranes.
3. The probe DNA fragment, part of the *GUS* gene, is labeled by PCR using DIG-dUTP following the supplier's instructions (*see* **Note 15**).
4. Hybridization, washing, and detection are performed using DIG Easy Hyb (hybridization solution) and DIG Wash and Block Buffer Set following the supplier's instructions.
5. Hybridization signals are detected with a cooled charge-coupled device (CCD) system.
6. Most marker-free transgenic plants have only a low copy number of transgenes (one or two genes) (*see* **Note 16**).

3.3. Single-Step Transformation Procedures (ipt-Type)

3.3.1. Agrobacterium Suspension Culture

1. The pMAT130HmGUS vector plasmid is introduced into the disarmed *A. tumefaciens* strain EHA105 by electroporation.
2. Transformants are selected on YEB agar medium containing 50 mg/L of hygromycin and 50 mg/L of kanamycin.
3. A fresh colony is inoculated onto YEB agar medium containing 50 mg/L of hygromycin and 100 mg/L of kanamycin and cultured over two nights at 27°C.
4. Transformants are cultured on YEB liquid medium overnight at 27°C.
5. The collected bacteria are suspended with N6 liquid medium containing 10 mg/L of acetosyringone.

3.3.2. Explant Preparation

1. Mature rice seeds are sterilized in 75% (v/v) sodium hypochlorite solution for 30 min and washed thoroughly in sterilized water.
2. The sterilized seeds are germinated on N6CL2 medium for 5 d at 30°C under continuous light (*see* **Note 17**).

3.3.3. Infection

1. The germinated seeds are immersed in diluted bacterial suspension ($OD_{630} = 0.15$) for 1.5 min and blotted dry with sterilized filter paper to remove liquid excess.
2. The germinated seeds are cocultured with *Agrobacterium* for 3 d at 28°C on Co-N6CL2 medium.

3. After cocultivation, the germinated seeds are washed with sterilized water containing 500 mg/L of carbenicillin and blotted dry with sterilized filter paper to remove liquid excess.
4. The seedlings are transferred to N6CL2 medium containing 2 mg/L of 2,4-D, 25 mg/L of hygromycin, and 500 mg/L of carbenicillin.
5. After 1 wk of cultivation, the scutellum tissues are aseptically excised from germinated seeds. The excised scutellum tissues are cultured on N6CL2 medium containing 4 mg/L of 2,4-D, 25 mg/L of hygromycin, and 500 mg/L of carbenicillin for a week.
6. After 1 wk, scutellum tissues are transferred to MSR medium containing 500 mg/L of carbenicillin.

3.3.4. Selection of Transgenic Shoots

1. After 1 wk of cultivation, the proliferated calluses are separated from scutellum tissues and transferred to MSR medium containing 500 mg/L of carbenicillin.
2. After 1 wk, regenerated shoots from these calluses are transferred to the same medium.
3. Genomic DNA is extracted from leaves of regenerated shoots using a DNAeasy Plant System, and used for PCR analysis.
4. PCR is performed under standard conditions with 1 min of denaturation, 1 min of annealing, and 2 min of extension at 94°C, 60°C, and 72°C, respectively, for 30 cycles (*see* **Note 18**).
5. Reaction products are resolved by electrophoresis in a 1.8% (w/v) agarose gel (*see* **Note 17**).
6. About 5% of the regenerated shoots are marker-free transgenic plants and 40% are transgenic plants with marker genes (ipt) (*see* **Notes 20** and **21**).

3.3.5. Induction of Marker-Free Transgenic Plants

1. Marker-free transgenic lines and transgenic lines with marker genes are divided into sublines with a single shoot and subcultured for 1 wk on MS medium containing 500 mg/L of carbenicillin.
2. After rooting (*see* **Note 22**), genomic DNA is extracted from leaves of each separated shoot using a DNAeasy Plant System, and used for PCR analysis.
3. From transgenic lines with marker genes, about 2–3% of the sublines are marker-free transgenic plants.
4. After 5–6 wk, these transgenic plants are transferred to a greenhouse at 25°C under an 11-h/13-h light/dark cycle.
5. After about 3 mo, seeds are obtained from both marker-free transgenic plants and transgenic plants with marker genes (*see* **Note 22**).
6. Seeds are germinated in water. DNA is isolated from seedlings and used for PCR analysis.
7. Marker-free transgenic plants are segregated from transgenic plants with marker genes in their progeny by crossing (*see* **Note 23**).

3.4. Two-Step Transformation Procedures (rol-Type)

3.4.1. Agrobacterium *Suspension Culture*

1. The pEXM120 vector plasmid is introduced into the disarmed *A. tumefaciens* strain EHA105 by electroporation.
2. Transformation procedures (*Agrobacterium* suspension culture, explant preparation, and infection) are the same as those described for the *ipt* type MAT vector, except that culture is done under continuous dark.

3.4.2. Selection of Transgenic Roots

1. Inoculated leaf segments are transferred to hormone-free MS agar medium containing 500 mg/L of carbenicillin every 2 wk under continuous dark (*see* **Notes 24** and **25**).
2. One month after *Agrobacterium* infection, the regenerated roots are separated from the leaf segments and transferred to the same medium containing 1 mg/L of BA and 0.1 mg/L of NAA under continuous light.

3.4.3. Selection of Marker-Free Transgenic Shoots

1. After 2 mo, the regenerated buds are separated from roots and transferred to hormone-free MS medium containing 500 mg/L of carbenicillin.
2. After 1 mo of cultivation, the developed shoots are classified into two groups according to phenotype: (a) normal shoots and (b) abnormal shoots (*see* **Note 26**), and maintained in a flask on hormone-free MS agar medium containing 500 mg/L of carbenicillin.
3. Genomic DNA is extracted from leaves of both the normal and abnormal shoots using a FastDNA™ Kit, and used for PCR analysis.
4. PCR is performed under standard conditions with 30 s of denaturation, 1 min of annealing, and 1.5 min of extension at 94°C, 60°C, and 72°C, respectively, for 30 cycles (*see* **Note 27**).
5. Reaction products are resolved by electrophoresis in a 1.8% (w/v) agarose gel (*see* **Note 28**).
6. About 20–30% of normal shoots are marker-free transgenic plants and 10–20% are chimeric transgenic plants.
7. These transgenic plants are transferred to a greenhouse. After about 2 mo, seeds are obtained from both marker-free transgenic plants and chimeric transgenic plants (*see* **Note 29**).
8. Seeds are germinated in pots and DNA is isolated from seedlings for PCR analysis.
9. Chimeric transgenic plants are segregated into marker-free transgenic plants in their progeny by crossing.

4. Notes

1. The *ipt* gene that codes for isopentenyl transferase ***(11)*** is used to induce shoot formation. Since the control of both cytokinin and auxin is needed to optimize the hormone levels in plant tissue and to regenerate transgenic shoots in many plant

species, we also constructed MAT vector plasmids that combine *ipt* genes with the *iaaM/H* genes to manipulate both the auxin and cytokinin levels (**Table 1**). The *iaaM/H* genes code for a tryptophan monooxygenase and an indoleacetamide hydrolase, which catalyze auxin synthesis ***(12)***.

2. The *rol* genes responsible for the proliferation of hairy roots ***(13)*** are used to induce root formation. We also constructed MAT vector plasmids that combine the *rol* genes with the *iaaH* gene to manipulate the auxin level (**Table 1**), as the auxin level in plant tissue must be increased to regenerate transgenic roots in many plant species.
3. We constructed MAT vector plasmids that combine the *rol* genes with the *ipt* gene to induce shoot formation from transgenic roots. The *rbcS* promoter is used to control the expression of the *ipt* gene. First, the *rol* genes induce transgenic roots under continuous dark, and then the *ipt* gene regenerates transgenic shoots under continuous light. Transgenic shoots with the *ipt* gene exhibit the *ipt*-shooty phenotype and marker-free transgenic plants are developed from these *ipt*-shooty lines.
4. PCR is performed using by PLATINUM™ *Taq* polymerase (Life Technologies) or KOD-Plus (Toyobo) according to the manufacturer's protocol. The pUC19 plasmid containing the gene of interest at LacZ' multicloning sites is used as a template. A forward primer (attB-P8) and a reverse primer (attB-P7) are used to amplify the gene of interest with terminal *attB1* and *attB2* sequences by PCR.

   The primer sequences are as follows:

   attB-P8: 5'-ggggacaagtttgtacaaaaaagcaggctgagcggataacaatttcacacagg-3';
   attB-P7: 5'-ggggaccactttgtacaagaaagctgggtcgacgttgtaaaacgacggccagt-3'.
5. The recombinase BP clonase is used to replace the *ccdB* gene of pTSattp plasmid with the gene of interest according to the protocol of the Gateway Cloning System (Life Technologies).
6. The reaction mixture is transformed into an *E. coli* DH5α strain (Takara Shuzo). Because the *ccdB* gene of pTSattp plasmid is lethal to this strain, only recombinants with replaced pTSattp plasmids can survive.
7. The GST-II promoter is less active in *A. tumefaciens* than the 35S promoter. However, the preparation of fresh bacterial culture is highly recommended to protect MAT cassettes from removal events in *A. tumefaciens.*
8. In the case of using *ipt*-type MAT vectors containing *iaaM/H* genes, explants are also placed on hormone-free MS agar medium. The *ipt*-type MAT vectors carrying only the *iaaH* gene are used with culture medium containing 0.04–0.2 mg/mL of NAM, as the *iaaH* gene codes for an indoleacetamide hydrolase, which can convert NAM into NAA.
9. Nontransgenic buds regenerate together with transgenic buds, as the overproduction of cytokinin by the *ipt* gene causes it to leak out from transgenic cells. We fused the *ipt* gene with several different promoters to optimize the cytokinin levels for the proliferation and differentiation of transgenic cells (**Table 1**). The *ipt* gene with the *rbcS* promoter is especially useful for increasing the percentage of transgenic shoots in hybrid aspen.

10. These abnormal shoots exhibit the *ipt*-shooty phenotype and lose apical dominance and rooting ability because of the overproduction of cytokinins.
11. Two pairs of PCR primers are designed to verify the presence of the *ipt* and *GUS* genes, and removal of the *R*/RS cassette.

    The primer sequences in **Fig. 1B** are as follows:

    IPT1: 5'-cttgcacaggaaagacgtcg-3';
    IPT2: 5'-aatgaagacaggtgtgacgc-3';
    GUS1: 5'-gtggaattgatcagcgttgg-3';
    GUS2: 5'-gcaccgaagttcatgccagt-3';
    Exb: 5'-agcctgaatggcgaatgcct-3';
    Exc: 5'-cgattaagtgggtaacgcc-3'.
12. In marker-free transgenic plants, the predicted 0.5-kb excision fragment and 1.7-kb GUS fragment are amplified by the primers Exb–Exc and GUS1–GUS2, respectively, and the predicted 0.8-kb ipt fragment is not amplified by IPT1-IPT2.
13. Owing to expression of the *R* gene in the callus, removal events occur early during regeneration and marker-free plants appear. Analysis of normal plants is highly recommended to obtain marker-free plants.
14. These *ipt*-shooty lines are chimeric transgenic plants that contain transgenic cells in which excision events do and do not occur. The predicted 0.5-kb excision fragment, 1.7-kb *GUS* fragment and 0.8kb ipt fragment are amplified by the primers Exb-Exc, GUS1–GUS2, and IPT1–IPT2, respectively.
15. The primer sequences are as follows:

    GUS1: 5'-gtggaattgatcagcgttgg-3';
    GUS2: 5'-gcaccgaagttcatgccagt-3'.
16. If more than one expressed copy of the *ipt* gene is inserted into the plant genome of the transgenic shoots, the elimination of one copy would not cause a loss of *ipt* function (*ipt*-shooty). This inference leads to the expectation that marker-free transgenic plants will be derived from low-copy-number transgenic plants.
17. The preculture period and hormone content of the preculture medium greatly affect the generation efficiency of marker-free transgenic rice plants. These conditions should be independently evaluated for each rice species.
18. Two pairs of PCR primers are designed to verify the presence of the *ipt* and *GUS* genes, and removal of the *R*/RS cassette.

    The primer sequences in **Fig. 1C** are as follows:

    IPT1: 5'-cttgcacaggaaagacgtcg-3';
    IPT2: 5'-aatgaagacaggtgtgacgc-3';
    GUS1: 5'-gtggaattgatcagcgttgg-3';
    GUS2: 5'-gcaccgaagttcatgccagt-3';
    PTEX1: 5'-cgtgccagctgcattaatgg-3';
    PRZEX12: 5'-ggagcccccgatttagagcttgac-3'.
19. In marker-free transgenic plants, the predicted 0.8-kb excision fragment and 1.7-kb *GUS* fragment are amplified by the primers PTEX1–PRZEX12 and GUS1–

GUS2, respectively, whereas the predicted 0.8-kb *ipt* fragment is not amplified by IPT1–IPT2.

20. Nontransgenic plants regenerate together with transgenic plants. About 40–50% of regenerated shoots (100–200 lines) from 1500 germinated seeds are transgenic plants.
21. These lines with marker genes are chimeric transgenic plants that contain transgenic cells in which excision events do and do not occur. The predicted 0.8-kb excision fragment, 1.7-kb *GUS* fragment and 0.8-kb *ipt* fragment are amplified by the primers PTEX1–PRZEX12, GUS1–GUS2, and IPT1–IPT2, respectively.
22. Transgenic shoots with marker genes (*ipt*) can root and produce seeds.
23. The *ipt* genes are removed from about half of chimeric transgenic plants by crossing.
24. Transgenic roots exhibit the hairy root phenotype that can grow rapidly and be maintained in long-term cultures. About 70% of the regenerated roots are transgenic roots.
25. The addition of 0.1 mg/L of IBA to MS agar medium increases the number of transgenic roots. Instead of IBA, 0.1 mg/L of, NAM is used for *rol*-type MAT vectors combined with the *iaaH* gene.
26. These abnormal shoots exhibit wrinkled leaves and shortened internodes due to the expression of the *rol* genes. About 30% of the regenerated shoots in a flask exhibit a normal phenotype.
27. Two pairs of PCR primers are designed to verify the presence of the *rolc* and *GUS* genes, and removal of the *R*/RS cassette.

    The primer sequences in **Fig. 1D** are as follows:

    rolC-P1:5'-ggtcacgaggtcatagtagtgg-3';
    rolC-P2:5'-gttagcaaagtaggaaataata-3';
    GUS1: 5'-gtggaattgatcagcgttgg-3';
    GUS2: 5'-gcaccgaagttcatgccagt-3';
    EX1 : 5'-ttgtcaagaccgacctgtcc-3';
    EX2 : 5'-tgcatcggcgaactgatcgt-3'.
28. In marker-free transgenic plants, the predicted 3.0-kb excision fragment and 1.7-kb *GUS* fragment are amplified by the primers EX1–EX2 and GUS1–GUS2, respectively, whereas the predicted 1.0-kb *rolc* fragment is not amplified by *rolC-P1–rolC-P2*. The chimeric transgenic plants contain transgenic cells in which excision events do and do not occur. The predicted 3.0-kb excision fragment, 1.7-kb *GUS* fragment and 1.0-kb *rolc* fragment are amplified by the primers EX1–EX2, GUS1–GUS2, and *rolC-P1–rolC-P2*, respectively.
29. Self-crossing is difficult for several chimeric transgenic plants and outcrossing with nontransgenic plants is necessary to produce seeds.

References

1. Ebinuma, H., Sugita, K., Matsunaga, E., and Yamakado, M. (1997) Selection of marker-free transgenic plants using the isopentenyl transferase gene as a selectable marker. *Proc. Natl. Acad. Sci. USA* **94,** 2117–2121.

2. Ebinuma, H., Sugita, K., Matsunaga, E., Endo, S., and Kasahara, T. (2000) Selection of marker-free transgenic plants using the oncogenes (IPT, ROL A,B,C) of *Agrobacterium* as selectable markers, in *Molecular Biology of Woody Plants II* (Jain, S. M. and Minocha, S. C., eds.), Kluwer Academic, Dordrecht, The Netherlands, pp. 25–46.
3. Ebinuma, H., Sugita, K., Matsunaga, E., Endo, S., Yamada, K., and Komamine, A. (2001) Systems for removal of a selection marker and their combination with a positive marker. *Plant Cell Rep.* **20,** 383–392.
4. Ebinuma, H. and Komamine A. (2001) MAT (Multi-Auto-Transformation) vector system. The oncogenes of *Agrobacterium* as positive markers for regeneration and selection of marker-free transgenic plants. *In Vitro Cell. Dev. Biol.* **37,** 103–113.
5. Ebinuma, H., Sugita, K., Matsunaga, E., Endo, S., and Yamada, K. (2002) GST-MAT vector for the efficient and practical removal of marker genes from transgenic plants, in *Molecular Methods of Plant Analysis. Vol. 22: Testing for Genetic Manipulation* (Jackson, J. F., Linskens, H. F., and Inman, R. B. eds.), Springer-Verlag, Heidelberg, Germany, pp. 95–117.
6. Gaudin, V., Vrain, T., and Jouanin, L. (1994) Bacterial genes modifying hormonal balances in plants. *Plant Physiol. Biochem.* **32,** 11–29.
7. Matsuzaki, H, Nakajima, R., Nishiyama J., Araki, H., and Oshima, Y. (1990) Chromosome engineering in *Saccharomyces cerevisiae* by using a site-specific recombination system of a yeast plasmid. *J. Bacteriol.* **172,** 610–618.
8. Onouchi, H., Yokoi, K., Machida, C., et al. (1991) Operation of an efficient site-specific recombination system of *Zygosaccharomyces rouxii* in tobacco cells. *Nucl. Acids Res.* **19,** 6373–6378.
9. Nagel, R., Elliott, A., Masel, A., Birch, R. G., and Manners, J. M. (1990) Electroporation of binary Ti plasmid vector into *Agrobacterium tumefaciens* and *Agrobacterium rhizogenes*. *FEMS Microbiol. Lett.* **67,** 325–328.
10. Doyle, J. J. and Doyle, J. L. (1989) Isolation of plant DNA from fresh tissue. *FOCUS* **12,** 13–15.
11. Akiyoshi, D. E., Klee, H., Amasino, R. M., Nester, E. W., and Gordon, M. P. (1984) T-DNA of *Agrobacterium tumefaciens* encodes an enzyme of cytokinin biosynthesis. *Proc. Natl. Acad. Sci. USA* **81,** 5994–5998.
12. Thomashow, L. S., Reeves, S., and Thomashow, M. F. (1984) Crown gall oncogenesis: evidence that a T-DNA gene from the *Agrobacterium* Ti plasmid pTiA6 encodes an enzyme that catalyses synthesis of indoleacetic acid. *Proc. Natl. Acad. Sci. USA* **81,** 5071–5075.
13. Tepfer, D. (1984) Transformation of several species of higher plants by *Agrobacterium rhizogenes*: sexual transmission of the transformed genotype and phenotype. *Cell* **37,** 959–967.
14. Matsunaga, E., Sugita, K., and Ebinuma, H. (2002) An asexual production of selectable marker-free transgenic woody plants, vegetatively propagated species. *Mol. Breed.* **10,** 95–106.
15. Endo, S., Kasahara, T.,Sugita, K., and Ebinuma, H. (2002) A new GST-MAT vector containing both the *ipt* gene and the *iaaM/H* genes can produce marker-free transgenic plants with high frequency. *Plant Cell Rep.* **20,** 923–928.

16. Endo, S., Sugita, K., Sakai, M., Tanaka, H., and Ebinuma, H. (2002) Single-step transformation for generating marker-free transgenic rice using the *ipt*-type MAT vector system. *Plant J.* **30,** 115–122.
17. Sugita, K., Matsunaga, E. and Ebinuma, H. (1999) Effective selection system for generating marker-free transgenic plants independent of sexual crossing. *Plant Cell Rep.* **18,** 941–947.
18. Sugita, K., Kasahara, T., Matsunaga, E., and Ebinuma, H. (2000) A transformation vector for the production of marker-free transgenic plants containing a single copy transgene at high frequency. *Plant J.* **22,** 461–469.
19. Sugita, K., Matsunaga, E., Kasahara, T., and Ebinuma, H. (2000) Transgene stacking in plants in the absence of sexual crossing. *Mol. Breed.* **6,** 529–536.

18

Simple and Efficient Removal of Marker Genes From Plastids by Homologous Recombination

Anil Day, Vasumathi Kode, Panagiotis Madesis, and Siriluck Iamtham

Summary

Removal of marker genes improves the design of transgenic plants. Homologous recombination between direct repeats provides a simple method for excising marker genes after transgenic cells and shoots have been isolated. Efficient implementation of the method requires high rates of homologous recombination relative to illegitimate recombination pathways. The procedure works well in plastids where homologous recombination predominates. Marker genes are flanked by engineered direct repeats. The number and length of direct repeats flanking a marker gene influence excision rate. Excision is automatic and loss of the marker gene is controlled by selection alone. After transgenic cells have been isolated selection is removed allowing loss of the marker gene. Excision is a unidirectional process resulting in the rapid accumulation of high levels of marker-free plastid genomes. Cytoplasmic sorting of marker-free plastids from marker-containing plastids leads to the isolation of marker free plants. Marker-free plants can be isolated following vegetative propagation or among the progeny of sexual crosses.

Key Words: Antibiotic resistance; gene targeting; homologous recombination; marker excision; plastid.

1. Introduction

Marker genes facilitate the identification of transformed cells and are important components of most plant transformation methods. Once transformed plants have been isolated, marker genes serve no useful purpose in a transgenic crop. Only foreign trait genes add value to a crop, and all excess foreign DNA associated with the transformation process, such as marker genes and bacterial vector sequences, are increasingly viewed as undesirable. Removing excess DNA sim-

From: *Methods in Molecular Biology, vol. 286: Transgenic Plants: Methods and Protocols*
Edited by: L. Peña © Humana Press Inc., Totowa, NJ

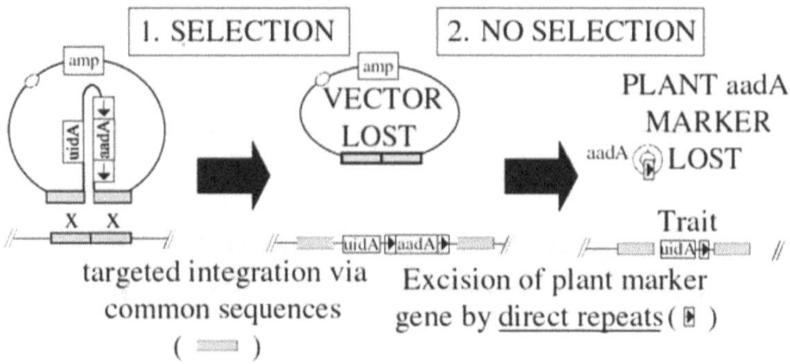

Fig. 1. Scheme showing targeted integration and excision of marker genes by homologous recombination. Vector sequences are excluded at integration. Integration of foreign genes is driven by selection for the plant marker gene. Once selection is removed the plant marker gene is excised by homologous recombination between flanking direct repeats.

plifies regulatory approval of a transgenic crop by limiting risk evaluation to the trait genes. Selectable marker genes based on antibiotic resistance, which might impact on human health and the environment, are particularly controversial *(1)* and European Union (EU) directive 2001/18/ EC requires their elimination from genetically manipulated organisms for commercial releases by 2004 and by 2008 for research purposes. Increasing the precision of transformation technologies to exclude all excess DNA will play an important role in enhancing the value of the next generation of transgenic crops *(2,3)*.

Exploitation of native homologous recombination pathways acting on DNA provides an attractive solution for increasing the precision of trait gene insertion into plant DNA and excluding vector and marker genes. The strategy uses native plant enzymes and is simple because it avoids the need for foreign site-specific DNA recombinases *(4,5)*. The overall scheme is shown in **Fig. 1**. Integration is based on homology between transforming DNA and its target site in plant DNA, allowing precise gene targeting. Homologous DNA flanking trait and marker genes promotes integration and excludes vector sequences, which lie outside regions of homology. The integrated marker gene enables selection of transformed cells. Once transgenic cells have been isolated the marker gene has served its purpose and will be excised by spontaneous homologous recombination events between direct DNA repeats flanking the marker gene. Excision of the marker gene can take place at any time during the transformation process but only cells that retain the marker will proliferate in the presence of selective agent.

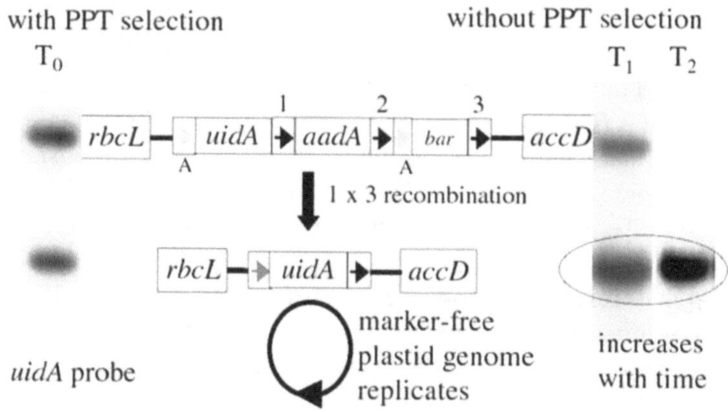

Fig. 2. DNA blot analysis showing loss of marker genes once PPT selection is removed. The construct contains three 418-bp direct repeats (labeled 1, 2, and 3) and two 174-bp repeats (labeled A). Recombination between 418-bp repeats 1 and 3 excises the *aadA* and *bar* gene resulting in a smaller band size. Note that excision takes place even in the presence of PPT. Excision is a unidirectional process, ensuring the rapid isolation of marker-free plants.

Once selection is removed, marker gene excision will lead to the accumulation of marker-free cells and eventually marker-free transgenic plants.

High rates of homologous recombination are required for efficient targeting and marker excision by flanking direct repeats. The procedure is not easily applied to foreign genes integrated into nuclear DNA owing to high rates of illegitimate recombination in the nucleus *(6)*. Excision of marker genes from transgenic plastids presents its own challenges because plastid DNA can be present in up to 10,000 copies per cell *(7)*. High rates of marker excision are needed and this can be achieved by increasing the number and sizes of direct repeats. This leads to the accumulation of marker-free plastid genomes and marker free plastids (**Fig. 2**). At first cells will be heteroplasmic because they contain a mixed population of two plastid types; marker-containing plastids and marker-free plastids. When two plastid types are found in the same cell, repeated cell divisions give rise to two populations of cells with a uniform content of either marker-containing plastids or marker-free plastids. Segregation of different plastid types during vegetative growth is known as cytoplasmic sorting. Marker excision by homologous recombination is an effective procedure in tobacco plastids *(8)*. Of the methods described to excise marker genes from plastid genomes *(4,5)* homologous recombination is the simplest to use. Unlike published methods based on herbicide resistance genes *(9)*, plant selectable markers *(10)*, and restoration of photosynthesis *(11)*, homologous recombination has allowed the isolation of transplastomic plants with an unselected trait gene (*uidA*) that is free of all selectable marker genes *(8)*. Conservation of plastid DNA recombination pathways resulting in homology-based excision is likely given the procedure also works in *Chlamydomonas reinhardtii (12)*. The procedure is likely to be applicable to all crops in which plastid transformation has been established.

Plastid transformation vectors are constructed by standard cloning techniques. The minimum requirement is two direct repeats flanking the marker gene to be excised (**Fig. 1**). Both the length and number of direct repeats influence excision frequency. Two direct repeats of 418 bp are not sufficient to promote high levels of excision *(8)*. Increasing the number of 418-bp direct repeats to three or increasing the length of two direct repeats to 650 bp raises excision frequencies to the levels needed to isolate marker-free plants. A variety of direct repeats promote excision indicating some flexibility in the choices of DNA sequences used to make direct repeats. The possibility that some DNA sequences might be more recombinogenic than others cannot be excluded because little is known on the substrate specificity of enzymes mediating homologous recombination in plastids. Another factor that might influence excision frequency is distance between direct repeats. We have observed recombination between direct repeats located 1 kbp, 3.6 kbp, and 5.7 kbp apart. The overall size of a foreign insert in plastid DNA, might also be a factor. If there are selective forces against large inserts in plastid DNA this would tend to favor marker gene excision.

Duplicating the 5' or 3' regulatory elements flanking a marker gene provides a simple method for making direct repeats. A particularly effective approach is to use two selectable marker genes and multiple direct repeats (**Fig. 2**), which allows stepwise selection of transformed plants on two different selective agents. This strategy enables the use of weak selectable markers such as the *bar* gene *(8)* to be used in combination with the efficient *aadA* marker gene *(13)* to select plastid transformants. Transplastomic plants containing *aadA* are first selected on spectinomycin plus streptomycin and then selected on phosphinothricin (PPT). With high rates of excision, PPT selection is needed to retain the *bar* gene. Once selection stops excision of *aadA* and *bar* results in the accumulation of high levels of marker-free genomes. The use of two series of direct repeats in a single construct allows multiple excision events from a single insertion event in plastid DNA. Excision events were observed between two 418-bp direct repeats and two 174-bp direct repeats in a construct containing three 418-bp direct repeats and two 174-bp direct repeats (**Fig. 2**). Recombination between the 418-bp repeats excised *aadA* and *bar* to leave *uidA* in the plastid genome, whereas recombination between the 174-bp repeats excised *uidA* and *aadA* leaving *bar* in the plastid genome.

2. Materials

2.1. Culture Media Components

1. Murashige and Skoog (MS) macrosalts (10X stock): 16.5 g of NH_4NO_3, 19 g of KNO_3, 4.4 g of $CaCl_2 \cdot 2H_2O$ (or 3.3 g of $CaCl_2$), 3.7 g of $MgSO_4 \cdot 7H_2O$, 1.7 g of KH_2PO_4. Dissolve salts separate in 100 mL of distilled water, mix solutions, and make up to 1 L. Store at 4°C (*see* **Notes 1** and **2**).
2. MS microsalts (1000X stock): 22.30 g of $MnSO_4 \cdot 4H_2O$ (or 16.90 g of $MnSO_4 \cdot H_2O$), 8.60 g of $ZnSO_4 \cdot 7H_2O$, 6.20 g of H_3BO_3, 0.83 g of KI, 0.250 g of $Na_2MoO_4 \cdot 2H_2O$, 0.025 g of $CuSO_4 \cdot 5H_2O$, 0.025 g of $CoCl_2 \cdot 6H_2O$. Dissolve in 1 L of distilled water and store at 4°C (*see* **Note 2**).
3. Ferrous sulfate-chelate (200X stock): 7.45 g of Na_2EDTA, 5.57 g of $FeSO_4 \cdot 7H_2O$. Place in distilled water separately, then mix slowly and heat to dissolve. Make up to 1 L with H_2O (*see* **Note 2**).
4. Spectinomycin dichloride pentahydrate (20 mg/mL stock): Check the purity of the powder, which can be as low as 60%. Dissolve in distilled water and make up to concentration, taking into account the purity of the powder. Sterilize through a 0.22-µm syringe filter unit (e.g., Millex GP, Millipore) and store as 10-mL aliquots at –20°C (*see* **Note 3**).
5. Streptomycin sulfate (20 mg/mL stock): Dissolve in distilled water to required concentration. Sterilize through a 0.22-µm syringe filter unit (e.g., Millex GP, Millipore) and store as 10-mL aliquots at –20°C (*see* **Notes 3** and **4**).
6. DL-PPT (10 mg/mL stock): Dissolve in distilled water to required concentration. Sterilize through a 0.22-µm syringe filter unit (e.g., Millex GP, Millipore) and store as 10-mL aliquots at –20°C (*see* **Note 3**).
7. 6-Benzylaminopurine (1 mg/mL stock): Dissolve 50 mg in a few drops of 1 *M* NaOH and make up to 50 mL with distilled H_2O. Store for 2 wk at 4°C. For longer periods store as 1-mL aliquots at –20°C.
8. α-Naphthyleneacetic acid (0.1 mg/mL stock). Dissolve 50 mg in a few drops of 1 *M* NaOH and make up to 0.5 L with H_2O. Store for 2 wk at 4°C. For longer periods store as 1-mL aliquots at –20°C.

2.2. Culture Media

1. MS plant salt mixture ***(14)*** with supplements: Fill a 1-L glass beaker (*see* **Note 1**) with 700 mL of distilled water. Add 100 mL of 10X macrosalts, 1 mL of 1000X microsalts, 5 mL Fe-chelate, 0.1 g of *myo*-inositol, 1 mL of vitamin B_1 (1 mg/mL stock), 0.5 g of 2-(*N*-morpholino)ethanesulfonic acid (MES), 30 g of sucrose. Add 1 *M* KOH to pH 5.8. Make up to 1 L with distilled water. For solid media add 7.5 g of agar per liter of media or 2.5 g of Phytagel (Sigma, St. Louis, MO) per liter of media (*see* **Note 5**). Autoclave 0.5- to 0.7-L volumes in 1 L Duran (Schott) bottles (*see* **Note 2**).

2. RMOP medium *(15)*: Make up MS plant salt mixture with supplements as described and add 1 mL of 6-benzylaminopurine (1 mg/mL stock) and 1.0 mL of α-Naphthyleneacetic acid (0.1 mg/mL stock) before autoclaving (*see* **Note 2**).

2.3. DNA Delivery and Selection

1. Gold particle suspension. Suspend 60 mg of gold powder (*see* **Note 6**) in 1 mL of 100% ethanol in a 1.5-mL microtube and vortex-mix at maximum setting for 2 min. Centrifuge the particles at maximum setting in a microfuge for 10 s and discard the supernatant. Repeat ethanol wash twice. Dislodge the gold pellet from the side of the tube with a yellow tip before vortexing. Finally, suspend the gold in 1 mL of sterile distilled water. Place aliquots of 50 µL of gold mixture in 1.5-mL microfuge tubes, vortex-mixing between aliquots to ensure an even suspension of gold particles. Store the mixture frozen at –20°C or –80°C.
2. Calcium chloride (2.5 *M* stock): The solution is made from $CaCl_2 \cdot 2H_2O$ and sterilized in an autoclave (*see* **Note 1**). Store aliquots of 1 mL frozen at –20°C or –80°C.
3. Spermidine solution (0.1 *M* stock): Dissolve spermidine (free base) in sterile water and store in 0.5-mL aliquots in 0.5 mL microfuge tubes. Store the tubes at –80°C for a maximum of 3 mo. Thaw each tube once and do not reuse.
4. Prepare plasmid DNA (1 mg/mL) using the Qiagen Purification Maxi Prep Kit and resuspend the DNA in sterile distilled water; store frozen at –20°C.

2.4. Histochemical β-Glucuronidase (GUS) Assay

1. Prepare 0.5 *M* Sodium phosphate buffer, pH 7.0, with separate 1 *M* stocks of Na_2HPO_4 and NaH_2PO_4. Mix 58 mL of 1 *M* Na_2HPO_4 and 42 mL of 1 *M* NaH_2PO_4 and make up to 200 mL with distilled water.
2. Make up 50 m*M* Potassium ferricyanide in distilled water and store as 20 mL stocks at –20ºC in 30-mL sterile universal containers.
3. Make up 50 m*M* potassium ferrocyanide in distilled water and store as 20-mL stocks at –20ºC in 30-mL universal containers.
4. Prepare X-Gluc buffer by dissolving 50 mg of X-Gluc (5-bromo-4-chloro-3-indolyl-β-D-glucuronide, cyclohexylammonium salt) in 1 mL of dimethylformamide. To this, add 20 mL of 0.5 *M* sodium phosphate, pH 7.0, 1 mL of 50 m*M* potassium ferricyanide, 1 mL of 50 m*M* potassium ferrocyanide, and 77 mL of distilled water. Filter-sterilize the solution using a 0.2-µm syringe filter unit and store at –20ºC in 5-mL aliquots.

2.5. Plant Material and Culture Conditions

1. Sterilize tobacco seeds (*see* **Note 7**) by placing in 100% ethanol for 30 s. Then replace the ethanol with sodium hypochlorite (5% active chlorine) for 10 min. Then wash seeds four times with sterile distilled water.
2. Germinate sterilized seeds on solid MS salts mixture (with *myo*-inositol, vitamin B_1, MES, and sucrose; *see* **Subheading 2.1.2.**) in 9-cm Petri dishes in dim light or in complete darkness by covering plates in aluminum foil.

3. In 3–7 d, transfer germinated seedlings to Magenta™ jars (Sigma) containing the same MS salt mixture with supplements and grow at 25°C in a 12-h/12-h day/night regimen at 10–50 μmol/m²/s (μE) light intensity for 6–8 wk.
4. Plants with three to five leaves are suitable for transformation. The top unexpanded leaf is not used for transformation.
5. Plants can be propagated in vitro by transfer of shoot tips to fresh Magenta jars (*see* **Note 8**). Cut each plant stem above the roots and excise all leaves below the apical meristem. Push the base of the bare stem into MS medium in a new Magenta jar and grow the plants at 25°C in a 12-h/12-h day/night regimen at 10–50 μE light intensity (*see* **Note 9**). After 4–8 wk, three to five leaves appear on the plants at which stage the leaves below the unexpanded top leaf are suitable for transformation.

3. Methods

3.1. Preparation of Leaf Explants and Particle Bombardment

Excise leaves from tobacco plants grown in vitro, and remove the midrib and petiole. Cut the remaining leaf sections into appropriate shapes to fit into a 4-cm diameter circle (*see* **Note 10**) in the center of a 9-cm plate containing RMOP medium (**Fig. 3**). Place leaf explants abaxial (bottom) side up on the medium. Place pieces from a single leaf on several plates such that each plate contains leaves from several different plants. Typically, 10–20 plates are prepared at a time. Once prepared, use the plates immediately or leave in dim light overnight at 25°C (*see* **Note 11**).

3.2. Particle Bombardment

3.2.1. DNA–Gold Precipitation

1. Allow 50-μL aliquot of gold mixture to thaw and place on a vortex mixer at maximum setting for 1 min.
2. Make sure the gold is evenly suspended before adding 5 μL of plasmid DNA. Add the plasmid DNA (1 mg/mL) slowly while moving the tip of the pipet in the gold mixture and finger-tapping the base of the microtube.
3. Add 50 μL of 2.5 *M* $CaCl_2$ quickly while moving the tip through the liquid and agitating the bottom of the tube with a finger to ensure good mixing.
4. Add 20 μL of 0.1 *M* spermidine quickly as the pipet tip moves through the liquid and finger-tap the bottom of the tube simultaneously to ensure good mixing. As soon as all the spermidine is added, vortex-mix at the maximum setting for 1 min.
5. Centrifuge the DNA-coated gold particles for 5 min at maximum setting in a microfuge. Discard the supernatant and add 250 μL of 100% ethanol. Dislodge and break up the pellet with a sterile pipet tip and suspend the particles by vortex-mixing briefly.
6. Centrifuge the DNA-coated gold particles for 2 min at maximum setting. Discard the supernatant and add 70 μL of 100% ethanol. The tube is kept on ice

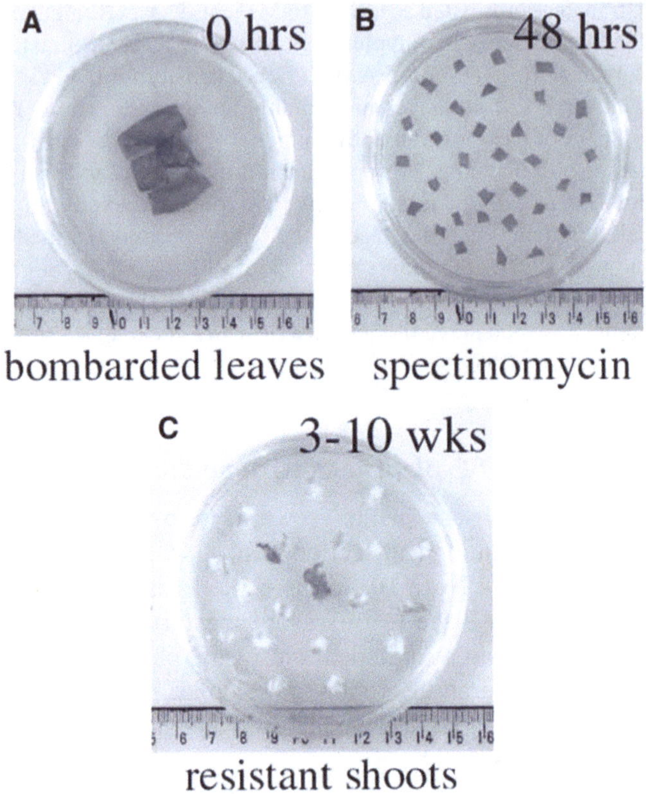

Fig. 3. Tobacco plastid transformation. **(A)** Bombarded leaves. **(B)** Cut leaves placed on RMOP medium with spectinomycin (0.5 mg/L). **(C)** Resistant shoots appear after 3–10 wk on this medium. On RMOP medium with spectinomycin and streptomycin resistant shoots appear after 8–20 wk (*see* **Note 13**).

without disturbing the pellet until it is needed. Just before use, the pellet is broken up with a micropipet tip and vortex-mixed very briefly by touching a mixer (about 1 s or less).

3.2.2. DNA Delivery

1. Set up Bio-Rad PDS 1000/He particle delivery system by placing the microcarrier launch assembly into the top groove (shelf position 1). This produces a 1-cm gap distance between rupture disk (1100 psi) and macrocarrier holder.
2. Two spacer rings separate the macrocarrier holder from the stopping screen, making a macrocarrier flight distance of 1.5 cm.

3. Place a 9-cm Petri dish containing the leaves placed on RMOP media in a Bio-Rad PDS 1000/He gun in the third groove from the top (shelf position 1). This results in a 6-cm target distance (distance from stopping screen to target plate).
4. Pipet 5 µL of the plasmid-coated gold suspension onto the center of a macrocarrier and allowed to dry in a laminar airflow hood.
5. Operate the device according to the instructions supplied with the Bio-Rad PDS 1000 He Particle Delivery system using a vacuum of 28 in. Hg. Leave the vacuum pump on for 2 min once it has reached 28 in. Hg. Then turn off vacuum pump and fire the device.
6. After the leaf explants have been bombarded, store them in dim light for 48 h at 25°C (*see* **Note 11**).
7. We routinely carry out ten bombardments per plasmid construct to ensure the isolation of 5–50 transplastomic plants for each plasmid.

3.2.3. Selection of Plastid Transformants

1. Cut microprojectile-bombarded leaves into 2- to 5-mm-long edges and place on RMOP solid medium containing spectinomycin (0.5 mg/L) and streptomycin (0.5 mg/L) in a 9 cm petridish (*see* **Note 12**). Use new scalpel blades (nos. 11 and 22) and change them before each set of plates. Sterilize scalpels and Waugh's forceps (15 cm, 20 cm) by rinsing in 70% (v/v) ethanol and placing in a dry bead sterilizer (STERI 350) for 30 s. Place the cut leaf sections from one bombarded plate abaxial side up on two RMOP selection plates (**Fig. 3**). Change the scalpel blade after cutting leaf pieces from four bombarded plates.
2. Seal plates with two strips of Parafilm and stack in groups of up to 10 plates. Incubate them at 25°C in a 12-h day/12-h dark cycle growth cabinet (*see* **Note 12**).
3. Green resistant shoots or clumps of cells appear after 8 wk and continue to appear after 20 wk postbombardment on RMOP medium containing spectinomycin and streptomycin (*see* **Fig. 4** and **Notes 13** and **14**). When selection is based on spectinomycin alone, green resistant cells are visible after 4–10 wk (*see* **Fig. 3** and **Note 13**).
4. Transfer green tissue to RMOP medium containing 5 mg/L of PPT for shoot regeneration. Any shoots appearing on spectinomyin/streptomycin plates are cut into small pieces before placing on the RMOP-PPT (5 mg/L) plates. Unselected marker genes such as *aadA* flanked by direct repeats are excised if material is propagated serially on PPT medium to maintain the *bar* gene (*see* **Note 15**).
5. Place regenerated shoots on MS medium containing 1 mg/L of PPT for rooting (*see* **Note 16**).
6. Combined spectinomycin plus streptomycin selection followed by PPT selection is very effective in driving homoplasmy of plastid transformants. Southern blots are used to verify homoplasmy using a plastid DNA probe (*see* **Note 17**).

3.3. GUS Expression Assay

1. Cut leaves or place whole seedlings in X-Gluc buffer with 0.1% (v/v) Triton X-100.

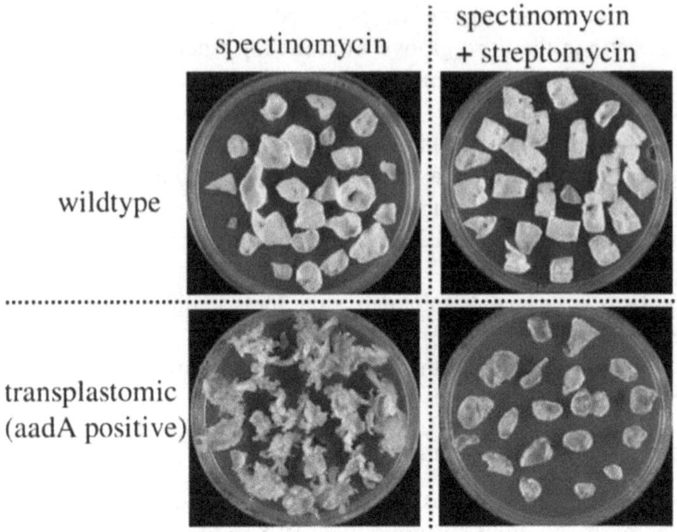

Fig. 4. Leaf pieces from wild-type and an *aadA* transplastomic tobacco plant placed on RMOP media with either spectinomycin (0.5 mg/L) alone or spectinomycin (0.5 mg/L) plus streptomycin (0.5 mg/L) for 22 d.

2. Incubate at 37°C for 16–40 h or until deep blue staining is apparent in the positive control, which contans a *uidA* transgene.
3. Remove X-Gluc buffer with a Pasteur pipet and replace with 70% (v/v) ethanol. Leave overnight and replace the liquid several times with 70% ethanol until all the chlorophyll has been removed from the explants.

3.4. Isolation of Marker-Free Transplastomic Shoots and Seedlings

1. Transfer transplastomic plants with roots to soil and grow without selection (*see* **Notes 17** and **18**). Flowers can be self-pollinated or pollinated with pollen from untransformed wild-type plants following excision of anthers (*see* **Note 19**).
2. For each plant, collect seeds from all the flower pods and store these separately (*see* **Note 20**).
3. Sterilize (100–200 seeds) from each flower pod and germinate on MS medium containing spectinomycin.
4. Very soon after germination, screen the plates for seedlings with all white cotyledons to identify seedlings that have lost the *aadA* gene.
5. Transfer the white seedlings to MS medium and propagate in dim light. *See* **Note 21** to enable greening.
6. After 2–4 wk move the seedlings with green true leaves to Magenta jars (Sigma).
7. Analyze the marker-free plants by DNA blot analysis and polymerase chain reaction (PCR).

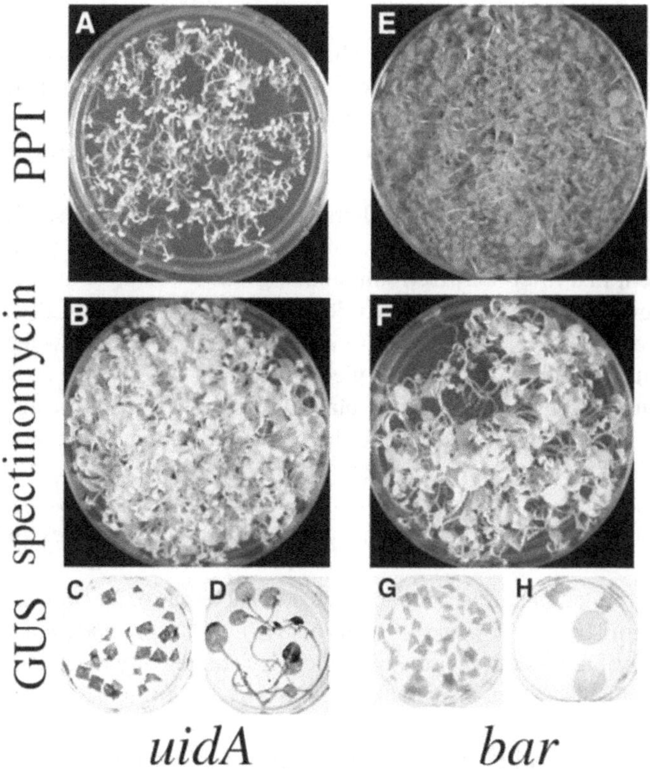

Fig. 5. Seedlings from *aadA*-free transplastomic plants containing either a single *uidA* gene **(A–D)** or single bar gene **(E–H)** in the plastid genome. The plants were isolated following homologous recombination between repeats 1 and 3, which excise *aadA* and *bar*, and between the A repeats, which remove *uidA* and *aadA* (**Fig. 2**). Seedlings on MS salts media (*see* **Subheading 2.1.2.**) with 10 mg/L PPT (*top*) or 500 mg/L spectinomycin (*middle*). Seeds were first germinated on MS salts medium (*see* **Subheading 2.1.2.**) for 3 d at 25°C before placing on the same medium with selective agents PPT or spectinomycin. Leaf pieces from seedlings **(C,G)** or whole seedlings **(D,H)** grown on MS medium and then stained with X-Gluc (*bottom*). White leaves **(A,B,F,G,H)**, green resistant leaves **(E)**, blue GUS-stained leaves **(C,D)**.

8. The instability resulting from direct repeats is specific for the marker genes and once a direct repeat is excised with the marker genes (**Fig. 1**), the remaining foreign genes inserted into plastids should be inherited stably (*see* **Fig. 5**). All progeny from *aadA*-free plants contain the foreign trait gene, either *uidA* or *bar*, but are sensitive to spectinomycin because of excision of *aadA* (*see* **Note 22**).

4. Notes

1. Keep all glassware used to make media separate from those used for general use. This avoids the inadvertent addition of contaminants such as detergents into media that might be deleterious to plant growth.
2. Autoclave solutions before storing at 4°C if kept longer than 2 wk. Autoclave settings are 121°C, 15 psi for 20 min.
3. The wetting agents used on some filters can be deleterious. Passing 10 mL of sterile water through a syringe filter unit before use will reduce the amount of wetting agent in a filter-sterilized solution.
4. Streptomycin stocks are usually discarded after 6 mo storage at –20°C.
5. We have not observed noticeable differences in our experiments when agar has been substituted with Phytagel. Leaf pieces placed on solid media are rarely flat and can curl with time. Leaf pieces placed on phytagel media tend to stay in closer contact with the solid surface for longer periods of time. Phytagel-medium tends to be better for rooting than agar solidified media.
6. We routinely use 1-μm gold for plastid transformation of tobacco. We have also obtained tobacco plastid transformants using 0.6-μm gold particles and 650 psi rupture discs.
7. Plastid transformation has been demonstrated in numerous of *N. tabacum* cultivars including Petit Havana ***(15)***, and Wisconsin 38 ***(8)***, and *N. plumbaginifolia* ***(16)***.
8. To avoid possible loss of regeneration efficiency and fertility resulting from prolonged growth, discard in vitro plants after 18 mo and germinate new plants from seeds.
9. We normally house all plants and explants grown in vitro in a Sanyo MLR350 illuminated plant growth chamber at light setting 4.
10. Leave a small circle of 0.5 cm clear of leaves at the center of the 4-cm circle. Experiments with *C. reinhardtii* plastid transformation have shown that most transformants are found in a ring located between 0.5 and 4 cm from the centre of the plate. It is not important to cover the entire target area with leaves. Spread the leaf pieces onto as many plates as is feasible, leaving gaps between the pieces. We have found that more bombardments with fewer leaf pieces per plate is a better strategy than fewer bombardments with more leaf pieces per plate. This is because of the large variation in numbers of plastid transformants obtained between different bombardments.
11. Place plates in stacks in a Sanyo MLR350 illuminated growth chamber at light setting 3 at least 12–15 cm away from the lights in a Sanyo MLR350 chamber at light setting 4. Place Petri dishes filled with 0.8% agar in 1 m*M* EDTA on top of the stacks to prevent condensation appearing on the top plate containing leaf explants.
12. Leaf pieces expand on RMOP solid medium. Make sure the leaf pieces are well separated. In some cases, it might be advisable to place the cut leaf pieces from one bombarded plate onto three 9-cm plates containing RMOP with antibiotics.
13. Resistant green clones appear later and grow more slowly with double selection using spectinomycin and streptomycin than with selection with spectinomycin

alone. Resistant shoots appear 3–10 wk post-bombardment with spectinomycin selection. Including streptomycin eliminates spontaneous spectinomycin-resistant mutants and allows *aadA* transformed clones to grow. For more rapid isolation of plastid transformants, we select first on spectinomycin (0.5 mg/L) and then select the green resistant cells on spectinomycin and streptomycin (0.5 mg/L).

14. In rare cases, only one or two green resistant clumps of cells are observed after 12 wk. In these cases, it is advisable to cut the leaf pieces again and transfer to fresh RMOP medium with spectinomycin (0.5 mg/L) and streptomycin (0.5 mg/L). If no green clones are obtained at 12 wk, it is better to repeat the transformation experiments.
15. Two out of 42 clones were PPT resistant and had lost the *aadA* marker gene as a result of recombination between the 174-bp repeats marked A in **Fig. 2**. The excision frequency would be expected to be higher if the sizes of the 174-bp direct repeats were increased.
16. The use of phytagel (0.25% w/v) rather than agar facilitates rooting.
17. To confirm integration into the plastid genome and the absence of wild-type plastid DNA, digest DNA from transplastomic plants with enzymes that cut outside the plastid DNA targeting regions in the plastid transformation vector. It is important that all wild-type plastid genomes have been replaced by recombinant plastid genomes before selection is stopped.
18. Retention of the *bar* gene in soil grown plants can be selected by spraying with a 1:1000 dilution of a herbicide such as Challenge (Hoescht) containing 15% (w/v) PPT.
19. Remove anthers from a transplastomic flower before pollen is shed and pollinate the flowers using a soft paintbrush loaded with pollen from a wild-type flower. Then cover the flower with a small waxed paper piece and tape it shut.
20. The segregation of marker-free plastids from *aadA* containing plastids gives rise to cells containing only marker free plastids. The spatial distribution of these marker free cells within a plant vary because cytoplasmic sorting is a stochastic process. This means the percentage of marker-free plants varies from flower to flower. It is important not to pool seeds from different flowers of an individual plant. Only egg cells need to contain marker-free plastids because pollen does not transmit plastid DNA to the zygote in tobacco. We have obtained 25% marker-free seedlings in tobacco. In species that inherit plastids from both parents, the frequency of marker-free plants is reduced. If 25% of eggs and pollen grains are marker-free, then 6% of seedlings ($0.25 \times 0.25 \times 100\%$) will be marker free. Any bottlenecks in sexual reproduction that reduce the number of copies of plastid DNA in egg cells will facilitate the isolation of marker-free plants.
21. Spectinomycin selection is not lethal and shoots will recover after they are placed on antibiotic-free medium. In tobacco, spectinomycin stops shoot growth ***(17)***. In other species such as *Brassica napus*, shoots bleach but continue to grow in the presence of spectinomycin; this can result in the irreversible loss of plastid ribosomes ***(17)***. In these species, screen *aadA*-free seedlings by PCR using *aadA* primers.

22. Plastids are inherited from the maternal parent in tobacco. Stable inheritance of trait genes inserted into plastids ensures a homoplasmic population of transgenic plastid genomes in *aadA*-free parents. All seedlings from these parents will express the trait gene if it is stable.

References

1. Day, A. (2003) Antibiotic resistance genes in transgenic plants: their origins, undesirability and technologies for their elimination from genetically modified crops, in *Transgenic Plants-Current Innovations and Future Trends* (Stewart, C. N., ed.) Horizon Scientific Press, Wymondham, Norfolk, UK, pp. 111–156.
2. Yoder, J. I. and Goldsbrough, A. P. (1994) Transformation systems for generating marker-free transgenic plants. *BioTechnology* **12,** 263–267.
3. Hare, P. D. and Chua, N. H. (2002) Excision of selectable marker genes from transgenic plants. *Nat. Biotechnol.* **20,** 575–580.
4. Hajdukiewicz, P. T. J., Gilbertson, L., and Staub, J. M. (2001) Multiple pathways for Cre/lox-mediated recombination in plastids. *Plant J.* **27,** 161–170.
5. Corneille, S., Lutz, K., Svab Z., and Maliga P. (2001) Efficient elimination of selectable marker genes from the plastid genome by the CRE-lox site-specific recombination system. *Plant J.* **27,** 171–178.
6. Zubko, E., Scutt, C., and Meyer, P. (2000) Intrachromosomal recombination between attP regions as a tool to remove selectable marker genes from tobacco transgenes. *Nat. Biotechnol.* **18,** 442–445.
7. Day, A. and Ellis, T. H. N. (1984) Chloroplast DNA deletions associated with wheat plants regenerated from pollen: possible basis for maternal inheritance of chloroplasts. *Cell* **39,** 359–368.
8. Iamtham, S. and Day, A. (2000) Removal of antibiotic resistance genes from transgenic tobacco plastids. *Nat. Biotechnol.* **18,** 1172–1176.
9. Ye, G. N., Colburn, S. M., Xu, C. W., Hajdukiewicz P. T. J., and Staub J. M. (2003) Persistence of unselected transgenic DNA during a plastid transformation and segregation approach to herbicide resistance. *Plant Physiol.* **133,** 402–410.
10. Daniell, H., Muthukumar, B., and Lee, S. B. (2001) Marker free transgenic plants: engineering the chloroplast genome without the use of antibiotic selection. *Curr. Genet.* **39,** 109–116.
11. Klaus, S. M. J., Huang, F. C., Eibl, C., Koop, H. U., and Golds, T. J. (2003) Rapid and proven production of transplastomic tobacco plants by restoration of pigmentation and photosynthesis. *Plant J.* **35,** 811–821.
12. Fischer, N., Stampacchia, O., Redding, K., and Rochaix, J. D. (1996) Selectable marker recycling in the chloroplast. *Mol. Gen. Genet.* **251,** 373–380.
13. Svab, Z. and Maliga, P. (1993) High frequency plastid transformation in tobacco by selection for a chimeric aadA gene. *Proc. Natl. Acad. Sci. USA* **90,** 913–917.
14. Murashige, T. and Skoog, F. (1962) A revised medium for rapid growth and bioassays with tobacco tissue cultures. *Physiol. Plant* **15,** 473–497.
15. Svab, Z., Hajdukiewicz, P., and Maliga, P. (1990) Stable transformation of plastids in higher plants. *Proc. Natl. Acad Sci. USA* **87,** 8526–8530.

16. Oneill, C., Horvath., G. V., Horvath, E., Dix, P. J., and Medgyesy, P. (1993) Chloroplast transformation in plants: polyethylene glycol (PEG) treatment of protoplasts is an alternative to biolistic delivery systems. *Plant J.* **3,** 729–738.
17. Zubko, M. K. and Day, A. (1998) Stable albinism induced without mutagenesis: a model for ribosome-free plastid inheritance. *Plant J.* **15,** 265–271.

V

Transgene Integration, Expression, and Localization

19

The Study of Transgene Copy Number and Organization

David J. Ingham

Summary

The development of efficient crop transformation systems has necessitated the development of efficient methods for detailed molecular characterization of putative events. This chapter details the routine use of quantitative real-time polymerase chain reaction to determine transgene copy number in putative transgenic events. This approach has allowed the analysis of plantlets in tissue culture prior to transfer to soil and greenhouse. Implementation of the TaqMan transgene copy assay permits the efficient utilization of limited resources and space to develop a highly efficient transgenic event production pipeline. Other applications for this assay within the biotechnology production pipeline are also discussed.

Key Words: Copy number; DNA; genomic DNA; polymerase chain reaction (PCR); quantitative PCR; real-time PCR; TaqMan; transformation; transgene.

1. Introduction

The development of efficient crop transformation systems has necessitated the development of efficient methods for detailed molecular characterization of putative transgenic events. Traditionally, standard polymerase chain reaction (PCR) methods have been utilized to detect the presence of recombinant DNA in transformed plants. But, for transformation methods in which the rate of escapes has dropped close to zero, these standard presence/absence type assays will become obsolete. In addition, researchers would like to be able to evaluate trait efficacy only for transgenic events with the desired molecular characteristics. Often, this translates to a desire for simple insertion events (single copy/locus) and the absence of vector backbone sequences.

Therefore, emphasis in recent years has focused on determining the number of inserted copies of a transgene that are present in a transgenic event as early as

From: *Methods in Molecular Biology, vol. 286: Transgenic Plants: Methods and Protocols*
Edited by: L. Peña © Humana Press Inc., Totowa, NJ

possible during event production. Traditionally, researchers have relied on nucleic acid blotting techniques (Southern blot *[1]*) to provide an estimate of transgene copy number and organization. Although Southern blotting techniques are critical for determining loci number and intactness of the insert, this method is difficult to apply in the high-throughput screening of putative transformants. For these reasons, many laboratories have expended significant resources to develop alternative methods to obtain estimates of transgene copy number.

Recently, the development of real-time quantitative PCR (Q-PCR) methods for determining transgene copy number has overcome the limitations of standard PCR and Southern analysis ***(2,3)***. There are several methods available to conduct Q-PCR. For this chapter, TaqMan assay is the method that is discussed. TaqMan is a real-time PCR detection technique in which the accumulation of PCR product is monitored by the accumulation of fluorescence in each reaction. Briefly, for TaqMan assay, the PCR contains standard forward and reverse PCR primers plus the addition of a dual-labeled (TaqMan) probe designed to hybridize to sequence between the primers. The probe is synthesized with a reporter fluorophore on the 5'-end and a quencher moiety on the 3'-end. While the probe is intact, emission from the reporter fluorophore is efficiently quenched by the quencher moiety, resulting in low detectable fluorescence. During each cycle of PCR, the polymerase extending from one of the PCR primers will encounter hybridized probe activitating the 5' to 3' exonuclease activity of the *Taq* polymerase. The hybridized probe is then degraded, releasing the fluorophore and decreasing the efficiency of quenching, resulting in an increase in relative fluorescence. The result is a detectable fluorescence emission from each reaction that is directly related to the accumulation of PCR product.

Q-PCR methods provide an accurate, quantitative and high-throughput approach for estimating transgene copy number from small amounts of sample. Because the required sample size is very small, these assays can be conducted while putative transgenics are still in tissue culture. This allows the selection of desirable transgenic events prior to expending the cost and resources required for transplantation to soil and propagation to maturity under greenhouse conditions. Therefore, utilization of quantitative PCR assays to determine transgene copy number for putative transgenic events allows greenhouse space to be utilized at higher efficiency, which will increase the capacity of the transgenic event production pipeline.

2. Materials

2.1. Genomic DNA Sample Preparation

2.1.1. Disposables and Equipment

1. 96-Well U-bottom plates.

2. Standard microplate Lid.
3. 1.2-mL deep-well block.
4. Capping mat for deep-well block.
5. 5/32-in. chrome steel ball bearings (grade 1000).
6. MagnaBot 96 magnetic separation device (Promega Corporation, Madison, WI).
7. MagnaBot spacers (Promega Corporation).
8. Model 4-96-A Tissue Pulverizer (Kinetic Laboratory Instrument Company, Visalia, CA).
9. Storage Mat Applicator (Corning Inc., Corning, NY).

2.1.2. Reagents

1. Lysis buffer A, plant (Wizard Magnesil Plant gDNA Kit, Promega Corporation).
2. Lysis buffer B, plant (Wizard Magnesil Plant gDNA Kit, Promega Corporation).
3. Magnesil paramagnetic particles (Wizard Magnesil Plant gDNA Kit, Promega Corporation).
4. Wash buffer, plant (Wizard Magnesil Plant gDNA Kit, Promega Corporation): wash buffer supplemented to 25% (v/v) final concentration each of 95–100% ethanol and isopropanol.
5. 1X Tris–EDTA (TE): 10 m*M* Tris-HCl, 1 m*M* EDTA, pH 8.0.

2.2. TaqMan Transgene Copy Assay

2.2.1. Disposables and Equipment

1. Clear adhesive seal.
2. 96-Well PCR plate.

2.2.2. Reagents

1. 2X Q-PCR mix: Jumpstart ReadyMix Taq (product number P2893, Sigma Aldrich, St. Louis, MO), consisting of 20 m*M* Tris-HCl, pH 8.3, 100 m*M* KCl, 4 m*M* $MgCl_2$, 0.002% gelatin, 0.4 m*M* each dATP, dCTP, dGTP, dTTP, 0.06 U/µL of *Taq* DNA Polymerase, Jumpstart *Taq* antibody), supplemented with an additional 11 m*M* $MgCl_2$ and 2X concentration of reference dye for Q-PCR (product no. R4526, Sigma Aldrich).
2. PCR primers.
3. Dual-labeled fluorescent probes.

3. Methods

3.1. Sampling Instructions

For best results, it is most important that samples are taken from healthy plant tissue. Young leaf tissue is best for these procedures, but we have also had success using various different tissue types including roots and callus. Severely necrotic or senescing tissue rarely produces gDNA of sufficient quality for successful application of TaqMan transgene copy assay. In general, the use of *smaller amounts of tissue* produces better results. Most plant tissues contain some level of contaminants that are inhibitory to PCR. Inhibitors are most abundant in tissues such as callus material where the amount of tissue must be kept very small to increase the chances for successful assay (*see* **step 2**). Alternatively, purified gDNA containing significant levels of PCR inhibitors may be serially diluted in an attempt to find a satisfactory concentration that yields no detectable inhibition, but sufficient levels of gDNA to perform TaqMan transgene copy determination. The procedures below have been systematically optimized for best performance across a wide variety of plant species.

1. Prepare a sufficient number of 1.2-mL deep-well blocks to contain the number of samples plus any control wells by placing a single 5/32-in. chrome steel ball bearing in each well (*see* **Note 1**).
2. Place a single sample in each well of the plate. Samples should be from young healthy tissue. The recommend size of each sample is <20 mg for leaf and other plant tissues and <2 mm^3 for callus. When possible, tissue should be folded in half lengthwise and placed in the well with the folded end toward the bottom of the well. This procedure increases the likelihood that each sample is efficiently ground (*see* **Note 2**).
3. Leave sufficient number of blank wells to allow addition of assay controls.
4. Securely seal each block with a capping mat using a storage mat applicator.
5. For shipment of samples from remote locations to a central facility, sample blocks may be shipped on wet ice or frozen on dry ice. Two to three days of storage on wet ice during shipment is not detrimental to the outcome of the majority of transgene copy number assay or standard PCR assays (unpublished observation). Alternatively, lyophilized samples may be shipped at room temperature (*see* **Note 3**).
6. Prior to sample preparation, samples should be stored in the laboratory at less than or equal to –20°C (*see* **Note 4**).

3.2. Sample Preparation

Genomic DNA (gDNA) may be prepared from plant samples via various of methods including published procedures (e.g., Dellaporta *[4]*) as well as using a variety of commercially available kits. We have found the commercial kits to be desirable because the manufacturer performs all necessary validation of kit components. Our laboratory can concentrate on processing samples rather than validating reagents we produce. Although many different procedures work well for the purification of genomic DNA from plants for TaqMan transgene copy assay, we have found the Magnesil Kit from Promega to be the most preferable for several reasons. First, the quality of the gDNA is excellent for all crops tested to date (including maize, soybean, wheat, barley, rice, cotton, canola, and *Arabidopsis*). Second, because the purification is solution based (magnetic particles), the purification can be scaled to meet the needs of the particular application. Lastly, this procedure is amenable to automation on a wide variety of automation platforms. We have successfully automated the procedure on the Biomek FX and Biomek 2000 workstations.

The following procedure is based on the preparation of two 96-well plates of plant leaf samples. This procedure has been modified from Promega Technical Bulletin 289. All steps are conducted at room temperature.

1. Freeze plant samples at minus 80°C for at least 15 min. Plant samples should already be in a well in a 96-deep-well block containing a single 5/32-in chrome steel ball bearing (*see* **Subheading 3.1.**).
2. Quickly transfer plates to a suitable grinding instrument and grind samples for at least 30 s. For optimal pulverizing of tissue, grinding must be completed prior to samples beginning to thaw. Several suitable grinding instruments exist on the market. Our preference for throughput and quality of instrumentation is the Model 4-96-A Tissue Pulverizer available through Kinetic Laboratory Equipment Company (KLECO). Lyophilization of tissue also works well for grinding, but ground tissue becomes prone to static charging, which generates high risk for cross contamination of samples (*see* **Note 3**). After grinding, sample may be stored on ice or frozen.
3. Spin samples briefly in a plate centrifuge with sufficient force to bring tissue to the bottom of the well.
4. Carefully remove capping mat and add 300 μL lysis buffer A to each well (*see* **Note 5**).
5. Replace capping mat and secure using Storage Mat Applicator.
6. To efficiently mix samples, place the block(s) back in the grinder and grind for 30 s. Alternatively, the blocks may be inverted several times until the samples are completely mixed.
7. Centrifuge sample blocks in a plate centrifuge with sufficient force to bring tissue debris to the bottom of the well.
8. Prepare paramagnetic particles (PMP) by combining 2.3 mL of completely resuspended Magnesil Paramagnetic Particles with 15 mL of lysis buffer B. This procedure will generate sufficient quantities of particles for two plates of samples, but creates a significant amount of leftover waste. Prepare only sufficient PMP/lysis buffer B mixture for the samples that will be processed that day. Any leftover reagent should be discarded.
9. Add 60 μL of PMP/lysis buffer B mixture to each well of two empty 96-well Greiner plates. These will be the plates used for the purification of gDNA from plant samples.
10. Remove capping mat from sample blocks and transfer 125 μL of the sample lysate to the appropriately labeled plate containing the freshly prepared PMP/lysis buffer B mixture. Care should be taken to avoid transfer of any tissue during this process. It is better to transfer less lysate than to risk transfer of tissue to the gDNA preparation plate.
11. Incubate samples for 5 min with continual shaking. Alternatively, mix samples completely by pipetting the mixture up and down several times and incubate mixture without shaking for 5 min. Our laboratory prefers use of orbital shakers for

all mixing steps. However, the instrument settings that provides sufficient mixing of samples while avoiding cross-contamination between wells must be determined empirically for each laboratory procedure.

12. Place each plate onto a MagnaBot 96 Magnetic Separation Device for 1 min. Discard the lysate mixture. Lysate is clear of PMPs in most cases in less than 20 s. When conducting procedure manually, the user need only wait until lysate is visually clear prior to aspiration of supernatant. For automated procedures, we recommend the user instruct the system to wait the full 1 min prior to aspirating the supernatant.
13. Remove the plate from the MagnaBot and add 150 µL of wash buffer that has been supplemented with ethanol and isopropanol as indicated in **Note 6**.
14. Incubate plates for 1 min with continual shaking. Alternatively, mix samples completely by pipetting the mixture up and down several times and incubate mixture without shaking for 1 min.
15. Place each plate onto a MagnaBot 96 Magnetic Separation Device for 30 s. Discard the wash buffer.
16. Remove the plate from the MagnaBot and add 100 µL of supplemented wash buffer (*see* **Note 6**).
17. Incubate the plates for 1 min with continual shaking. Alternatively, mix samples completely by pipetting the mixture up and down several times and incubate mixture without shaking for 1 min.
18. Place each plate onto a MagnaBot 96 Magnetic Separation Device for 30 s. Discard the wash buffer. At this step, to remove as much liquid as possible, repeat the aspiration for each well to remove as much residual wash buffer as possible.
19. Remove the plate from the MagnaBot and allow the plates to air-dry uncovered for 5 min.
20. Add 60 µL of 1X TE to each well and incubate plates for 5 min with continual shaking. Alternatively, mix samples completely by pipetting the mixture up and down several times and incubate mixture without shaking for 5 min.
21. Place each plate onto a MagnaBot 96 Magnetic Separation Device for 1 min. For this step, we recommend incubating plates for the full minute to ensure complete clearing of each sample.
22. Transfer 40–50 µL of the purified DNA to an appropriately labeled 96-well U-bottom plate. To minimize the carry over of PMP we recommend leaving some of the purified DNA behind. Purified gDNA may be stored for several days at 4°C and indefinitely at –20°C (*see* **Note 7**).

3.3. Assay Design Considerations

The design considerations for TaqMan assays to be used for transgene copy number are basically the same as those used for other types of TaqMan assays (i.e., Quantitative RT-PCR). The reader is referred to the Primer Express Software User's Manual (ABI User Manual 303014D) for details of the design considerations. A summary of the most relevant factors is outlined in the following.

1. Target sequence for transgene assay can be any sequence contained within the DNA used for transformation. Sequences that do not share known homology to endogenous sequences should be used, when possible. When this is not possible, targeting assays to the junctions between two components has proven successful in the past.
2. Target sequences for the endogenous control assay should be to known genomic sequences from the plant species of interest, when possible. The endogenous copy number of the assay target should be low, but does not have to be single copy. Because the copy number calculation normalizes to a known copy sample, the actual copy of the endogenous gene is irrelevant. However, to avoid potential competition in duplexed reactions, the copy number of the endogenous gene should be low. In addition, the choice of sequences that are well conserved among related species or ecotypes would also simplify assay development because the same endogenous control assay can be used for the majority of sample sources.
3. Amplicon length should be kept as short as possible. Rather than attempt to target specific sequences within the transgenic DNA, unless necessary, we would recommend the design be left as flexible as possible to ensure the most efficient assay is designed. This should allow the selection of assays with very short amplicon sizes preferably in the range of 70–100 basepairs.
4. The choice of fluorophores is quite flexible. For transgene copy determination we recommend the use of fluorescein (FAM) and tetrachloro-6-carboxy-fluoroscein (TET) for the GOI and endogenous assay probes, respectively, each in combination with a nonfluorescent quencher (Black Hole Quencher). We have not detected any advantage to the use of VIC (or related dyes) in direct comparisons to use of TET (unpublished observation). In addition, we have found both fluorescent (tetramethylrhodamine [TAMRA]) and nonfluorescent (4[dimethyaminoazo] benzene-4-carboxylic acid [DABCYL] or Black Hole Quencher [BHQ]) quenched probes to work equally well. The use of the dark quencher gives the added flexibility to triplex transgene copy number assays (*see* **Note 8**).

3.4. Assay Validation

Prior to the use of each new assay on unknown samples, we recommend performing basic validation of the assay's performance. Because of the fact that the assays are all run at maximal primer concentrations, very few assays fail this test. The test is conducted by running the duplexed assay on one or more known positive samples. The actual copy number of the test sample does not need to be known. In the absence of a confirmed positive sample, vector DNA "spiked" into nontransgenic genomic DNA can serve as a test target. The vector DNA should be present at a molar equivalent level approximating a 1- to 10-copy sample.

1. Prepare a set of eight samples for testing purposes. Samples 1–7 should consist of a twofold dilution series of a genomic DNA prepared from a confirmed positive transgenic event. The eighth sample should be left blank to serve as a negative control.

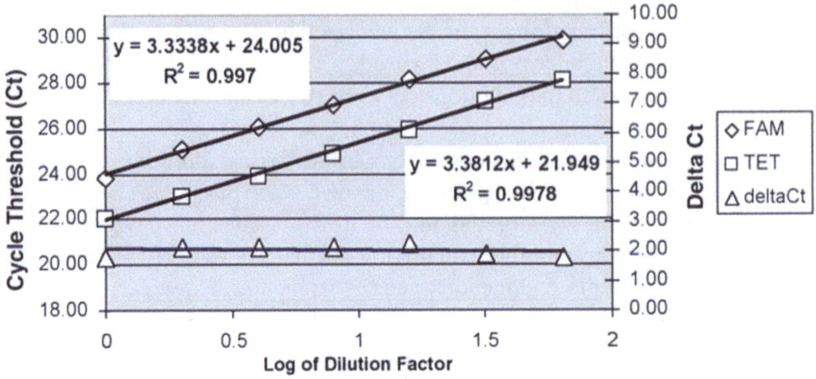

Fig. 1. Example validation experiment for TaqMan transgene copy number assays. Two fold dilutions of a known positive transgenic corn plant were run in duplexed TaqMan assays as described in the text. The average cycle threshold (C_t) value for three replicate reactions was plotted for each assay vs the log of the dilution factor. In addition, the delta C_t (ΔC_t) was calculated from each set of FAM and TET average C_ts and plotted vs the log of each dilution factor. Logarithmic curves were fitted to each C_t data set and the line equation and R^2 value for each curve is displayed on the graph (FAM or TET data sets are shown in diamonds or squares, respectively).

2. Set up a 70-µL reaction mixture for each sample as described in **Subheading 3.5.** Setup standard duplex reactions containing both GOI and endogenous control TaqMan sets.
3. Thoroughly mix each reaction and split the mixture into three 20-µL replicate reactions.
4. Setup and run the plate on the Q-PCR instrument as described in **Subheading 3.5.** and export the cycle threshold (C_t) values to a standard spreadsheet program.
5. Calculate the average and standard deviation for each set of triplicate C_t values.
6. Plot the average C_t value for each dilution versus the log of the dilution factor and fit a logarithmic curve to the data set. The result should be a straight line (**Fig. 1**).
7. Calculate the efficiency of the reaction using the slope of the fitted curve (efficiency = $10^{1/\text{slope}}$). Using this approach, 100% efficiency should yield a slope of approx 3.3.
8. Evaluate the result (*see* **Fig. 1** and **Notes 9** and **10**).

3.5. Quantitative PCR Reaction Setup

The following procedure describes the setup of quantitative PCR reactions for TaqMan assay basically as described by Ingham et al. *(2)*. For the purposes of this procedure, the reactions are run in 0.2-mL PCR tubes on an ABI7700 (or equivalent) instrument with a total reaction volume of 20 µL. In addition, the procedure is written for 96 samples (including controls) that are assayed for the same transgene (GOI) and endogenous gene. For different numbers of samples or assay approaches, volumes will need to be adjusted appropriately. Quantitation of genomic DNA samples is unnecessary, but the sample preparation method should be designed so that reactions setup as described below will deliver endogenous control C_ts in the range of 22–28 cycles for the majority of samples (*see* **Note 7**). Variations in DNA concentrations of two- to fourfold within a sample set are not deleterious to the outcome of the assay (*see* **step 5**). For a thorough review of quantitative PCR applications, the reader is directed to reviews by Stephen A. Bustin and David G. Ginzinger and references therein *(5–7)*.

1. Prepare 2X Jumpstart Mix for Q-PCR, by supplementing the supplied mixture with an additional 11 m*M* $MgCl_2$ and reference dye for Q-PCR to a 2X final concentration. For example, for every 10 mL of 2X Jumpstart Mix for Q-PCR, add 110 µL of 1 *M* $MgCl_2$ and 200 µL of 100X reference dye for Q-PCR. This results in final 2X stock concentrations of 15 m*M* $MgCl_2$ (4 m*M* from the 2X Jumpstart Mix for Q-PCR plus 11 m*M* of additional $MgCl_2$) and 2X concentration of the reference dye for Q-PCR. The additional magnesium is required for efficient TaqMan assay (unpublished observation).
2. Prepare 50X TaqMan primer and probe mixtures by generating solutions containing 45 µ*M* of each primer (forward and reverse) and 5 µ*M* of the correct TaqMan probe. When diluted to 1X concentration in the Q-PCR reactions, the primers will each be a 900 n*M* and the probes at 100 n*M* final concentrations. Recommended procedure for preparing stocks is to combine 45 µL of each 1 m*M* primer stock (forward and reverse) and 50 µL of a 100 µM TaqMan probe stock with 860 µL of 1X TE. These 50X TaqMan primer and probe mixtures can be stored for at least 1 mo at 4°C and indefinitely at –20°C.
3. For each set of samples, it is recommended that a reaction cocktail is prepared containing all constituents that are common. Although this approach will generate larger volumes of waste, savings are realized through efficiency of labor and reaction consistency. For 96 samples run in single 20-µL reactions, prepare a cocktail by combining 1100 µL of 2X Jumpstart Mix for Q-PCR (supplemented as described in **step 1**), 44 µL each the 50X FAM and 50X TET TaqMan primer and probe mixtures and 500 µL of molecular biology grade sterile water. This results in a an approx 1.3X reaction mixture. Replicate reactions are not necessary (*see* **Notes 11** and **12**).
4. Dispense 15 µL of 1.3X reaction cocktail into each well of the reaction plate. Dispensing can be conducted using various techniques including automated delivery or manual dispensing utilizing single-channel, multichannel, or repeat pipettors. All of these methods deliver sufficient levels of accuracy. By dispensing the reaction mixture into the plates prior to addition of DNA, the entire plate can be dispensed using the same pipet tips, resulting in time and material savings.

5. Add 5 µL of genomic DNA samples to each wells of the reaction plate using single or multichannel manual or automated pipettors. It is not necessary to mix the reactions. For maize, we find approx 200 ng gDNA per 20 µL reaction will yield desirable results (C_ts approx 25 cycles; *see* **Note 7**). Two- to fourfold increases or decreases in DNA concentration will not affect the outcome of the assay. Each laboratory should determine empirically what levels of gDNA work best for their particular systems (*see* **Notes 13–16**).
6. Immediately seal plates using clear sealing films or caps. Sealing films can be either heat or adhesive seals. We have not observed any significant differences in the performance of clear sealing films from various sources. We have detected no benefit justifying the added expense of sealing films marketed as validated for use in Q-PCR. Plates that will not be run immediately may be stored overnight at room temperature, for at least several days at 4°C and for at least 1 mo at –20°C. Although exposure to standard laboratory lighting for up to 2 d appears to have no detrimental effect, we recommend avoiding long-term exposure to intense sources of light.
7. Set up the reaction plate template in instrument software following manufacturer's instructions.
8. Run plate on ABI7700 under the following cycling conditions: hot start at 95°C for 5 min followed by 35 cycles of 95°C for 15 s and 60°C for 60 s. Addition of a step for uracyl *N*-glycolase digestion (usually 2 min at 50°C prior to hot start) is not necessary because the reaction mixture used here does not contain these components. Also, we use a shorter hot start time because the Sigma Aldrich jumpstart mixtures allow rapid activation at 95°C.

3.6. Postrun Manipulations

During the course of the Q-PCR, the ABI7700 (as well as other Q-PCR instruments) detect the accumulation of PCR product by the increase of fluorescence derived from TaqMan probe degradation (for review, *see* Bustin *[6]* and references therein). The plotting of the fluorescence value for each well versus cycle number yields the amplification plot, which represents the accumulation of PCR product in each reaction. After completion of the run, follow the manufacturer's instructions to generate C_t (cycle threshold) values for each sample. We perform all transgene copy number calculations in standard spreadsheet software utilizing C_t values for each sample exported from the instrument software.

1. The baseline for the Q-PCR run is set to cycles 3–15 by default for each run on the ABI7700. The baseline is used by the Sequence Detection System (SDS) software to determine relative normalized fluorescence for each well at each reading. The default baseline usually does not need to be modified. However, if amplification above background occurs for some wells prior to cycle 15, the high-end baseline cycle should be adjusted appropriately.
2. The cycle threshold setting is used by the SDS software to determine the C_t value for each well. Set the threshold so that the majority of reactions are intersected

late in the log phase of amplification. We have observed that lower well to well variability is observed with thresholds set late, rather than early, in the log phase of amplification.

3. The previous steps should be carried out for both dye layers (FAM and TET).
4. Export the C_t values for both the FAM and TET layers for each reaction to a "comma separated value" (csv) file. This file will be imported into a spreadsheet program to perform transgene copy number calculations.

3.7. Transgene Copy Number Calculation

Relative quantitation ($\Delta\ C_t$ method) is used to calculate transgene copy number as originally described by Ingham et al. *(2)* . Using this method, each reaction is first normalized relative to the endogenous control. This step removes variability resulting from variations in DNA concentration between wells and obviates the requirement for accurate quantitation and normalization of each purified sample. Next, each unknown sample is normalized to a control sample of known copy number, if available. This value is then used to calculate an estimate of the actual transgene copy number of each sample.

1. Import the C_t values for both the FAM and TET assays into a standard spreadsheet program for further manipulations.
2. Calculate ΔC_t by subtracting the TET C_t from the FAM C_t value.
3. Calculate $\Delta\ C_t$ by subtracting the ΔC_t of the known copy control from the ΔC_t value for each unknown sample.
4. Calculate transgene copy number using the Eq. $2^{(-\Delta\ C_t)}$. Use of the negative value in the exponent of the calculation is dependent on the setup of the ΔC_t calculation and its requirement in the copy number calculation needs to be determined empirically.
5. In the absence of a control sample of known copy number, transgene copy number can still be estimated using the TaqMan results. The accuracy of this approach is dependent on several factors including the size of the sample set and the distribution of copy number.
   a. Provided the availability of a sufficiently large sample set and a significant proportion of samples of one, two, and three transgene copies, one can select an arbitrary control from the population. By evaluating the set of data in a similar fashion as is routinely conducted during genotyping studies, one can identify populations of samples corresponding to various copy numbers.
   b. We recommend this be visualized by graphing the FAM C_t values (*y*-axis) vs the TET C_t values (*x*-axis) on a standard scatter plot. Variation along the *x*-axis is due to variation in DNA concentration. Variation along the *y*-axis is due to variation in relative transgene copy numbers.
   c. Because of the logarithmic nature of PCR amplification, one-cycle differences are observed between samples that vary in concentrations by twofold. Therefore, one-copy and two-copy sample sets should be separated by one cycle, while two-copy and three-copy samples would be separated by approx 0.67 cycles. This predictable progression can be used to identify likely

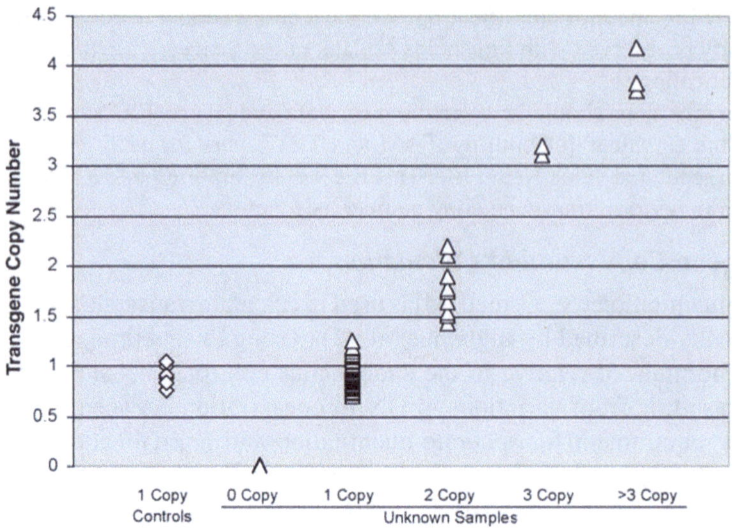

Fig. 2. Example data from a transgene copy number assay. Copy numbers were calculated from TaqMan copy assays as described in the text. The resulting data was grouped into zero, one, two, or three, and more than three data sets and plotted as shown. One copy control or unknown samples are displayed as diamonds or triangles, respectively.

single copy populations. Once the putative single copy population is identified, either a single representative sample can be used as the arbitrary copy control or an average ΔC_t can be calculated from several samples and used for $\Delta\ C_t$ calculations.

3.8. Interpretation of Results

The calculated transgene copy number is a *quantitative* estimate of the transgene copy number for each sample. The data produced by this assay can be used to answer various questions regarding transgenic plants at various stages of production from putative transgenic to a deregulated event in a commercial product.

1. Transgene copy determination on putative transgenics (T_0): The resulting population of single copy samples will be distributed over a range (e.g., 0.7–1.3 copies, **Fig. 2**). This does not mean that there are 1.3 copies of the transgene in a particular sample. This is simply a manifestation of the inherent assay variability. "Copy calls" consisting of whole number values can be generated from the quantitative estimates by many different techniques. The simplest approach is to set arbitrary cutoffs for each group (one-copy, two-copy, etc.) by visual delineation of groups from graphed data (**Fig. 2**).

2. Use of transgene copy number data in zygosity determination: Assays developed to determine transgene copy number in putative transgenics may be used to determine transgene zygosity during breeding programs. Traditionally in maize breeding programs, zygosity of individual transgenic lines selected for backcrossing is determined by screening the offspring derived from that individual. This requires pollination of each individual in the population and the planting and screening of several offspring from each individual in order to determine the zygosity of the parent. Owing to the specificity and quantitative nature of the transgene copy number assay described here, breeders can determine zygosity of each individual line as soon as the seedlings are large enough to sacrifice a small sample for analysis. This way, zygosity of each individual in a segregating population may be known prior to conducting pollinations. This allows the elimination of nulls and hemizygous plants from the population so that labor is expended for pollination of only homozygous individuals which will make up only approx 25% of the population (*see* **Notes 17** and **18**).
3. Discriminating true integration from *Agrobacterium* contamination: The chances are remote that the level of contaminating recombinant binary *Agrobacterium* DNA mimics the level of a low copy transgenic event in a TaqMan assay. Therefore, contaminating *Agrobacterium* is easily discriminated from a true transgenic integration event. In addition, the level of *Agrobacterium* detected in contaminated plants decreases over time, so in the rare occurrence that the contaminating level does mimic a true integration event, subsequent plant samples taken later in plant development will reveal the true molecular characteristics of the event (unpublished observation).

4. Notes

1. We have found chrome steel ball bearings to be the best choice in terms of efficient grinding and reduced cost. Other viable alternatives include glass, stainless steel and titanium balls of various sizes. The metal balls appear to be the most efficient at grinding (especially for difficult tissue) presumably owing to their high mass-to-volume ratio. Each laboratory should determine which material works best for their application.
2. It is not necessary to rinse or decontaminate sampling utensils (i.e., Leaf Punch) in between each sample. The TaqMan transgene copy assay is designed to be unaffected by small amounts of cross contamination between samples (*see* **Note 19**).
3. The benefits of shipment at room temperature should be weighed against the risks of cross contamination created by use of lyophilized sample material. Lyophilized tissue tends to become statically charged during the grinding process causing some tissue to "leap" from the well when the plate is uncapped.
4. Samples can be taken on ice or at room temperature. Freezing of samples in liquid nitrogen or dry ice is unnecessary. The reason for freezing samples is to avoid degradation of gDNA in samples. For PCR assays, some degree of degradation of gDNA is acceptable and usually beneficial (unpublished observation).

5. Dispensing of reagents in 96-well plates can be accomplished efficiently in a semi-automated fashion by utilizing instruments such as the Q-Fill2 (Genetix, cat. no. X3001). Alternatively, a more cost effective approach would utilize manual reagent dispensers such as the Stat-Matic Dispenser available from Sigma Aldrich (product no. S1061)
6. Supplement the supplied wash buffer by adding the appropriate amount of 95–100% Ethanol and Isopropanol. Typically, the alcohols are each added to a final concentration of 25% (v/v). For example, for the two plate kits from the manufacturer (Promega, product no. FF3760), add 20 mL each of 95–100% Ethanol and Isopropanol to the supplied 40 mL of wash buffer.
7. For maize, the average yield of gDNA prepared from approx 20 mg leaf samples using this procedure is approx 30–40 ng/μL.
8. Often in plant biotechnology, it is desirable to assay for two different transformation targets in addition to the endogenous control for any particular plant sample. For example, it is common to assay for both the plant selectable marker and the trait gene of interest. To increase the efficiency of the assay for these applications, we have successfully triplexed our TaqMan transgene copy number assay (unpublished data). This was accomplished using TaqMan probes that were all dark quenched (using DABCYL or BHQ as opposed to quenching by TAMRA). The reporter fluorophores should be selected to minimize the level of potential "crosstalk" between the various probes in the triplex reaction. As with the duplexed assay described above, no primer optimization is necessary and we recommend using the same concentrations described for duplexed assays (*see* **Subheading 3.5.**).
9. By using nonlimiting primer concentrations, we are promoting the highest efficiency for each reaction. By ensuring that both reactions are highly efficient, we are also ensuring that the reaction efficiencies for the duplexed assays are closely matched. This is a simpler and more accurate approach to matching efficiency than trying to adjust limiting primer concentrations to produce matched efficiencies ***(2,3)***.
10. Using nonlimiting primer concentrations, we have been unable to define a threshold that would indicate an assay that will be nonfunctional. In the great majority of assays, the validation results showed very high (and closely matched) efficiencies for both reactions (**Fig. 1**). Failed assays were obvious through qualitative analysis of the data.
11. We have found replicate reactions to provide little or no benefit to the TaqMan transgene copy number assay. This results from the fact that the variability is very low, we screen large number of samples and we do not require nor expect 100% accuracy. Inclusion of replicate reactions did not produce any detectable increase in the accuracy of the assay results.
12. Alternatively, we do recommend the use of replicate samples. This way, sample preparation variability is detected. Any failed preparations will also have a backup sample that will likely yield viable results. When using this approach, we recommend calculating a separate copy number for each replicate sample.

This provides multiple independent copy determinations for each transgenic event, which increases the confidence of the estimation of the transgene copy number. Because the assay normalizes to an endogenous control (DC_t calculation, **Subheading 3.7.**), it is not appropriate to average the C_t values of replicate samples.

13. We recommend optimizing the sample preparation procedure by evaluating performance of the purified samples in the transgene copy number assay rather than analyzing purified DNA by ultraviolet absorbance or using fluorescent intercalating dyes. We have found using the latter methods to determine quality of purification methods to be misleading and frequently incorrect (unpublished data). In contrast, we prefer testing sample preparation methods in the same type of experiment as used for assay validation (*see* **Subheading 3.4.**).
14. Using an assay validation experiment to test a sample preparation method will yield valuable information. First, the yield of DNA is determined empirically by the assay and is an indication of the amount of *amplifiable* DNA that is recovered. This is indicated by the C_t values generated by the endogenous control assay. For this assay, the undiluted sample should yield a C_t in the mid to low 20s. If this is not the case, the sample preparation method should be adjusted to increase or decrease the gDNA yield appropriately.
15. The assay validation experiment will detect the presence and relative abundance of any inhibitors of the assay that may be copurified. If inhibitors are present, the endogenous control C_t values will not increase by one cycle over the entire twofold dilution range. Instead, the C_ts for the initial dilutions will remain about the same or actually drop with each successive dilution until the inhibitors are diluted to a level where their effect is no longer observed (**Fig. 3**). Owing to the large linear dynamic range of the assay, samples my be recovered from the presence of inhibitors to yield reliable copy number information by diluting each sample appropriately as indicated by the dilution at which the affect of the inhibitor is no longer observed.
16. Using the approaches described in **Notes 8–10**, the researcher will have more specific information to allow them to efficiently troubleshoot the sample preparation procedure.
    a. If DNA yield is a problem, the researcher can adjusted amounts and/or elution volumes to compensate for a low or high yield.
    b. If inhibitors are a problem, the solution can be various of approaches.
        i. Several references exist purporting the use of polyvinyl pyrrolidone complexes (PVP or PVPP) for the removal of inhibitors of PCR during sample preparation *(8–10)*. Using the trouble shooting approach suggested here, we observe very low yields of amplifiable DNA when procedures incorporate the use of PVP or PVPP (unpublished observation). Therefore, the use of PVP and PVPP were found to be functionally equivalent to dilution of samples prepared without the use of these compounds.
        ii. As an alternative to the use of PVP or PVPP, it has been reported that the addition of *nonacetylated* bovine serum albumin (BSA) to PCR effectively

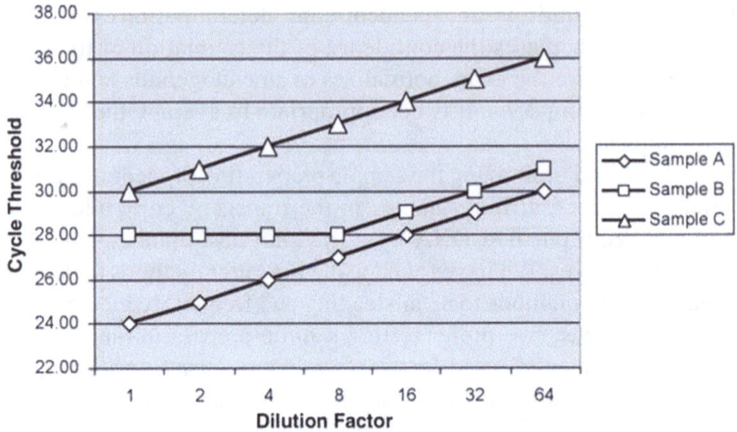

Fig. 3. Testing of sample preparation methods. Hypothetic data was plotted to represent results from TaqMan transgene copy number assay run on samples representing different levels of genomic DNA (gDNA) concentration and purity. Sample A (*diamonds*) represents a dilution series of a hypothetical sample with high relative yield of gDNA and no detectable presence of PCR inhibitors. Sample B (*squares*) represents a dilution series of a hypothetical sample with high relative yield of gDNA, but contains detectable levels of PCR inhibitory compounds. Sample C (*triangles*) represents a dilution series of a hypothetical sample with low relative yield of gDNA and no detectable presence of PCR inhibitors.

protects the polymerase from the effects of inhibitors that may be present *(11)*. This approach effectively dilutes the level of inhibitors in the reaction by the BSA protein acting as a sink for the inhibiting compounds, thus protecting the polymerase but not affecting the level of amplifiable gDNA. Very high levels of BSA can be tolerated in PCR and TaqMan reactions. We recommend adding BSA to a final concentration of 800 ng/μL in the PCR. The use of acetylated BSA as opposed to nonacetylated BSA may result in a failed reaction because of the carryover of the acetylation compounds into the PCR.

17. For zygosity analysis of a one-copy event, the segregating population should be comprised of approx 25% zero copy (null), 50% one-copy (hemizygous), and 25% two-copy (homozygous) genotypes. Similarly, a two-copy, *single locus* line would segregate as approx 25% zero copy (null), 50% two-copy (hemizygous), and 25% four-copy (homozygous) genotypes. Therefore, the TaqMan transgene copy assay is quite useful for zygosity determination for any low copy transgenic events segregating as a single locus. Higher copy (greater than approx four copies), single locus events have proven difficult presumably due to recombination occurring within the locus (unpublished observation).

18. The number of independent loci containing insertions of recombinant DNA may be easily inferred from the results of the zygosity assay described above. A single locus segregating in Mendelian fashion will yield approx 25% nulls, while a multilocus event would generate less than approx 6% null siblings. This can be reliably observed in populations as small as 20 siblings.

Acknowledgments

I would like to acknowledge the hard work, dedication, and creativity contributed by Mary Fielder, Carla Thomas, Jamie Huang, Philip Rivers, Leslie Ireland, Sandra Beer, Stephanie Money, Wenjin Yu, and Qing Zhou. Their combined efforts were essential to the development, implementation, and refining of the TaqMan transgene copy number assay.

References

1. Southern, E. M. (1975) Detection of specific sequences among DNA fragments separated by gel electrophoresis. *J. Mol. Biol.* **98,** 503–517.
2. Ingham, D. J., Beer, S., Money, S., and Hansen, G. (2001) Quantitative real-time PCR assay for determining transgene copy number in transformed plants. *BioTechniques* **31,** 132–140.
3. Mason, G., Provero, P., Vaira, A. M., and Accotto, G. P. (2002) Estimating the number of integrations in transformed plants by quantitative real-time PCR. *BMC Biotechnol.* **2,** 20.
4. Dellaporta, S. L. (1993) Plant DNA miniprep and microprep: version 2.1–2.3, in *The Maize Handbook* (Freeling, M. and Walbot, V., eds.), Springer-Verlag, New York, NY, pp. 522–525.
5. Bustin, S. A. (2002) Quantification of mRNA using real-time reverse transcription PCR (RT-PCR): trends and problems. *J. Mol. Endocrinol.* **29,** 23–39.
6. Bustin, S. A. (2000) Absolute quantification of mRNA using real-time reverse transcription polymerase chain reaction assays. *J. Mol. Endocrinol.* **25,** 169–193.
7. Ginzinger, D. G. (2002) Gene quantification using real-time quantitative PCR: an emerging technology hits the mainstream. *Exp. Hematol.* **30,** 503–512.
8. John, M. E. (1992) An efficient method for isolation of RNA and DNA from plants containing polyphenolics. *Nucl. Acids Res.* **20,** 2381.
9. Kim, C. S., Lee, C. H., Shin, J. S., Chung, Y. S., and Hyung, N. I. (1997) A simple and rapid method for isolation of high quality genomic DNA from fruit trees and conifers using PVP. *Nucl. Acids Res.* **25,** 1085–1086.
10. Pich, U. and Schubert, I. (1993) Midiprep method for isolation of DNA from plants with a high content of polyphenolics. *Nucl. Acids Res.* **21,** 3328.
11. Kreader, C. A. (1996) Relief of amplification inhibition in PCR with bovine serum albumin or T4 gene 32 protein. *Appl. Environ. Microbiol.* **62,** 1102–1106.

20

Analysis of Gene Expression in Transgenic Plants

Andrew F. Page and Subhash C. Minocha

Summary

Recent years have seen a huge increase and improvement in techniques for analysis of transgene expression in plants. The analysis of RNA frequently provides a vital link between changes in enzyme levels and/or metabolites and the phenotype. This chapter focuses on RNA-based techniques for the analysis of transgene expression, beginning with the extraction of RNA and its evaluation in terms of purity and integrity by spectrophotometry and gel electrophoresis, respectively. Common methods of transcript analysis by Northern and dot-blot hybridizations using nonradioactive probing methods are described. A protocol for reverse-transcriptase polymerase chain reaction (RT-PCR) as a method of establishing transgene expression qualitatively, as well as a procedure for quantitative RT-PCR for comparing relative abundance of transcript levels of two or more genes are described. Lastly, a protocol for localization of RNA transcripts within tissues by *in situ* hybridization is included.

Key Words: Digoxigenin; *in situ* hybridization; Northern blot hybridization; probe; quantitative RT-PCR; real-time PCR; reverse-transcriptase (RT) PCR; RNA; spectrophotometry.

1. Introduction

The study of transgene expression is of vital importance whenever transgenic plants are produced. Transgene expression levels are influenced by many factors, in particular the site of integration of the transgene within the plant genome, gene silencing, and the promoter attached. Although some of these factors can be circumvented to some degree in the experimental design, it is still necessary to correlate phenotypic differences between transgenic and control plants with transgene expression. To examine thoroughly the ultimate effect of transgenes, enzymes and metabolites must also be studied. Many techniques exist for the

From: *Methods in Molecular Biology, vol. 286: Transgenic Plants: Methods and Protocols*
Edited by: L. Peña © Humana Press Inc., Totowa, NJ

analysis of transgenic proteins (e.g., Western blotting, enzyme-linked immunosorbent assay [ELISA]), and specific techniques must be used for analysis of protein (enzyme) activity; however, this chapter focuses on the analysis of gene expression at RNA level, in particular the quantification and localization of mRNA transcripts. Numerous medium-to-high throughput analytical techniques are available to quantify the levels of mRNA transcripts of large numbers of genes *(1–4)*. Kuhn *(1)* published a comparison of these techniques based upon specific goals, facilities available, genome sequence information availability, and the ease of employment. However, these techniques are not suitable for examination of the expression of specific genes in small amounts of tissues, nor do they have the precision and power to allow localization of the expression of a particular gene in specific cells or tissues.

The following protocols for the analysis of transgene expression focus primarily on the analysis of one or more transgene transcripts in the presence or absence of native genes of high homology. Protocols are described for the extraction of RNA, establishing RNA purity and integrity, the use of Northern blotting, reverse transcriptase-polymerase chain reaction (RT-PCR) and quantitative or real-time RT-PCR for quantifying RNA, and *in situ* hybridization for studying tissue-level expression patterns. More details on the selection and suitability of and a comparison of various techniques for analysis of gene expression in general can be found in Jones *(5)* and Bartlett *(6)*.

1.1. RNA Extraction and Analysis

RNA extraction is frequently the stage at which the researcher most influences the success of an experiment. A well-organized and patient scientist can easily extract a large amount of good quality RNA, whereas less prepared individuals or those who are hurried are often disappointed. Careful planning is paramount, not only in preparing equipment and reagents ahead of time but also in maintaining an RNase free environment (*see* **Note 1**). The extraction method in **Subheading 3.1.1.** involves grinding tissues/cells in liquid nitrogen to break cells open, partitioning with chloroform to remove protein, and precipitation of RNA from the resulting solution using lithium chloride. **Subheading 3.1.2.** describes the quantification of RNA by spectrophotometry, and **Subheading 3.1.3.** contains a protocol for gel electrophoresis of RNA to establish integrity.

1.2. Transcript (mRNA) Analysis

The presence of a transgenic transcript can be detected as well as quantified in different tissues using numerous techniques that rely on the use of a labeled probe. The probe can be either single-stranded complementary RNA or double-stranded DNA; the latter being preferred because of its ease of use. The technique most often used for detection of the transcript is Northern blot hybridization, which employs a transgene-specific labeled probe and a variety of detection mechanisms depending on the label used. Although this approach does not distinguish between translationally active and inactive messages, it often is used reliably to study the expression levels of various transcripts. In brief, in this method, total RNA is isolated and separated by gel electrophoresis on 1% agarose containing formaldehyde. The RNA is transferred to a nylon membrane and hybridized with a labeled gene-specific probe. One of the several methods of detection is used for quantification of the specific band by densitometry. When comparing transgene expression in different tissues or plants, equal loading of the RNA is verified by ethidium bromide staining of a parallel gel. The protocol in **Subheading 3.2.** uses a nonradioactive labeling approach and a colorimetric or a chemiluminescent detection system. Nonradioactive detection systems, which have recently become as sensitive as the traditional ^{32}P-labeling systems, provide the advantage of convenience of use and minimal regulatory burdens. Various of nonradioactive labeling and detection systems are described in Kricka *(7,8)*, and are available from various vendors in the form of easy-to-use kits.

As an alternate to the Northern blot hybridization described above, a quick method of detecting the presence of a transgene transcript is dot-blot or slot-blot hybridization. This technique works reasonably well for transgenes that have little homology with a native gene in the plant, as the technique does not allow size analysis of the transcript. The method is very efficient for screening of transgene expression in large numbers of samples or for comparing the expression of a transgene in different organs, and in a population of transgenic plants. The procedure is quite similar to the standard Northern blot hybridization, except a gel is not run. This saves time as well as reduces the chances of RNA degradation.

1.3. RT-PCR

RT-PCR enables researchers to quickly identify plants that are expressing transgene(s), particularly if the transgene sequence shows low homology to the native gene(s). The availability of kits has made RT-PCR a fast and reliable technique for transgene expression analysis, circumventing the need for Northern analysis (*see* **Subheading 3.2.**). It is common for a population of putatively transgenic plants to be first screened by PCR for integration of the transgene(s), and then by RT-PCR for expression of these genes. As both methods are qualitative, other techniques may subsequently be used for quantification of transgene integration and expression, for example Southern blot analysis and Northern blot analysis or quantitative (real-time) RT-PCR (*see* **Subheading 3.4.**) (*see also* Chapter 19). RT-PCR is identical to conventional PCR, except that the template DNA is created by reverse transcription of RNA. This means that positive or negative PCR results (established by gel electrophoresis) represent the presence or absence of a RNA transcript in the original RNA sample and, therefore, is an indicator of specific gene expression.

1.4. Quantitative or Real-Time PCR

Quantitative or real-time PCR (QRT-PCR) is a highly sensitive technique for quantifying mRNA copy numbers of specific genes *(9)*. The method permits a direct measurement of products during the log-linear phase of the PCR reaction via the incorporation of a fluorescent probe in the PCR reaction mix and the use of a thermocycler equipped with an optical sensor for fluorescence quantification. Because sequence specific primers can be designed to amplify DNA from a target gene of known sequence, transcripts of genes that are closely related (e.g., a transgene with high homology to a native gene) can be independently quantified *(10,11)*. Using different dyes, up to four genes can be analyzed in the same PCR reaction. Thermocyclers formatted for 96-well plates allow rapid sample throughput; and because the reaction is monitored *in situ*, gel electrophoresis is avoided. This results in savings of time and money; at the same time accuracy is increased. The principle of the method is that as the sequence-specific PCR product accumulates, at a certain cycle number, the fluorescence of labeled product exceeds a threshold set by the operator. This is called the threshold cycle (C_t), which is inversely proportional to the initial number of cDNA target molecules. Thus, the smaller the C_t value (i.e., fewer number of cycles to reach detectable product), the greater the number of initial target molecules. By comparing C_t value for the target gene to that of an internal control gene (a constitutively expressed gene, for example, actin or glycerol-3-phosphate dehydrogenase—*g3pdh*), it is possible to determine its relative abundance in a starting population of mRNA.

Several approaches that use different chemistries are available for real-time PCR *(9,12,13)*. The probe is often a short linear molecule that matches a region in the PCR product, and is attached to both a fluorescent dye and a quencher molecule (e.g., Taqman chemistry of Applied Biosystems; ***11,13***). The close proximity of the quencher moiety to the fluorescent dye results in most of the fluorescence being quenched following optical excitation (**Fig. 1**). The probe binds to a segment of DNA between the two PCR primers at the PCR extension temperature. During DNA extension, the 5' exonuclease activity of *Taq* polymerase cleaves the dye from the quencher, allowing the free dye molecule to fluoresce when excited. At each PCR cycle, the free dye concentration increases, causing increased fluorescence proportional to PCR product formation.

Other techniques rely on PCR probes that are designed to fold into a hairpin configuration. In their linear configuration, these "molecular beacons" have a fluorescent dye at one end and a quencher at the other (**Fig. 2**). When free and folded the hairpin configuration, the quencher is close to the dye; thus prevent-

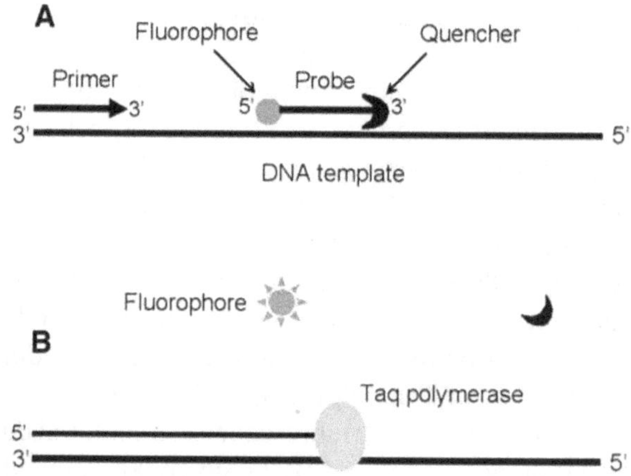

Fig. 1. The Taqman-labeled probes for QRT-PCR. The probe bears a fluorophore at the 5' end and a quencher at the 3' end. Probe cannot fluoresce when bound to DNA template because of the proximity of the fluorophore to the quencher (**A**). During amplification, the 5'- to 3'-exonuclease activity of the *Taq* polymerase (gray oval) cleaves the probe and releases the fluorophore into solution (**B**), where it is free to fluoresce.

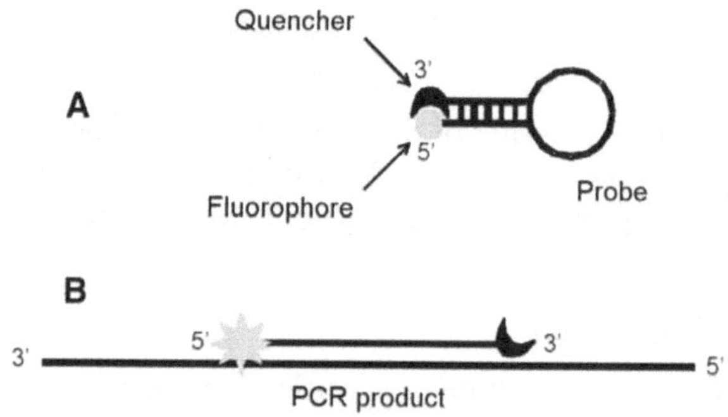

Fig. 2. The hairpin probe procedure of QRT-PCR. The probe consists of a sequence-specific region (loop) flanked by short complementary sequences that form a hairpin with the fluorophore and the quencher at the ends, keeping the fluorophore in close proximity to the quencher (**A**). When the molecule anneals to its target DNA, it unfolds and causes the fluorophore to be further from quencher, resulting in fluorescence.

ing fluorescence. However, when the molecule hybridizes to the PCR product of interest, the fluorescent dye and the quencher are separated; the fluorescence can be detected and is proportional to the amount of PCR product. Commercially available probes cost approx $1–2 per PCR reaction (*see* Website: http://molecular-beacons.org).

The QRT-PCR procedure described in **Subheading 3.4.** is modified from Carleton and Kocher ***(11)*** and has been used successfully in our laboratory for quantification of mRNA for *S*-adenosylmethionine decarboxylase transcripts in carrot ***(14)***. Specific primers and probes are designed for each gene using Primer Express, ver. 1.5 (Applied Biosystems, Foster City, CA) to amplify short (60–90 bp) fragments based on available sequences. To prevent false signals from genomic DNA, primers are designed to overlap exon–exon boundaries wherever possible, however, DNase treatment of RNA can also be used. The probes used here are 5'-labeled with 6'-FAM (6-carboxyfluorescine) and 3'-labeled with TAMRA (6-carboxy-tetramethyl rhodamine). In principle, during PCR, the 5'→3' exonuclease activity of Taq polymerase releases the 3'-labeled TAMRA dye molecule. The relative TAMRA fluorescence is monitored during 40 cycles of PCR on a detection system (e.g., ABI Prism 5700, Applied Biosystems). Critical cycle number C_t is determined when the fluorescence exceeds a threshold set close to the background fluorescence. Relative gene expression for a gene of interest is determined with respect to a normalization gene such as g3pdh according to the following equation:

$$\frac{T_i}{T_{norm}} = \frac{(1 + E_{norm})^{C_{t_{norm}}}}{(1 + E_i)^{C_{ti}}} \tag{1}$$

where T_i/T_{norm} is the expression of the gene of interest (the template amount) corrected by the expression of the normalization gene, E is the PCR efficiency for each gene and its primers, and C_t is the critical cycle number for each gene (*see* **Note 2**).

1.5. In Situ *Hybridization*

In addition to the quantification of specific transcripts in different organs of the plant, semiquantitative localization of transcripts in different tissues can be accomplished by *in situ* hybridization of labeled single-stranded, antisense probes to specific mRNA sequences in thin sections of plant tissues/organs. This technique is especially valuable when a developmentally regulated and/or tissue-specific promoter is used to regulate transgene expression. Tissue sections from treated and control transgenic plants are cryo-immobilized by fast-freezing and freeze substitution, the tissue embedded in methacrylate resin and sections hybridized with a DIG-labeled single-stranded (complementary to the mRNA) probe of the respective gene ***(15)***. An alternate technique of two-color *in situ* hybridization using two gene-specific RNA probes labeled with different tags provides an extremely powerful tool for comparing the spatial expression patterns of two genes in a specific tissue/organ; for example, expression of the selective marker gene and the gene of interest ***(16)***. The technique involves the production of two probes, one labeled with digoxigenin (DIG) and the other with dinitrophenol (DNP), performing standard *in situ* hybridization, and detection of the two probes independently using alkaline phosphatase (AP)-conjugated anti-DIG antibodies with BM purple as substrate for AP, and *p*-iodonitrotetrazolium violet/X-Gal 4 toluidine (INT RED/BCIP; Sigma, St. Louis, MO) for DNP. The INT RED/BCIP staining allows a clear single cell localization of the transcript. This method has been shown to be superior to similar methods using fluorescein or biotin as a second label ***(17)***. The protocol described here is based on that of Angerer et al. ***(15)***.

2. Materials

Important: Read **Notes 1** and **2** before preparing solutions.

2.1. RNA Isolation and Analysis

2.1.1. RNA Isolation

1. Liquid nitrogen.
2. Extraction buffer: 2% sodium dodecyl sulfate (SDS), 1% polyvinylpyrrolidone-10 (PVP-10), 1.5 *M* NaCl, 25 mM ethylenediamine tetraacetic acid (EDTA), 0.2 *M* Tris-HCl, pH 8.0, 2% β-mercaptoethanol (added just before use).
3. 30-mL centrifuge tubes (polytetrafluoroethylene [PTFE], Nalgene).
4. Chloroform–isoamyl alcohol (24:1).
5. 10 *M* Lithium chloride.
6. 80% Ethanol.
7. Diethylpyrocarbonate (DEPC)-treated water.

2.1.2. Determination of RNA Purity and Quantity by Spectrophotometry

1. UV/VIS spectrophotometer.
2. Quartz cuvets.
3. DEPC-treated water.

2.1.3. Determination of RNA Integrity by Gel Electrophoresis

1. 95% Ethanol.
2. RNase Zap (Ambion, Austin, TX).
3. DEPC-treated water.
4. Agarose.
5. 12.3 *M* Formaldehyde.
6. Formamide.
7. 10X MOPS–EDTA buffer: 0.5 *M* 3-(*N*-morpholino)propaneslufonic acid (MOPS), 10 m*M* EDTA, pH 7.0, store at 4°C.
8. Gel running buffer: 1X MOPS–EDTA buffer (from above, diluted with DEPC-treated water) containing 2.2 *M* formaldehyde (final conc.).

9. 10X Loading dye: 50% (v/v) glycerol, 1 m*M* EDTA, pH 8.0, 0.25% (w/v) xylene cyanol FF, 0.25% (w/v) bromophenol blue.
10. RNA molecular weight markers of appropriate size.
11. 0.5 mg/mL of ethidium bromide.
12. UV light box (transilluminator).

2.2. Transcript (mRNA) Analysis

2.2.1. Denaturing Gel Electrophoresis

See **Subheading 2.1.3.**

2.2.2. Blotting and Transfer

1. DEPC-treated water.
2. 20X Saline sodium citrate (SSC): 175.3 g/L of NaCl, 88.2 g/L of trisodium citrate, pH 7.0.
3. Nylon membranes (0.22 µm).
4. 95% Ethanol.
5. RNase Zap.
6. Filter paper (e.g., Whatman no. 1) or UV crosslinker.

2.2.3. Nonradioactive Probe Preparation

1. Probe labeling kit (e.g., Digoxigenin DNA Labeling and Detection Kit; Roche Applied Science, Indianapolis, IN).
2. PCR product of 600–1200 bp for gene of interest.
3. Agarose.
4. DEPC-treated water.
5. Gel purification kit (e.g., QIAquick Gel Extraction Kit [Qiagen Inc., Valencia, CA] or GFX™ PCR and Gel Band Purification Kit [Amersham Biosciences Corp., Piscataway, NJ]).
6. Either DIG High Prime Labeling Mix from the kit or hexanucleotide mix, dUTP labeling mix, and Klenow enzyme.
7. 0.2 *M* EDTA pH 8.0.

2.2.4. Probe Quantification

1. Control DIG-labeled probe (e.g., from Roche).
2. Nylon membrane (0.22 µm).
3. Filter paper (e.g., Whatman no. 1) or UV crosslinker.
4. Washing buffer 3: maleic acid buffer, 0.3% Tween-20.
5. Blocking solution (e.g., from Roche).
6. Antibody solution: anti-DIG-alkaline phosphatase conjugate in fresh 1X blocking solution in a ratio of 1:10,000, i.e., 4 µL in 40 mL of blocking solution.
7. Detection buffer: 0.1 *M* Tris-HCl, 0.1 *M* NaCl, pH 9.5.
8. Color substrate solution: nitroblue tetrazolium (NBT)–5-bromo-4-chloro-3-indolyl phosphate (BCIP) or chemiluminescent detection kit.

2.2.5. Hybridization

1. Hybridization oven and bottles.
2. Prehybridization solution: 5X SSC, 0.1% Sarkosyl, 0.2% SDS, 1X blocking solution.
3. Hybridization solution: probe diluted to appropriate concentration (e.g., 1:10,000) in prehybridization solution.
4. Washing buffer 1: 2X SSC, 0.1% SDS.
5. Washing buffer 2: 0.1X SSC, 0.1% SDS.
6. Washing buffer 3: maleic acid buffer, 0.3% Tween-20.
7. Maleic acid buffer: 0.1 *M* maleic acid, 0.15 M NaCl, pH 7.5 (use approx 8 g/L of NaOH pellets).
8. Blocking solution (e.g., from Roche).
9. Antibody solution: anti-DIG-AP conjugate in fresh 1X blocking solution in a ratio of 1:10,000, i.e., 4 µL in 40 mL of blocking solution.
10. Detection buffer: 0.1 *M* Tris-HCl, pH 9.5, 0.1 *M* NaCl.
11. Color substrate solution: NBT–BCIP.

2.2.6. Dot-Blot Hybridization

1. 10X MOPS–EDTA buffer: 0.5 *M* 3-(*N*-morpholino)propaneslufonic acid (MOPS), 10 m*M* EDTA, pH 7.0, store at 4°C.
2. Denaturing solution: 500 µL of 100% formamide, 162 µL of 37% formaldehyde, and 100 µL of MOPS–EDTA buffer.
3. 20X SSC: 175.3 g/L of NaCl, 88.2 g/L of trisodium citrate, pH 7.0.
4. Blotting equipment (e.g., Vacuum Manifold [Schleicher & Schuell, Keene, NH]).
5. Parafilm or Nescofilm.

2.3. RT-PCR

1. DNase (e.g., RQ1 RNase-free DNase from Promega, Madison, WI), 10X reaction buffer, stop solution.
2. Oligo-$(dT)_{15}$ solution (e.g., from Promega).
3. 10 m*M* dNTP mix.
4. 5X First-strand buffer: 250 m*M* Tris-HCl, pH 8.3, 375 m*M* KCl, 15 m*M* $MgCl_2$.
5. 0.1 *M* Dithiothreitol (DTT) (make fresh).
6. RNasin (Promega).
7. Reverse transcriptase (e.g., SUPERSCRIPT™ II RNase H$^-$ Reverse transcriptase; Invitrogen, Carlsbad, CA).

2.4. Quantitative or Real-Time PCR

1. Sample of complementary DNA (cDNA) (*see* **Subheading 3.3.** for protocol).
2. Primers and probes for gene of interest and a gene for normalization. These may be designed using Primer Express, ver. 1.5 (Applied Biosystems). Probes should be 5'-labeled with 6'-FAM and 3'-labeled with TAMRA.
3. TaqMan® Universal PCR Master Mix (Applied Biosystems).
4. Appropriate equipment and tubes.

2.5. In Situ *Hybridization*

2.5.1. *Preparation of the Riboprobe*

1. Appropriate restriction enzyme to linearize DNA if in plasmid form.
2. Phenol–chloroform (1:1).
3. 100% and 70% ethanol.
4. T3 or T7 polymerase with buffer and riboNTPs (e.g., from Stratagene, La Jolla, CA).
5. DIG-11-UTP (e.g., from Stratagene).
6. Ribonuclease inhibitor (e.g., RNasin from Promega).
7. Yeast tRNA (e.g., from GIBCO-BRL Grand Island, NY).
8. DNase (e.g., RQ1 RNase-free DNase from Promega), buffer, and stop solution.
9. 4 *M* Ammonium acetate.

2.5.2. *Preparation of Plant Material*

1. Phosphate-buffered saline (PBS) buffer: 130 m*M* NaCl, 70 m*M* Na_2HPO_4, 30 m*M* NaH_2PO_4, pH 7.0.
2. Paraformaldehyde (PFA) fixative solution: 4% PFA in PBS buffer.
3. Ethanol: 30%, 40%, 50%, 60%, 70%, 85%, 95%, and 100%.
4. EosinY (Sigma).
5. Hemo-De (Fisher Scientific, Pittsburgh, PA): 25%, 50%, and 75% in ethanol.
6. Paraplast chips (Oxford Labware, St. Louis, MO).
7. Aluminum weighing boats.
8. Hot plate.
9. Plastic cassette mold for sectioning.
10. Fisher ProbeOn Plus slides (Fisher Scientific).

2.5.3. In Situ *Pretreatment*

1. Hemo-De.
2. 100%, 95%, 90%, 80%, 60%, and 30% ethanol.
3. DEPC-treated water.
4. 20X SSC: 175.3 g/L of NaCl, 88.2 g/L of trisodium citrate, pH 7.0.
5. Proteinase K: 1 µg/mL in 100 m*M* Tris-HCl, pH 8, 50 m*M* EDTA.
6. Glycine: 2 mg/mL in PBS.
7. Triethanolamine–acetic anhydride (prepare 0.1 *M* triethanolamine in DEPC-treated water, adjust to pH 8.0 using NaOH, add 0.5% [v/v] acetic anhydride just before use).

2.5.4. In Situ *Hybridization*

1. Formamide.
2. 10X *In situ* salt solution: 3 *M* NaCl, 100 m*M* Tris-HCl, pH 8.0, 100 m*M* sodium phosphate, 50 m*M* EDTA, pH 8.0, store at –20°C.
3. Denhardt's solution 50X (Sigma).
4. Hybridization solution (200 µL): 80 µL of formamide, 10X *in situ* salt solution (20 µL), dextran sulfate (10%, [w/v]—20 mg), Denhardt's solution (1X; 4 µL),

yeast transfer RNA (tRNA) (10 µL of 20 mg/mL), denatured RNA probe (40 µL), DEPC-treated water (46 µL).

2.5.5. In Situ *Posthybridization*

1. 20X SSC: 175.3 g/L of NaCl, 88.2 g/L of trisodium citrate, pH 7.0.
2. NTE: 0.5 *M* NaCl, 10 m*M* Tris-HCl, pH 8.0, 1 m*M* EDTA.
3. 20 µg/mL of RNase A in NTE.
4. Bovine serum albumin (BSA) solution: 1% BSA in 100 m*M* Tris HCl, pH 7.5, 150 m*M* NaCl, 0.3% Triton X-100.
5. Anti-DIG antibody in BSA solution (1:1250).
6. Detection buffer: 100 m*M* Tris HCl, pH 9.5, 100 m*M* NaCl, 50 m*M* $MgCl_2$.
7. Western Blue substrate solution (Promega).
8. TE buffer: 10 m*M* Tris-HCl, pH 8.0, 1 m*M* EDTA ***(8)***.
9. Cytoseal 60 (Stephens Scientific, Riverdale, NJ).
10. 1% Blocking reagent (Roche) in 100 m*M* Tris HCl, pH 7.5, 150 m*M* NaCl.
11. 30%, 50%, 70%, 95%, and 100% ethanol.
12. Hemo-De.

3. Methods

3.1. RNA Isolation and Analysis

3.1.1. RNA Isolation

The following protocol was modified from Mason and Schmidt ***(18)*** and has been used successfully used in our laboratory for isolation of total RNA from poplar cell suspension cultures. The technique is equally suitable for RNA extraction from leaf and other tissues. Typical yields are 400–500 µg/g FW of cells.

1. Collect 0.5–1 g of plant tissue/cells and freeze in liquid nitrogen.
2. Grind to powder with a prechilled mortar and pestle, adding liquid nitrogen as necessary to prevent the sample from thawing.
3. Add the sample to 5 mL of heated (60°C) Extraction buffer in a 30-mL PTFE centrifuge tube. Vortex-mix vigorously for 30 s.
4. Add an equal volume of chloroform–isoamyl alcohol (24:1), vortex-mix vigorously for 1 min. The sample can now remain at room temperature while other samples are brought to this stage, provided the waiting sample is vortex-mixed every few minutes.
5. Centrifuge at 4°C for 5 min at 10,000*g*. (At this point, RNase-free supplies and solutions must be used; *see* **Notes 1** and **3**).
6. Transfer the aqueous (upper) layer to a new PTFE tube and repeat the chloroform: isoamyl alcohol twice, vortex-mixing and centrifugation each time.
7. Transfer the aqueous layer to a new tube and add 0.25 volumes of 10 *M* LiCl. Incubate overnight at 4°C.
8. Centrifuge at 4°C for 20 min at 5000*g*, and discard the supernatant.

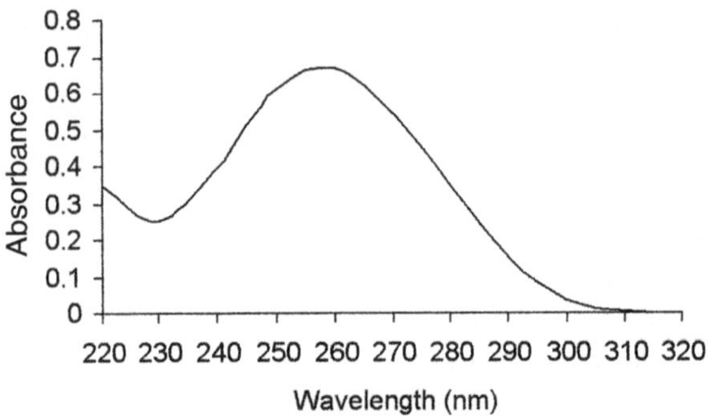

Fig. 3. Typical absorbance spectrum of nucleic acids, featuring a trough at approx 230 nm, a peak at 260 nm, and a tail approaching zero by about 310 nm.

9. Wash the pellet with 1 mL of 80% ethanol, centrifuge at 4°C for 5 min, and again discard the supernatant.
10. Dry the pellet in air or in a vacuum centrifuge, and resuspend in 50 µL of DEPC-treated water (*see* **Note 4** for storage temperatures).

3.1.2. Determination of RNA Purity and Quantity by Spectrophotometry

1. Zero the spectrophotometer using DEPC-treated water. Mix 4 µL of RNA sample with 996 µL DEPC-treated water in a 1-mL quartz cuvet. A smaller cuvet may be used with appropriate adjustment of volume.
2. Measure the absorbance at 230, 260, and 280 nm, or if possible scan within a range of 220–320 nm.
3. Nucleic acids should give a classic absorbance spectrum, featuring a trough at approx 230 nm, a peak at 260 nm, and a tail approaching zero by approx 310 nm (**Fig. 3**).
4. As the A_{260} of a 40 µg/mL of RNA solution is 1 ***(19)***, the concentration of RNA in µg/µL can be calculated by multiplying the A_{260} by 10.
5. As proteins absorb maximum at 280 nm, protein contamination can be detected by comparing the A_{260}/A_{280} ratio. Values of 1.7–2.1 indicate clean RNA.
6. As carbohydrates absorb maximum at 230 nm, carbohydrate contamination can be detected by comparing the A_{260}/A_{230} ratio. Values of about 2.0 indicate clean RNA.

3.1.3. Determination of RNA Integrity by Gel Electrophoresis

The following protocol is based on that in Sambrook et al. ***(19)*** and is suitable for a 7 × 8.5 cm gel.

1. Wash the gel box with detergent, rinse with 95% ethanol, air-dry, treat with RNase Zap, and then wash with DEPC-treated water.
2. Prepare the gel in a fume hood as follows: Mix 0.36 g of agarose with 3 mL of 10X MOPS–EDTA buffer in a 125-mL flask, add 21.6 mL of DEPC-treated water, and

heat in the microwave until agarose is dissolved. Allow to cool to 70°C, add 5.4 mL of 12.3 *M* formaldehyde stock solution (final concentration of formaldehyde is 2.2 *M*), and pour the gel immediately.

3. Prepare the RNA samples for electrophoresis by mixing the following components: 4.5 µL of total RNA (up to 30 µg), 10 µL of formamide (final concentration of formamide is 50% v/v), 2 µL of 10X MOPS–EDTA buffer, and 3.5 µL of 12.3 *M* formaldehyde.
4. Denature the samples for 15 min at 65°C, then cool on ice. Add 2 µL of 10X loading dye to each sample (total volume of each sample loaded is 22 µL).
5. When the gel is set, cover to a depth of approx 2 mm in 1X running buffer, and run the gel for 5 min at 5 V/cm.
6. Turn off the current and load the gel, alongside RNA markers of known size. Run the gel for about 60 min at 4–5 V/cm. Stain for 15 min in 0.5 mg/mL of ethidium bromide, destain in water for 5 min, and visualize under UV light.
7. The 28S ribosomal subunit should be approximately twice as bright as the 18S subunit. Degradation may have occurred if this is not observed, or if smears are present.

3.2. Transcript (mRNA) Analysis

3.2.1. Denaturing Gel Electrophoresis

1. Denaturing gel electrophoresis should be performed as described in **Subheading 3.1.3.**, except that duplicates of each sample should be loaded in separate lanes in the two halves of the gel, so that half can be cut and stained and the other used for transfer to a membrane.
2. Run the gel at 3–4 V/cm for 2–3 h.
3. Cut the gel in half and stain half for 30 min in ethidium bromide to check for RNA integrity. Take a picture of the gel with a fluorescent ruler next to the ladder to mark the size of the ribosomal bands and to determine the transcript size of the samples.

3.2.2. Blotting and Transfer

1. Wash the other half of the gel three times (15 min per wash) in 150 mL each of DEPC-treated water under constant gentle agitation.
2. Equilibrate the gel for 45 min in 250 mL of 10X SSC with gentle but constant agitation.
3. Transfer the RNA to a 0.22-µm nylon membrane overnight (e.g., using the Turbo blotter kit—Schleicher and Schuell) and 10X SSC. Prior to setup, wash the blotter with detergent, treat with 95% ethanol, air dry, treat with RNase Zap and rinse with DEPC-treated water to make it RNase free (*see* **Note 4**).
4. Following transfer, fix the RNA to the membrane by baking at 80°C for 1 h between two clean sheets of filter paper. Alternatively, use a UV crosslinker. Perform prehybridization and hybridization as described in **Subheadings 3.2.5.** and **3.2.6.**; *see also* **Note 6**.

3.2.3. Nonradioactive Probe Preparation

The non-radioactive probe described here is prepared by using the Digoxigenin (DIG) DNA Labeling and Detection Kit (Roche Applied Science). The PCR product (0.6–1.2 kb) for the appropriate gene is labeled by random-primed incorporation of DIG-labeled deoxyuridine-triphosphate (DIG-11-dUTP) according to the protocol suggested by the manufacturer. Numerous of other labeling and detection techniques are described in Kricka *(7,8)*, and are available from various vendors.

1. Pool PCR product of three or four tubes, and run a 1% agarose gel (wider slot to accommodate larger volume).
2. Cut out the desired band and purify the DNA, following the procedure suggested by the kit manufacturer.
3. Dissolve the DNA in water and quantify the amount. Probe making requires 0.5–3.0 μg of DNA.
4. For making the probe, denature the DNA (1–3 μg) in a boiling water bath or thermocycler for 10 min and chill on ice for 5 min.
5. To a 0.5 mL microfuge tube on ice, add the following reagents: DNA template from **step 4** (15 μL), 2 μL of hexanucleotide mix, 2 μL of dUTP labeling mix, and 1 μL of Klenow DNA polymerase. Alternatively, 16 μL of DNA template and 4 μL of DIG high prime labeling mix may be used.
6. Vortex and briefly centrifuge, then incubate at 37°C for 20 h.
7. Stop the reaction by adding 2 μL of 0.2 *M* EDTA and/or by heating at 65°C for 10 min.

3.2.4. Probe Quantification

1. Add 1.0 μL of the DIG-labeled probe to 9.0 μl of water to make a 10^{-1} (1:10) dilution.
2. Sequentially prepare three other dilutions (10^{-2}, 10^{-3}, and 10^{-4}) by adding 1.0 μL of the previous dilution to 9.0 μL of H_2O. Make similar dilutions of the control provided in the kit.
3. Spot 1.0 μL of the probe dilutions and the control dilutions (1 ng/μL, 100 pg/μL, 10 pg/μL, and 1 pg/μL) on a 2.5 × 5.0 cm, 0.22-μm nylon membrane.
4. Fix DNA to the membrane by baking the membrane between two clean sheets of Whatman filter paper at 80°C for 30 min or using a UV crosslinker.
5. Wet the membrane in washing buffer 3, agitate in blocking solution for 10 min, and transfer to antibody solution for 10 min.
6. Wash the membrane twice with washing buffer 3 for 5 min each and incubate in 20 mL detection buffer to equilibrate the membrane.
7. Place the membrane in detection buffer containing the color substrate (NBT–BCIP) (40 μL of substrate in 2 mL of detection buffer) in the dark without shaking until color develops. Alternatively, a chemiluminescent detection system can be used.

8. The probe concentration is estimated by exposing photographic film to the membrane (in case of chemiluminescent detection) and comparing the spot intensities with those of the control DNA provided in the kit. Various image analysis software can be used to assist in quantification of the resulting spots. The probe of appropriate dilution is added to the hybridization solution for Southern and Northern blot hybridizations and reused three to four times.

3.2.5. *Hybridization* (see **Note 7**)

1. Place the membrane in a 50 mL of hybridization bottle (with the side to which the RNA was transferred facing inward).
2. Add 20 mL of prehybridization solution and incubate at 65°C for 2–3 h.
3. Replace the prehybridization solution with 10 mL of hybridization solution.
4. Incubate on a roller overnight at 65°C.
5. Wash the membrane twice for 5 min each at room temperature in washing buffer 1.
6. Wash the membrane twice for 5 min each at 68°C in washing buffer 2 (prewarm buffer before use).
7. Remove the membrane from the hybridization bottle and transfer it to a polypropylene (e.g., Tupperware) tray containing washing buffer 3; incubate for 2–5 min.
8. Replace washing buffer 3 with 40 mL of blocking solution.
9. Replace the blocking solution with antibody solution and incubate for 30 min.
10. Wash the membrane twice for 15 min each in 50 mL of washing buffer 3.
11. Incubate the membrane for 2–5 min in 20 mL of detection buffer.
12. Finally, incubate the membrane in 10 mL of color substrate solution (200 μL of NBT–BCIP per 10 mL of detection buffer). Leave it undisturbed in the dark to allow color to develop until desired bands appear.
13. Take picture of the membrane and store the membrane in clear plastic wrap after air-drying.

3.2.6. *Dot-Blot Hybridization*

The following procedure uses a Schleicher & Schuell Vacuum Minifold and is modified from the manufacturer's instructions.

1. Mix 10–20 μg of (DNase-treated) total RNA with three volumes of denaturing solution.
2. Incubate the samples at 65°C for 15 min and chill the samples on ice.
3. Add two volumes of ice-cold 20X SSC to each of the denatured samples.
4. Assemble the dot-blot or slot-blot apparatus according to the manufacturer's instructions.
5. Load the samples on the dot-blot apparatus.
6. Cover the slots that are not used with Parafilm and seal them tight.
7. Apply a vacuum and wash the samples three times with 1 mL of 10X SSC each.
8. Disassemble the apparatus, remove the membrane and bake it at 80°C for 1 h.
9. From this point on, treat the membrane the same way as described earlier for Northern blot hybridization.

3.3. RT-PCR

The following procedure uses SUPERSCRIPT™ II RNase H$^-$ Reverse Transcriptase (Invitrogen) for making cDNA from total RNA:

1. Sequentially add 1 µL of 10X reaction buffer, 1 µL of RQ1 RNase-free DNase, 1–5 µg of total RNA, and DEPC-treated water to a final volume of 10 µL.
2. Incubate at 37°C for 30 min.
3. Add 1 µL of RQ1 DNase stop solution; incubate at 65°C for 10 min.
4. Add 1 µL oligo-$(dT)_{15}$ solution and 1 µL of dNTP mix, heat to 65°C for 5 min in a thermocycler, and then place on ice.
5. Add 4 µL of 5X first-strand buffer, 2 µL of DTT, and 1 µL of RNasin; incubate for 2 min at 42°C.
6. Add 1 µL (200 U) of SUPERSCRIPT™ II RNase H$^-$ Reverse Transcriptase, incubate for 50 min at 42°C and inactivate at 70°C for 15 min.
7. Use 2 µL of the reaction mix for a standard PCR reaction.
8. Run a 1% agarose gel and stain with ethidium bromide to detect the band of expected size. For storage, *see* **Note 3**.

3.4. QRT-PCR (see **Notes 2 and 8**)

1. Make cDNA from RNA as described in **Subheading 3.3.**
2. In a 0.2-mL optical microfuge tube, sequentially mix 1.0 µL of cDNA, 9 pmol each of both the forward and the reverse primers (1–2 µL), 6 pmol of the probe (1–2 µL), 15 µL of TaqMan® Universal PCR Master Mix, and sterile-distilled H_2O to a final volume of β L.
3. Set the PCR profile in the thermocycler, for example, 2 min at 50°C, 10 min at 95°C (for activation of AmpliTaq enzyme), followed by 40 cycles of 15 s at 95°C, 30 s at 55°C and 1 min at 65°C.
4. Determine the threshold level using an amplification plot, which is cycle number (*x*-axis) plotted against fluorescence signal intensity (ΔR_n). The threshold level (usually 0.05–0.1) is set at a ΔR_n within the linear range for all template concentrations. The C_t values are then calculated from the threshold level.
5. Determine the efficiency of the primers using a dilution series for each gene (1, 0.1, 0.01, 0.001, 0.0001, and 0.00001 pg), performing the PCR as described above, and then graphing the log (concentration) (*x*-axis) against the C_t value (*y*-axis). The efficiency is then calculated using the slope of the line in the following equation:

$$E = 10^{\frac{-1}{\text{slope}}} \tag{2}$$

This compensates for differences between primer and probe efficiencies for different genes, which would otherwise result in inaccurate template estimation. For example, the equation to calculate the ratio of *samdc* to *g3pdh* is:

$$\frac{T_{sam}}{T_{g3p}} = \frac{E^{C_{t_{g3p}}}}{E^{C_{t_{sam}}}} \tag{3}$$

where T = template amount, E = efficiency, and C_t = cycle number at threshold.

6. Use these values to quantify levels of gene expression as described in **Subheading 1.4.**

3.5. In Situ *Hybridization*

3.5.1. Preparation of the Riboprobe

1. Linearize the transgene DNA, if in plasmid form, by digestion with a restriction enzyme, and purify DNA by phenol–chloroform (1:1) treatment followed by ethanol precipitation.
2. Transcribe DNA in a 25-µL reaction containing 1 µL of buffer, 0.5 m*M* each of riboCTP, riboATP and riboGTP, 0.25 m*M* riboUTP and DIG-11-UTP, 2 µg of linearized template DNA, 25 units RNasin ribonuclease inhibitor, 0.4 U T3 or T7 RNA polymerase. Incubate the reaction mix for 45 min at 37°C, add an additional 0.2 µL of the appropriate RNA polymerase, and incubate for an additional 45 min at 37°C. Dilute reaction with 75 µL of distilled water and add 100 µg of yeast tRNA.
3. Add 5 U of RNase-free DNase and incubate at 37°C for 10 min to remove DNA template.
4. Add 100 µL of 4 *M* ammonium acetate and 400 µL of 100% ethanol, and centrifuge at 16,000 µg for 10 min.
5. Wash pellet with 70% ethanol, dry in a vacuum centrifuge, resuspend in 100 µL of distilled water and store at –20°C.
6. Perform denaturing gel electrophoresis as in **Subheading 3.1.3.**, then a Northern transfer and process the resultant membrane as in **Subheading 3.2.2.** to confirm the identity of the probe.
7. Determine probe concentration by spotting 1 µL of serial dilutions (1:10–1:10^4) of probe onto a nitrocellulose or nylon membrane, followed by 1-µL dilutions of a DIG-labeled standard (1 ng/µL–0.1 pg/µL). Fix the membrane by baking at 80°C for 60 min or using a UV crosslinker and determine the probe concentration by chemiluminesence and exposure of photographic film as described in **Subheading 3.2.4.**
8. Determine probe specificity by probing digested DNA containing the gene of interest, and other DNA samples, preferably containing similar sequences.

3.5.2. Preparation of Plant Material

1. Place tissues or whole seedlings in vials containing 10 mL paraformaldehyde (PFA) fixative solution; incubate on a rotary shaker at 100 rpm, 4°C overnight. This fixation step will crosslink proteins and preserve tissue structure.
2. Process the tissue through the following washes: two washes of 30 min each in PBS at 4°C, followed by 1 h each in 30, 40, 50, 60, 70, and 85% ethanol, all at 4°C. Subsequent washes should be performed at room temperature on a rotary shaker at 100 rpm. Overnight wash in 95% ethanol, 0.2% (w/v) EosinY. Four treatments with 100% ethanol, 0.2% EosinY (2 × 30 min, 2 × 60 min), followed by washes in 25, 50, and 75% Hemo-De in ethanol. Wash samples twice with

100% Hemo-De for 60 min each. After the second wash, add 0.25 vol of Paraplast chips and incubate at 37°C overnight.

3. Add an additional 0.25 vol of Paraplast chips and incubate at 37°C for several hours until the wax has completely melted. Replace wax with fresh liquid wax and incubate samples at 60°C, replacing wax 6–12 times over 3–8 d.
4. Pour material in liquid wax into aluminum weighing boats on a 55°C hot plate. Orientate samples before removing from hot plate and allowing samples to cool. Either trim and mount immediately or store at 4°C.
5. Following trimming with a razor blade, mount sample onto a plastic cassette mold and cut 8-μm sections. Place sections "shiny side" down on a Fisher ProbeOn Plus slide, add a few drops of DEPC-treated water and incubate at 42°C on a slide warmer overnight.

3.5.3. In Situ *Pretreatment*

Place slides in a slide rack and incubate with Hemo-De (two washes, 10 min each), ethanol 100, 100, 95, 90, 80, 60, and 30% (1–2 min each), DEPC-treated water (1–2 min), 2X SSC (15–20 min), proteinase K (37°C, 30 min), glycine (2 mg/mL in PBS, 2 min), PBS (two washes, 2 min each), PFA fixative solution (10 min), PBS (two washes, 5 min each), triethanolamine/acetic anhydride (stir during 10 min incubation), PBS (two washes, 5 min each), ethanol: 30, 60, 80, 90, 95, 100, and 100% (30 s each). Store slides at 4°C in a closed plastic container with a small amount of 100% ethanol at the bottom for several hours to overnight.

3.5.4. In Situ *Hybridization*

1. Dry the slides on a sheet of filter paper while preparing hybridization solution. Dilute the probe in DEPC-treated water such that a 100-μL contains 0.5 ng/μL/kb. For newly synthesized probes, higher and lower concentrations may also be tested. Add 100 μL of formamide to the probe solution and incubate at 80°C to denature the probe, then chill on ice for 2 min, and add to hybridization solution.
2. Add 100 μL of hybridization solution to the slide, and another 100 μL to a fresh slide. Slowly sandwich the two slides together, and place them upright in a closed plastic container, over wet paper towels, and incubate at 55°C overnight.

3.5.5. In Situ *Posthybridization*

1. Remove the slide from the plastic container and place it in a slide rack on a rotary shaker at 100 rpm and incubate with 0.2X SSC (two washes at 55°C, 60 min each), NTE (two washes at 37°C, 5 min each), RNase A in NTE (37°C, 30 min), NTE (two washes, 5 min each), 0.2X SSC (55°C, 60 min), 1X PBS (room temperature, 5 min or 4°C, overnight).
2. Place slide in a large plastic container on a rocking platform set at maximum speed, and immerse in 1% blocking reagent at room temperature (45 min). Replace solution with BSA solution and incubate for 45 min.

3. Sandwich slide with a second slide, and draw up a solution of 1:1250 anti-DIG antibody in BSA by capillary action, incubate for 2 h, then separate slides and immerse in 1% BSA (four washes, 15 min each) in a plastic container on a rocking platform. Replace this solution with detection buffer and incubate for 10 min.
4. Sandwich the slide with a second slide, and draw up Western Blue substrate solution by capillary action and incubate for 2 d. Separate the slides and immerse them in TE buffer for 5 s each in a series of ethanol concentrations (30, 50, 70, 95, 100, and 100%), followed by two washes of 5 s each in Hemo-De.
5. Mount slides with Cytoseal 60, cover with a no. 1 cover slip, and place in a holder overnight to drain while medium hardens. Examine using a microscope and photograph.

4. Notes

1. RNase-free environment: Care should be taken to prevent ribonucleases coming into contact with RNA samples. Glassware should be baked at 200°C overnight, as should other equipment capable of withstanding such a temperature. Plasticware should be filled with 0.1% DEPC and left overnight before autoclaving to remove DEPC. Many companies supply DNase- and RNase-free plasticware.
2. Parallel RT-PCR reactions should be set up for each gene using the same master mix such that each 30 µL of RT-PCR reaction contains equal amounts of the real-time cDNA mix.
3. DEPC treatment of water: All water used when working with RNA (reagents, etc.) should be rendered RNase free by making a 0.1% solution of DEPC and incubating overnight, before autoclaving to remove DEPC.
4. RNA may be stored at –70°C, cDNA can be stored at 4°C or –20°C.
5. The gel can be stained with ethidium bromide before discarding to ensure that no bands were seen and efficient transfer had occurred.
6. Alternatively, the protocol for Southern blot hybridization can be used, except that all solutions used for Northern blot hybridization up to the step of hybridization must be RNase free.
7. During all prehybridization, hybridization, and posthybridization washes, the membrane should be constantly agitated on a shaker.
8. An important issue in mRNA quantification using any technique is data replication and verification ***(2,17)***. A sufficient number of reactions from the same mRNA stock spanning a range of 10^3- to10^4-fold variations must be run to ensure reliable quantitation. Thus, PCR efficiencies should be determined from a dilution series (over at least three orders of magnitude) for each gene. Data should be collected for 2–10 replicates per sample.

Acknowledgment

This study is scientific contribution no. 2201 from the New Hampshire Agricultural Experiment Station.

References

1. Kuhn, E. (2001) From library screening to microarray technology: strategies to determine gene expression profiles and to identify differentially regulated genes in plants. *Ann. Bot.* **87,** 139–155.
2. Wu, S. H., Ramonell, K., Gollub, J., and Somerville, S. (2001) Plant gene expression profiling with DNA microarrays. *Plant Physiol. Biochem.* **39,** 917–926.
3. Seki, M., Narusaka, M., Abe, H., et al. (2001) Monitoring the expression pattern of 1300 *Arabidopsis* genes under drought and cold stresses by using a full-length cDNA microarray. *Plant Cell* **13,** 61–72.
4. Donson, J., Fang, Y., Espiritu-Santo, G., et al. (2002) Comprehensive gene expression analysis by transcript profiling. *Plant Mol. Biol.* **48,** 75–97.
5. Jones, H., ed. (1995) *Plant Gene Transfer and Expression Protocols*, Humana Press, Totowa, NJ.
6. Bartlett, J. M. S. (2002) Approaches to the analysis of gene expression using mRNA—a technical overview. *Mol. Biotechnol.* **21,** 149–160.
7. Kricka, L. J. (1995) Chemiluminescence and bioluminescence. *Anal. Chem.* **67,** R499–R502.
8. Kricka, L. J. (2002) Stains, labels and detection strategies for nucleic acids assays. *Ann. Clin. Biochem.* **39,** 114–129.
9. Freeman, W. M., Walker, S. J., and Vrana, K. E. (1999) Quantitative RT-PCR: Pitfalls and potential. *BioTechniques* **26,** 112–125.
10. Gause, W. C. and Adamovicz, J. (1994) The use of PCR to quantitate gene-expression. *PCR Methods Applicat.* **3,** S123–S135.
11. Carleton K. L. and Kocher T. D. (2001) Cone opsin genes of African cichlid fishes: tuning spectral sensitivity by differential gene expression. *Mol. Biol. Evol.* **18,** 1540–1550.
12. Sellner, L. N. and Turbett, G. R. (1998) Comparison of three RT-PCR methods. *BioTechniques* **25,** 230–234.
13. Heid, C. A., Stevens, J., Livak, K. J., and Williams, P. M. (1996) Real-time quantitative PCR. *Genome Res.* **6,** 986–994.
14. Chretien, R. L. (2003) Characterization of *S*-adenosylmethionine decarboxylase (*samdc*) and arginine decarboxylase (*adc*) genes from *Daucus carota* and their expression during somatic embryogenesis. MS Thesis, University of New Hampshire, Durham, NH.
15. Angerer, L. M., Cox, K. H., and Angerer, R. C. (1987) Demonstration of tissue-specific gene expression by *in situ* hybridization. *Meth. Enzymol.* **152,** 649–661.
16. Long, S. and Rebagliati, M. (2002) Sensitive two-color whole-mount *in situ* hybridizations using digoxygenin- and dinitrophenol-labeled RNA probes. *BioTechniques* **32,** 494–500.
17. Lee, M. L. T., Kuo, F. C., Whitmore, G. A., and Sklar, J. (2000) Importance of replication in microarray gene expression studies: statistical methods and evidence from repetitive cDNA hybridizations. *Proc. Natl. Acad. Sci. USA* **97,** 9834–9839.

18. Mason, M. G. and Schmidt, S. (2002) Rapid isolation of total RNA and genomic DNA from *Hakea Actities. Funct. Plant Biol.* **29,** 1013–1016.
19. Sambrook, J., Fritsch, E. F., and Maniatis, T. (1989) *Molecular Cloning: A Laboratory Manual.* Cold Spring Harbor Laboratory Press, Cold Spring Harbor, NY.

21

Transgene Integration

Use of Matrix Attachment Regions

George C. Allen, Steven Spiker, and William F. Thompson

Summary

Matrix attachment regions (MARs) are operationally defined as DNA elements that bind specifically to the nuclear matrix in vitro. When MARs are positioned at the 5'- and 3'-ends of a transgene higher more predictable expression of the transgene results. MARs are increasingly being applied to prevent unwanted transgene silencing, which is especially common when direct DNA transformation methods are used. This chapter describes methods for the isolation of MARs and the subsequent methods allowing the investigator to incorporate MARS into transformation strategies that can both improve transformation frequency and result in predictable, stable expression of the transgenic trait.

Key Words: Matrix attachment regions; nuclear matrix; nuclear scaffold; MARs; SARs; posttranscriptional gene silencing; transcriptional gene silencing; transformation.

1. Introduction

Transgenic crops have the potential to be an inexpensive source of specialized products, and to be a part of the solution for feeding the increasing world population. However, for transgenic technology to be useful, transgenes must have predictable and stable expression. Extensive testing, which leads to higher production costs, is no guarantee that an apparently stable trait will remain stable in subsequent generations *(1)* under all environmental conditions *(2)*. Thus, technologies have been sought that would enhance our ability to create transgenic plants with the desired expression characteristics. One of these technologies involves the use of matrix attachment regions (MARs). MARs are DNA sequences that bind specifically to a network of proteinaceous fibers, called the nuclear matrix, which permeates the nucleus. These MAR-matrix

From: *Methods in Molecular Biology, vol. 286: Transgenic Plants: Methods and Protocols*
Edited by: L. Peña © Humana Press Inc., Totowa, NJ

interactions are thought to organize chromatin into a series of independent loop domains. Several laboratories have found that MARs enhance and stabilize transgene expression to varying extents. Most results are consistent with the hypothesis that MARs function by reducing or eliminating some forms of gene silencing. These properties make MARs a promising tool for combating unpredictable transgene expression ***(3)***.

When the results of numerous studies involving MARs are interpreted, the following important experimental differences in these studies must be considered: (a) The methods by which MARs are isolated. MARs have been isolated by two primary methods in which "high salt" ***(4)*** or lithium diiodosalicylate (LIS) ***(5)*** is used to remove histones and other soluble proteins from the nucleus. The LIS extraction method, which we predominantly use in our work, was originally developed to address criticisms that the use of 2 *M* NaCl causes precipitation artifacts. (b) Nature of the biological system. We have typically found the use of MARs leads to much greater increases in gene expression in cells in culture than in whole plants ***(3)***. The greater MAR effect may be related to the greater number of rapidly dividing cells in cell culture than in whole plants ***(6)***. However, in plants, we and others have shown that MARs greatly increase the transmission of a trait into subsequent generations ***(7–9)***. (c) The specific MAR sequence used. Different MARs when used in otherwise identical experimental conditions can lead to different results ***(3)***. (d) Transformation procedure. Direct DNA transformation protocols, such as the gene gun or electroporation, may yield results different than those resulting from *Agrobacterium*-mediated transformation ***(8,10)***. Direct DNA transformation typically leads to the integration of large numbers of transgenes that are localized within the same region. This transgene arrangement has been speculated to lead to silencing. Although the mechanism by which MARs increase and stabilize transgene expression is unknown, we have speculated that they prevent transcriptional silencing by shielding the transgene locus from the effects of the surrounding chromatin and indirectly prevent posttranscriptional silencing by changing the structure of the transgene locus in a manner that decreases the likelihood of dsRNA formation ***(3)***.

Much of our gene expression work has been done using the Rb7 MAR, which was originally isolated from the *Rb7* gene of tobacco ***(11)***. The Rb7 MAR is known to both increase and stabilize transgene expression when it flanks the transgene (*see* Table 1 in **ref. 3**). This effect has now been seen in a number of different crop species and is presently being incorporated into an increasing number of transformation strategies.

We routinely use NT1 (or BY2) tobacco suspension cells because they are easily transformed using either direct gene transfer or *Agrobacterium*. NT1 cells are easy to grow and offer a convenient system in which to test many construct designs. Additionally, NT1 cells are an excellent source for obtaining pure nuclei. This is important because clean nuclei are necessary for the isolation of nuclear matrix and the associated MARs. Clean nuclei can be easily isolated from suspension cultures if one wants to isolate a MAR from a particular plant, it would be advisable to use a suspension culture for that plant if one is available.

Because the nuclear matrix is highly conserved, nuclear matrix from tobacco suspension cells can be used to test the binding of MARs from animals, fungi, or plants. Although all MARs may bind to the matrix in an in vitro binding assay, the same MARs may have varying effects on gene expression. For example, the A element MARs from the chicken lysozyme gene can be used to normalize transgene expression in plants ***(12,13)*** but seems to have little effect on gene expression levels. The methods that we describe are based on tobacco suspension cells but can be modified for isolating MARs from a variety of different plants.

The detailed methods presented here are designed to provide readers with the tools for isolating MARs, developing a transformation vector for their particular needs, and producing a transgenic plant or cell line. Two major options are presented that give readers the choice of using MARs isolated from their experimental system or using well-characterized MARs known to have an impact on transgene silencing in numerous different plants.

2. Materials

1. 4-D NT1 or BY2 tobacco suspension cells.
2. Rotary shaker and growth chamber.
3. NT1 media (per liter: 30 g of sucrose, 180 mg of potassium monophosphate, 100 mg of inositol, 10 mg of thiamine-HCl, one packet of Murashige and Skoog (MS) salt mixture [Invitrogen, Carlsbad, CA]; adjust to final pH 5.7 with KOH).
4. Protoplasting enzyme solution: 0.4 *M* mannitol, 10 m*M* *N*-(2-hydroxyethly) piperazine-*N*'-2-ethanesulfonic acid), (MES), pH 5.5, 1% Cellulase RS (Onozuka, Yakult Honsha Co., Ltd., distributed by Yakult Pharmaceutical Ind. Co., Tokyo, Japan), 0.1% Pectolyase Y23 (Onozuka).
5. Cell wash buffer: 0.4 *M* mannitol, 10 m*M* MES pH5.5.
6. 0.4 *M* Mannitol solution in water.
7. Refrigerated swinging bucket centrifuge.
8. Brightfield and fluorescence microscope.
9. UV spectrophotometer.
10. 5.5 *M* Urea solution in water.
11. Percoll (Sigma, St. Louis, MO).
12. Nuclear isolation buffer 1 (NIB-1): 1X Nuclear Isolation Buffer 2 and 0.5% Triton X-100; A mixture of Percoll and NIB-1 at a final concentration (v/v) of 85% NIB-1 and 15% Percoll.

13. NIB-2: 0.5 *M* hexylene glycol (FW 118.18; d = 0.985), 20 m*M* HEPES, pH 7.4, 20 m*M* KCl, 1% thiodiglycol, 0.5 m*M* NaEDTA, pH 7.4, 1 m*M* PMSF (PMSF solution can be made as at 10 m*M* (1.74 mg/mL) in isopropanol and stored in aliquots at –20°C. PMSF rapidly undergoes inactivation in aqueous solutions, particularly at room temperature), 4 μ*M* apoproteinin, 50 μ*M* spermine, 125 μ*M* spermidine.
14. NIB-3: 0.5 *M* hexylene glycol, 20 m*M* HEPES, pH 7.4, 20 m*M* KCl, 1% thiodiglycol, 50 μ*M* spermine, 125 μ*M* spermidine, 0.5 m*M* phenylmethylsulfonylfluoride (PMSF), 2 μg/mL aprotinin.
15. Halo isolation buffer (HIB): 10 m*M* lithium diiodosalisylic acid, 0.1 *M* lithium acetate, 20 m*M* HEPES, pH 7.4, 2 m*M* NaEDTA pH 7.4, 0.1% digitonin, 0.5 m*M* PMSF, 2 μg/mL of Aprotinin.
16. Digestion binding buffer: 20 m*M* HEPES, pH 7.4, 20 m*M* KCl, 10 m*M* $MgCl_2$, 1% thiodiglycol, 0.2 m*M* PMSF, 5 mg/mL aprotinin, 10 μ*M* E-64 (Sigma Aldrich).
17. 100% Glycerol (FW 92.09).
18. 100-μm nylon mesh.
19. 50 m*M* $CuSO_4$ stock solution in water.
20. 10 mg/mL of proteinase K stock solution in water.
21. 7% (w/v) Trichloroacetic acid (TCA) in water.
22. Restriction enzymes (New England Biolabs, Beverly, MA).
23. T4 ligation kit (Roche Applied Science, Indianapolis, IN).
24. Competent *Escherichia coli* strain DH5α (Invitrogen).
25. Luria broth for growing bacteria and agar for plating transformants (Difco, BD, Franklin Lakes, NJ).
26. Antibiotic for selecting appropriate bacterial transformants containing the desired plasmid (typically ampicillin).
27. Plasmids that include first, a high-copy cloning plasmid (pBluescript from Stratagene, La Jolla, CA), and second, a plant expression plasmid for cloning isolated MARs into containing desired gene cassette with a plant promoter, gene and polyadenylation signal. The plant expression plasmid could either be a high-copy plasmid if direct DNA transformation is to be used or a binary vector if *Agrobacterium* is to be used.
28. Incubator and shaker at 37°C.
29. Electrophoresis equipment including gel trays and power supplies.
30. Molecular biology grade agarose (Sigma Aldrich) and either TAE or TBE buffers. TAE: Concentrated 50X TAE stock solution contains 242 g of Tris base, 57.1 mL of glacial acetic acid, and 100 mL of 0.5 *M* EDTA. The TAE mixture is brought to 1.0 L with water. TBE buffer: 10X concentrated TBE stock solution contains 121.1 g of Tris base, 55 g of boric acid, 7.4 g of Na_2EDTA. The TBE is brought to 1.0 L with water.
31. Ethidium bromide (Sigma) or alternative DNA (Molecular Probes, Eugene, OR) stain and a ultraviolet or blue light source for visualizing DNA in gels.
32. Equipment for transforming plant cells such PDS-1000 Gene Gun (Bio-Rad, Hercules, CA) or a BTX electroporator.
33. NT1 media in 0.8% Phytagar with the appropriate selection agent (kanamycin, hygromycin, gluphosinate ammonium) at a concentration for selecting resistant

plants or plant cells.
34. Plant growth chamber for regenerating plants or plant cells. It is preferable that the chambers have lighting and temperature control (Percival Scientific, Perry, IA).
35. PCR machine for screening plants containing the transgene (Applied Biosystems, Foster City, CA).
36. Materials for plant DNA extraction.
37. Buffers for denaturing DNA in an agarose gel.
38. Hybridization membranes for transferring electrophoresed DNA.
39. Hybridization solutions and hybridization bottles.
40. Hybridization ovens capable of holding a range of temperatures from 37°C to 70°C.

3. Methods

Several MARs have been shown to work well in various plants and can be used. However, in certain cases, researchers may desire to use MARs from the plant that they are planning on transforming. For these investigators we have included a section on isolating and cloning MARs. We have found that nuclear matrix preparations from various sources can be used for binding assays, but the important point is that the nuclear matrix used should be easy to obtain and free of nonmatrix proteins. For these reasons, we use matrix preparations from the tobacco suspension line NT1 that serves as our standard matrix preparation ***(14)***. In tobacco NT1 suspension cultures the cells are very fine, divide rapidly, and can be easily synchronized ***(15,16)***. This culture system has been used as a model system for approx 25 yr for studies ranging from *Agrobacterium* infection ***(17)*** to signal transduction ***(18,19)***. The methods outlined below outline (a) the growth of an NT1 tobacco culture, (b) the isolation of nuclei, (c) the isolation of nuclear matrix, (d) the construction of MAR flanked gene expression plasmid, (e) the transformation of NT1 plant cells, and (f) screening the putative transgenic cells and final analysis of the transformed cells for integration patterns and gene expression.

3.1. Growing Cells for Isolation of Nuclear Matrix

3.1.1. Growth of Cells

To isolate tobacco nuclear matrices we use a tobacco cell culture (NT1) originally provided to us by G. An ***(20)***. We culture the cells as follows:

1. Prepare sterile NT1 medium by dispensing 100 mL into a 500-mL culture flask. The shape of the flask is important because the characteristics of the culture can change in different volumes of media and different types of flasks.
2. Inoculate 3 mL of a 1-wk NT1 suspension culture into the 100 mL of NT1 medium.
3. Prior to harvest prepare the protoplasting enzyme solution, cell wash buffer, and mannitol solution. The enzymes need to be stirred 30–45 min to go into solution. Enzymes should not be stirred too rapidly or denaturation may occur.

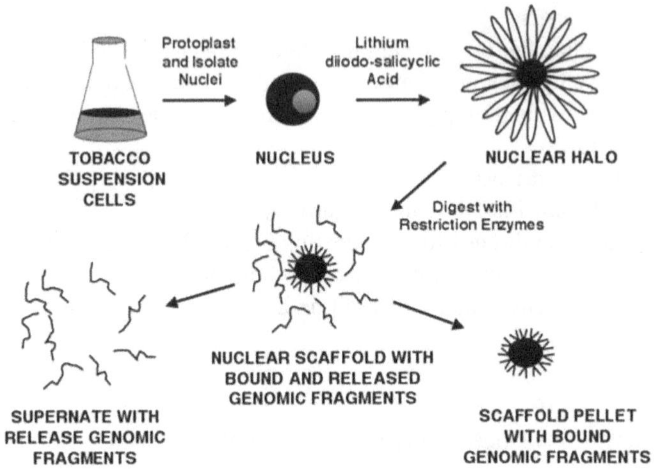

Fig. 1. Preparation of nuclear matrix and isolation of MAPs from tobacco suspension cells. The cells are grown to mid-log and treated with protoplasting enzymes. The resulting protoplasts are lysed and the nuclei are isolated, treated with LIS to remove histones. The resulting nuclear halos are digested with restriction enzymes releasing nonbound DNA fragments, which can be separated from bound DNA (MAR DNA) by centrifugation. The purified matrix can be used for in vitro binding assays or digested to allow cloning of the MAR sequences.

4. The enzymes can be filter sterilized by using a 45-µm filter (Corning, Corning, NY) to remove the larger particulates followed by a 22-µm filter for sterilization.

3.1.2. Making Protoplasts

A major barrier to isolating nuclei is the cell wall. We gently isolate nuclei from NT1 cells by first making protoplasts (**Fig. 1**).

1. Grow NT1 tobacco suspension cells 4 d in two 500-mL growth flasks containing 100 mL NT1 medium incubated at 27°C on a rotary shaker at 120 rpm. Keep the sterile growth flask for use later.
2. Harvest the cells by centrifuging four 50-mL Falcon tubes containing 45 mL of cells each at 450*g* for 10 min on a swinging bucket centrifuge with the brake on low.
3. The cell yield is approx 10 mL packed volume. Too many cells can cause problems during the protoplasting step.
4. Gently pour off the supernatant and replace with a 45 mL of cell wash buffer per Falcon tube. The cell wash buffer removes calcium present in the MS, which is known to inhibit cellulases.
5. Mix to suspend the cells in cell wash buffer and harvest the cells by again centrifuging four 50-mL Falcon tubes containing 45 mL of cells each at 450*g* for 10 min on a swinging bucket centrifuge with the brake on low.

6. Gently pour off the supernatant and re-suspend the cells in 45 mL of enzyme solution per Falcon tube.
7. Mix and transfer the cells to 500-mL sterile growth flask (from above) for incubation at 27°C on the rotary shaker at 100 rpm for 30 min.
8. Remove a small sample for microscopic examination. Cells that retain cell walls are not round and may appear in files, whereas protoplasts will appear as rounded cells with some individuals. After the cells appear to be protoplasts, harvest the protoplasts by again centrifuging four 50-mL Falcon tubes containing 45 mL of cells each at 450*g* for 10 minutes on a swinging bucket centrifuge with the brake on low (*see* **Note 1**).
9. Gently remove the supernatant and gently resuspend the protoplasts in an equal volume of ice-cold 0.4 *M* mannitol to remove the residual enzyme mixture.
10. Repeat the centrifugation of the four 50-mL Falcon tubes containing 45 mL of protoplasts at 450*g* for 10 min on a swinging bucket centrifuge with the brake on low.
11. Repeat the resuspension of the protoplasts in an equal volume of ice-cold 0.4 *M* mannitol to remove the residual enzyme mixture.
12. Repeat the centrifugation of the four 50-mL Falcon tubes containing 45 mL of protoplasts each at 450*g* for 10 min.

3.1.3. Isolation of Nuclei

The protoplasts are now lysed to release the nuclei. The nuclei are cleaned to remove cellular debris and treated to remove histones and other soluble proteins. The remaining nuclear halo is digested with the appropriate restriction enzyme to remove the DNA not bound to the nuclear matrix (**Fig. 1**). The proteinaceous nuclear matrix is then digested with proteases, which allows the isolation and cloning of the DNA fragments that were bound to the nuclear matrix.

1. After the protoplasts have been washed with mannitol in **Subheading 3.1.3., step 13** discard the supernatant.
2. To lyse the protoplasts gently resuspend the protoplasts in NIB-1. The final volume of NIB-1 used should be half the original culture volume (i.e., for a 100 mL culture of cells use 50 mL of NIB-1 solution).
3. Dispense 10 mL of the Percoll–NIB-1 solution per 50-mL Falcon tube and layer 25 mL of the filtrate onto the Percoll layer. Typically, two gradients are required per 100 mL of cell culture. An alternative is to aliquot the filtrate first and then layer the Percoll under the filtrate.
4. Centrifuge the nuclei at 450*g* at 4°C for 10 min in a swinging bucket centrifuge rotor with the brake on low.
5. Resuspend the nuclei pellet into the NIB-2 solution. The nuclei pellets from four Percoll gradients may be combined into two tubes at this stage. Wash the nuclei in each tube with 30 mL of NIB-2.
6. Centrifuge the nuclei at 450g at 4°C for 10 min on a swinging bucket centrifuge with the brake on low and resuspend in a total of 1–5 mL of NIB-2.

7. To estimate yield measure the A_{260} using 5 µL of the nuclei preparation in 500 µL of 5.5 *M* urea solution using the urea solution as the blank.
8. Add cold NIB-2 solution to bring the total volume to 30 mL and centrifuge the nuclei at 450*g* at 4°C for 10 min in a swinging bucket centrifuge rotor with the brake on low.
9. Resuspend in a volume of NIB-2 solution sufficient to give a final concentration of 500 µg nuclei/mL and store at –70°C (*see* **Note 2**).
10. Carefully remove the supernatant and resuspend the nuclear pellet in 200 µL of NIB-3.
11. Add 4 mL of 50 m*M* $CuSO_4$.

3.1.4. Preparation of Nuclear Matrix

1. Thaw the 1-mL aliquot of nuclei on ice and add 9.8 ml of NIB-3 and centrifuge the nuclei at 450*g* at 4°C for 10 min in a swinging bucket centrifuge rotor with the brake on low.
2. Stabilization of the nuclear matrix of NT1 cells requires heat shock stabilization at 42°C for 15 min.
3. Add 10 mL of Halo Isolation Buffer (HIB) and incubate at room temperature for 15 min. Mix by inverting the tube every 5 min.
4. Centrifuge the resulting nuclear halos at 2500*g* for 10 min at room temperature and discard the supernatant.
5. Resuspend the nuclear halo pellet into 10 mL of digestion and binding buffer-I (DBB-I) which contains protease inhibitors. It is important to mix by gently rocking the tube back and forth.
6. Centrifuge the nuclear halos at 250*g* for 10 min at room temperature and discard the supernatant.
7. Repeat the resuspension of the nuclear halo pellet into 10 mL of DBB-I, which contains protease inhibitors. The halos can be frozen at this stage by resuspending the pellet in 1 mL of DBB-IM containing glycerol (this results in a 1:1 dilution of 2X DBB-IM with glycerol). Store at –70°C.
8. To continue, if nuclear halos were frozen, thaw them nuclear halos on ice and wash the pellet with 10 mL of DBB-IM (containing protease inhibitors with $MgCl_2$).
9. Centrifuge the nuclear halos at 2500*g* for 10 min at room temperature and discard the supernatant.
10. Resuspend the nuclear halos in DBB-IM and add 250 U of each restriction enzyme and incubate according to the manufacturer's instructions.
11. Replenish the digestion mixture with an equal amount of enzyme and incubate an additional 1 h. A precipitate falling out of the solution indicates complete digestion. The matrix can be directly used for cloning or can be frozen at this point.

3.2. Cloning the MARs

The restriction enzyme digestion releases unbound DNA and the fragments remaining associated with the purified nuclear matrixes are considered MARs by operational definition (*see* **Note 3**).

1. Centrifuge the nuclear halos at 2500*g* for 10 min in a microfuge and discard the supernatant. The MAR fragments remain bound to the matrix, which is the pellet.
2. Add SDS to 1% and add proteinase K to 500 mg/mL. Incubate overnight at 37°C.
3. Add an equal volume of phenol–chloroform and gently mix the microfuge tube.
4. Add 10% (v/v) 3 *M* sodium acetate and an equal volume of ethanol and precipitate the MARs overnight at –20°C.
5. Digest the plasmid vector (pBluescript II SK+, Stratagene) with the same enzymes used for releasing the halos (*see* **Note 4**).
6. Ligate the fragments into the vector using T4 ligase and transform *Escherichia coli* DH5α (Invitrogen) according to the manufacturer's instructions. The resulting clones contain putative MARs that can be tested by using the in vitro binding assay. It is beyond the scope of this chapter to explain the details but the procedure is outlined in detail in Hall and Spiker ***(14)***.

3.3. Plant Transformation

3.3.1. Biolistic (Direct) of Transformation Tobacco Cells

The following procedures should be performed under a laminar flow hood to prevent the possibility of contamination (*see* **Note 5**).

1. Grow NT1 cells for 4 d as described earlier and centrifuge in 50-mL Falcon tubes at 450*g* for 10 min at room temperature.
2. Determine wet weight and resuspend the cells NT1 medium at 1.0 g/mL.
3. Pipet the resuspended cells (0.5 mL) onto a sterile piece of microscope lens paper, cut to fit the plate and previously placed on 0.8% NT1 agar medium. A wide-bore pipet is recommended for pipetting the cells at this stage.
4. Shake the cells gently by hand until a thin film of cells covers the plate.
5. Place the plated cells in a 25°C in the dark for 2–3 h prior to transformation.
6. Coat the microcarriers (gold beads) with the plasmid mix containing the plasmid and the desired gene (**Fig. 2**) and the selection plasmid according to the manufacturer's instructions (Bio-Rad) (*see* **Note 6**).
7. Follow manufactures instructions for using the Biolistic PDS-1000/He Particle Delivery System (Bio-Rad) (*see* **Note 7**).
8. Incubate the cells without selection for 24 h at 25°C.
9. Transfer the microscope lens paper containing the cells onto 0.8% NT1 Phytagar medium containing the selective agent at a concentration high enough to kill untransformed cells.
10. Wrap the plates in Parafilm (micropore tape) and incubate in the dark at 25°C for 12 d.
11. Check for transformants, which will be noted as small bumps on a flat lawn of cells. The lawn represents the remnants of nontransformed cells whereas the dividing cells, are seen as bumps or microcalli.

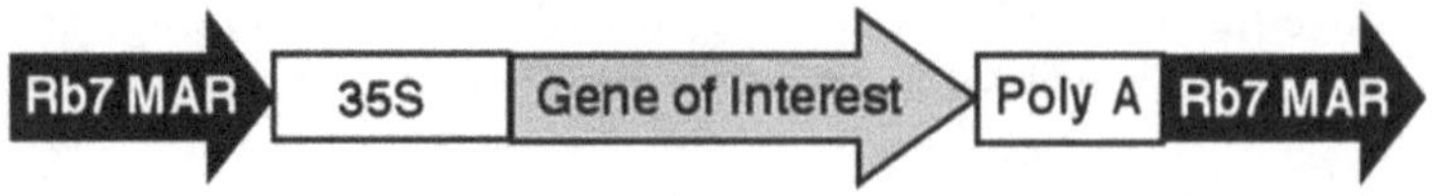

Fig. 2. A transformation construct containing a transgene flanked by MARs. The transgene cassette consists of a promoter (CaMV35S) a gene of interest and a polyadenylation signal (pA). The Rb7 MAR, which was isolated from tobacco, flanks the transgene cassette.

3.3.2. Agrobacterium *Transformation*

The *Agrobacterium* strain used for transformation and the selectable marker may vary depending on the type of plant being transformed. We suggest that using a strain that is deficient in recombination (RecA$^-$) be used because we and others have found that the MAR elements can recombine in *Agrobacterium* (Allen, unpublished) (*see* **Note 8**).

1. Grow NT1 cells for 4 d as described earlier.
2. Grow the *Agrobacterium* strain containing the MAR-flanked desired gene in a binary vector for approx 36 h at 27–30°C on a rotary shaker and shake at 225 rpm. The culture should be in late-log or early stationary phase for the transformation step.
3. Pipet 4 mL of the NT1 suspension into a 120-mm Petri plate and mix with 100 µL of the late-log *Agrobacterium* and gently mix.
4. Seal the plate with Parafilm and incubate at 27°C for 48 h (*see* **Note 9**).
5. Prepare 0.8% Phytagar (Invitrogen) NT1 medium plates containing 100 mg/mL of timentin (PhytoTechnology Laboratories, Shawnee Mission, KS) and the plant cell selective agent (*see* **Note 10**).

3.4. Identifying the Transformed Plants

3.4.1. Identification With Visual Markers or Selectable Markers

Typically, the transgenic plant can be identified by its ability to grow in the presence of a selectable agent such as basta or kanamycin if such selectable markers were included in the transformation. Because researchers are trying to avoid including the selectable marker in the transgenic plant there has been a shift towards the use of visible markers such as green fluorescent protein (GFP). Visual markers are especially advantageous because they do not require killing nontransformed cells to identify the transformed cell. In addition, the use of a screenable marker instead of a selectable marker may be less stressful, perhaps leading to higher numbers of transgenic events. We have found that the soluble modified red-shifted GFP from the *Arabidopsis* stock center gives excellent fluorescence and does not require damaging ultraviolet light to visualize.

3.4.2. Screening With Polymerase Chain Reaction

Polymerase chain reaction (PCR) used in combination with a high-throughput DNA isolation protocol can also be used to identify transgenic plants. The advantage of using this method is that if the transformation is robust it may not be necessary to use a selectable or visual marker gene. Increasingly in transgenic plants, it is desirable to reduce the amount of transgenic DNA to a minimum for public acceptance (*see* **Note 11**).

4. Notes

1. If the culture is too old or the density is too high, good protoplasts may not appear within 60 min. In this case, it is advised that continuation of the procedure will yield poor nuclei.
2. Set up a series of three funnels such that the funnel below is receiving the filtrate from the funnel at top in a cold room or refrigerator for column chromatography. The topmost funnel should be lined with a 100-µm nylon mesh, followed by 50-µm and a 30-µm mesh (Tetko, Lancaster, NY).
3. It is advisable that the chosen restriction enzymes be active on methylated genomic DNA and allow directional cloning into the shuttle vector. Examples of common restriction enzymes that work well to digest genomic DNA are *Hin*dIII and *Eco*RI.
4. The vector design for creating a transgene that is to be protected from transgene silencing must be carefully considered. For example, the desired gene (reporter gene) and the selection gene can both be flanked by the MARs. An advantage to this approach is that by when MARS flank both genes the likelihood is increased that any plant cell that is resistant to selection will also contain a complete copy of the desired gene.
5. Direct gene transfer is the direct introduction of DNA into the cell through physical means. We typically use biolistic transformation but other direct methods such as electroporation or whiskers can be applied. It is advisable to test the plasmid constructs on a well-characterized system before attempting any difficult plants. We use tobacco suspension cell lines to test plasmids to be used to transform dicots and Black Mexican Sweet maize suspension cells to test plasmids that are to be used for monocot transformation. Plasmid DNA, produced in *E. coli*, is introduced the cell by cotransformation with the plasmid containing the desired gene flanked by MARs, and the appropriate plant selectable marker. We use a 4:1 molar ratio of the desired gene plasmid to selectable marker plasmid with 500 ng of total DNA per shot. This assures us that all cells that survive are likely to have integrated the desired gene as well as the selectable marker. The amount of DNA used per shot can be also decreased to achieve lower gene copy number as shown by Brouwer et al. ***(21)***.
6. Our initial constructs were designed for use in biolistic cotransformation experiments in which the reporter gene, or gene of interest, and the selectable marker were on separate plasmids. We use a molar ratio of four reporter plasmids per selection plasmid to insure that the resistant cells are likely to contain the reporter

gene. Before stable transformation experiments are attempted with any new construct it is wise to first test the construct in a transient assay using either biolistics or electroporation. If *Agrobacterium*-mediated transformation is used for transient expression assays, it is desirable to include an appropriate plant intron within the gene to ensure that the expression results from the plant cells and not the agrobacteria.

7. We use a PDS1000 gene gun from Bio-Rad placed in a Baker laminar flow hood (The Baker Company, Sanford, ME) for our transformation and follow the manufacturer's instructions. Depending on the plant or plant cell to be transformed several parameters can be optimized, which include pressure, distance of shooting, and bead size. We use 1100 psi, which is controlled by the type of rupture disc used, 9-cm shooting distance, which is the distance of the plate, and 1.0-μm gold beads, which contain the precipitated DNA to be used for transformation. Prior to the transformation, all of the PDS1000 parts that could introduce contamination are sterilized in 70% ethanol and allowed to dry in a laminar flow hood.
8. *Agrobacterium* transformation to introduce a binary vector is widely used because it is generally assumed the complete desired gene will be integrated with a lower copy number than direct gene transfer. Lower transgene copy numbers may be important because multiple gene copies often have been associated with gene silencing. Numerous protocols are now available for *Agrobacterium*-mediated transformation, which include leaf disk (tobacco), vacuum infiltration and floral dip (*Arabidopsis*), and embryogenic callus (maize) or suspension cell cultures (NT1). Although a detailed transformation for each protocol is beyond the scope of this manuscript, the NT1 protocol is included for a comparison with the other procedures. Additionally, there are several general points that should be considered for all transformation protocols where *Agrobacterium* is used to introduce MAR-flanked desired gene cassettes. When MAR-flanked DNA is introduced into *Agrobacterium* there is the possibility that the directly repeated MARs may recombine to remove the gene of interest. To circumvent this problem we use *Agrobacterium tumefaciens* strains that are RecA$^-$.
9. Although micropore tape has been used instead of Parafilm, we have found that Parafilm keeps the mixture from drying during the incubation period.
10. We find that timentin works extremely well for killing *Agrobacterium*; however, other antibiotics can be used such as carbenicillin (Sigma Aldrich).
11. Putative transgenic plants must be analyzed to determine the presence and structure of the transgene locus. It is important to avoid the use of transgenic plants with multiple transgenes or partial inserts of transgenes, because these plants are more likely to have unpredictable expression of the transgene. Simple, low copy transformation events are likely to pass the transgene to progeny that will reflect the same expression patterns as the parent. To determine whether the transgenic plant in question has low gene copy numbers and complete integrations it is necessary to do Southern blots. Although a detailed description of the Southern blotting procedure is beyond the scope of our review, we refer you to Sambrook et al. ***(22)***, which is one of several excellent methods books that describes the procedure in detail.

Acknowledgments

We thank the past and present members of George Allen's, Steve Spiker's, and W.F. Thompson's laboratories for their valuable contributions to this work.

References

1. Finnegan, J. and McElroy, D. (1994) Transgene inactivation: Plants fight back! *BioTechnology* **12,** 883–888.
2. Meyer, P., Linn, F., Heidmann, I., Meyer, H. Z. A., Niedenhof, I., and Saedler, H. (1992) Endogenous and environmental factors influence 35S promoter methylation of a maize A1 gene construct in transgenic petunia and its colour phenotype. *Mol. Gen. Genet.* **231,** 345–352.
3. Allen, G. C., Spiker, S., and Thompson, W. F. (2000) Use of matrix attachment regions (MARs) to minimize transgene silencing. *Plant Mol. Biol.* **43,** 361–376.
4. Berezney, R. and Coffey, D. S. (1975) Nuclear protein matrix: association with newly synthesized DNA *Science* **189,** 291–293.
5. Mirkovitch, J., Mirault, M.-E., and Laemmli, U. K. (1984). Organization of the higher-order chromatin loop: specific DNA attachment sites on nuclear scaffold. *Cell* **39,** 223–232.
6. Mitsuhara, I., Shirasawa-Seo, N., Iwai, T., Nakamura, S., Honkura, R., and Ohashi, Y. (2002) Release from post-transcriptional gene silencing by cell proliferation in transgenic tobacco plants: possible mechanism for noninheritance of the silencing. *Genetics* **160,** 343–352.
7. Vain, P., Worland, B., Kohli, A., et al. (1999) Matrix attachment regions increase transgene expression levels and stability in transgenic rice plants and their progeny. *Plant J.* **18,** 233–242.
8. Ulker, B., Allen, G. C., Thompson, W. F., Spiker, S., and Weissinger, A. K. (1999) A tobacco matrix attachment region reduces the loss of transgene expression in the progeny of transgenic tobacco plants. *Plant J.* **18,** 253–263.
9. Ascenzi, R., Ulker, B., Todd, J. J., et al. (2003) Analysis of trans-silencing interactions using transcriptional silencers of varying strength and targets with and without flanking nuclear matrix attachment regions. *Transgen. Res.* **12,** 305–318.
10. Vaucheret, H., Elmayan, T.,Thierry, D., et al. (1998) Flank matrix attachment regions (MARs) from chicken, bean, yeast or tobacco do not prevent homology-dependent trans-silencing in transgenic tobacco plants. *Mol. Gen. Genet.* **259,** 388–392.
11. Hall, G. E., Jr., Allen, G. C., Loer, D. S.,Thompson, W. F., and Spiker, S. (1991) Nuclear scaffolds and scaffold-attachment regions in higher plants. *Proc. Natl. Acad. Sci. USA* **88,** 9320–9324.
12. Mlynarova, L., Loonen, A., Heldens, J., et al. (1994) Reduced position effect in mature transgenic plants conferred by the chicken lysozyme matrix-associated region. *Plant Cell* **6,** 417–426.
13. Mlynarova, L., Keizer, L. C. P., Stiekema, W. J., and Nap, J. P. (1996) Approaching the lower limits of transgene variability. *Plant Cell* **8,** 1589–1599.

14. Hall, G. E., Jr. and Spiker, S. (1994) Isolation and characterization of nuclear scaffolds in *Plant Molecular Biology Manual*, Vol. D2, Kluwer Academic, Belgium, The Netherlands, p. 1.
15. Yang, S. W., Jin, E., Chung, I. K., and Kim, W. T. (2002) Cell cycle-dependent regulation of telomerase activity by auxin, abscisic acid and protein phosphorylation in tobacco BY-2 suspension culture cells. *Plant J.* **29,** 617–626.
16. Criqui, M. C., Parmentier, Y., Derevier, A., Shen, W. H., Dong, A., and Genschik, P. (2000) Cell cycle-dependent proteolysis and ectopic overexpression of cyclin B1 in tobacco BY2 cells. *Plant J.* **24,** 763–773.
17. Narasimhulu, S. B., Deng, X., Sarria, R., and Gelvin, S. B. (1996) Early transcription of *Agrobacterium* T-DNA genes in tobacco and maize. *Plant Cell* **8,** 873–886.
18. Boniotti, M. B. and Gutierrez, C. (2001) A cell-cycle-regulated kinase activity phosphorylates plant retinoblastoma protein and contains, in *Arabidopsis*, a CDKA/cyclin D complex. *Plant J.* **28,** 341–350.
19. Porceddu, A., Stals, H., Reichheld, J. P., et al. (2001) A plant-specific cyclin-dependent kinase is involved in the control of G2/M progression in plants. *J. Biol. Chem.* **276,** 36,354–36,360.
20. An, G. (1985) High efficiency transformation of cultured tobacco cells. *Plant Physiol.* **79,** 568–570.
21. Brouwer, C., Bruce, W., Maddock, S., Avramova, Z., and Bowen, B. (2002) Suppression of transgene silencing by matrix attachment regions in maize: a dual role for the maize 5' ADH1 matrix attachment region. *Plant Cell* **14,** 2251–2264.
22. Sambrook, J., Frisch, E. F., and Maniatis, T. (1989) *Molecular Cloning: A Laboratory Manual*, 2nd Ed., Cold Spring Harbor Laboratory Press, Cold Spring Harbor, NY.

22

Fluorescence *In Situ* Hybridization to Localize Transgenes in Plant Chromosomes

Wendy A. Harwood, Lorelei J. Bilham, Silvia Travella, Haroldo Salvo-Garrido, and John W. Snape

Summary

Production of transgenic plants is now routine for many of our crop species. Methods for the detailed molecular analysis of transgenic plants are available, but often the exact location of the transgene within the crop genome is poorly understood. As a starting point to understanding more about the site of transgene insertion, transgenes can be physically located using fluorescence *in situ* hybridization (FISH). This technique allows transgenes to be located to specific chromosome regions following the hybridization of a fluorescent labelled probe to a chromosome spread. The technique is sensitive enough to detect single transgene copies and can reveal information about the complexity of a transgene insertion site as well as identifying plants homozygous for the transgene. A FISH method is described that has been used successfully to detect single-transgene copies in mitotic metaphase chromosome preparations of wheat and barley.

Key Words: Barley; fluorescence *in situ* hybridization (FISH); genetic modification; plant chromosomes; transgene location; wheat.

1. Introduction

Methods for the production of transgenic plants are continually being improved, increasing the chances of obtaining plants with desirable low numbers of transgene copies and the required transgene expression profiles. One aspect of the genetic transformation process that is at present very difficult or impossible to control, is the exact insertion site of the transgene in the host genome. As the transgene insertion site may play an important role in determining the stability and expression characteristics of the introduced gene, an understanding of the location of transgenes within the host genome is often desirable.

From: *Methods in Molecular Biology, vol. 286: Transgenic Plants: Methods and Protocols*
Edited by: L. Peña © Humana Press Inc., Totowa, NJ

Fluorescence *in situ* hybridization (FISH) is a powerful technique that enables transgenes to be located to specific chromosomes and to specific chromosome regions. The technique involves the preparation of a labeled probe, homologous to the target sequence, hybridization of the probe to a suitable chromosome spread, and then detection of the hybridized probe. The method provides a good starting point for more detailed analysis of the exact transgene insertion site using either genetic mapping techniques or a range of molecular techniques.

The first *in situ* hybridizations of nucleic acid probes to chromosomes preparations used radioactive probes ***(1)***. These methods have now been superseded by nonradioactive techniques including the use of fluorescence. The FISH technique offers numerous advantages including more rapid and sensitive detection that has in turn allowed low- and single-copy transgene sequences to be detected in plant chromosome preparations. Other advantages include improved safety and the ability to combine several probes, labeled with different fluorophores, in a single experiment.

There have now been many reports of the use of FISH to detect the location of transgenes in a range of plants, for example, petunia ***(2)***, tobacco ***(3)***, rice ***(4)***, *Vicia faba* ***(5)***, barley ***(6–8)***, wheat ***(6,9,10)***, triticale ***(6)***, and oat ***(11,12)***. In oat and barley, the FISH technique was able to reveal complex transgene integration sites where linked transgene insertions were interspersed with genomic DNA ***(7,13)***. In addition, the technique may be used to identify plants homozygous for the transgene at an early stage ***(10)***. Recent reviews give excellent summaries of the use of the technique to localize transgenes ***(14,15)***. Schwarzacher and Heslop-Harrison ***(16)*** provide a comprehensive practical guide to all aspects of *in situ* hybridization and this should be referred to for further details. This chapter concentrates on providing details of a protocol that has been successfully used to detect transgenes in wheat and barley and that is sensitive enough to detect single-transgene copies.

2. Materials

2.1. Mitotic Metaphase Chromosome Preparation

1. Fixative solution: three parts 100% ethanol to 1 part glacial acetic acid.
2. 1X Enzyme buffer: 4 m*M* citric acid, 6 m*M* sodium citrate.
3. Cellulase (Sigma, St. Louis, MO).
4. Pectinase (Sigma).
5. 45% acetic acid.

2.2. Probe Preparation

1. 1X TAE buffer: 0.04 *M* Tris-acetate, 1 m*M* EDTA. A 1X working solution is made by dilution of a 50x concentrated stock solution in water (50X TAE: 242 g of Tris, 57.1 mL of glacial acetic acid, 100 mL of 0.5 *M* EDTA at pH 8.0).

2. QIAquick polymerase chain reaction (PCR) Purification Kit (Qiagen, Valencia, CA).
3. 10X Nick translation buffer: 0.5 *M* Tris-HCl pH 7.8, 0.05 *M* $MgCl_2$, 5 mg/mL bovine serum albumin (BSA).
4. Unlabeled nucleotide mix: 1:1:1 ratio of 0.5 m*M* dATP, dCTP, dGTP in 100 m*M* Tris-HCl, pH 7.5.
5. 0.1 *M* dithiothreitol (DTT).
6. DNA polymerase I/ DNase I (Gibco BRL, Invitrogen, Carlsbad, CA).
7. 1 m*M* Biotin-11-dUTP (Sigma).
8. Digoxigenin-11-dUTP 1 m*M* (Roche Diagnostics, Nutley, NY).
9. dTTP 1 m*M* (in 100 m*M* Tris-HCl, pH 7.5).
10. QIAquick Nucleotide Removal Kit (Qiagen).
11. 10X PCR buffer (Amersham, Piscataway, NJ).
12. dNTP stock containing 2 m*M* each of dATP, dTTP, dCTP, dGTP.
13. 0.2 m*M* M13 17-bp reverse primer (Amersham).
14. 0.2 m*M* M13 single-stranded 17-bp primer (Amersham).
15. 1X TE: 10 m*M* Tris-HCl, 1 m*M* EDTA, pH 8.
16. *Taq* DNA polymerase (5 U/µL) (Pharmacia, Pfizer, New York, NY).
17. Buffer 1: 0.1 *M* Tris-HCl, pH 7.5, 0.15 *M* NaCl.
18. Buffer 2: 0.5% (w/v) blocking reagent (Roche) in buffer 1.
19. Buffer 3: 0.1 *M* Tris-HCl, pH 9.5, 0.1 *M* NaCl, 0.05 *M* $MgCl_2$.
20. Anti-biotin-AP Fab fragment (Roche).
21. Anti-digoxigenin-AP Fab fragment (Roche).
22. Detection solution: 22.5 µL of nitroblue tetrazolium (NBT) 50 mg/mL of (Promega, Madison, WI), 17.5 µL of 5-bromo-4-chloro-3-indolyl-phosphate (BCIP) 50 mg/mL of (Promega), 4.96 mL of buffer 3. (Prepare immediately prior to use).

2.3. Hybridization

1. 100 µg/mL of RNase A (Sigma).
2. 2X Saline sodium citrate (SSC): 0.3 *M* NaCl, 30 m*M* sodium citrate at pH 7.0).
3. Pepsin (25 µg/µL) (Sigma) in 0.01 *M* HCl.
4. Depolymerized paraformaldehyde (Sigma) 4% in water. To prepare a 50-mL solution: add 2 g of paraformaldehyde to 30 mL of sterile water and heat to 50°C, add 10 mL of 0.1 *M* NaOH to dissolve. Make up to 50 mL with sterile water.
5. 100% Formamide (Sigma).
6. 50% Dextran sulfate (Sigma).
7. 20X SSC: 3 *M* NaCl, 0.3 *M* sodium citrate at pH 7.0.
8. 10% Sodium dodecyl sulfate (SDS).
9. Sheared salmon sperm DNA (5 µg/µL).
10. 20% (v/v) Formamide in 0.1X SSC.

2.4. Probe Detection

1. 4X SSC, 0.2% Tween-20 (Sigma).
2. 5% BSA (Sigma) in 4X SSC, 0.2% Tween-20.

3. Extra-avidin conjugated to Cy3 (Sigma).
4. Anti-digoxigenin conjugated to fluorescein (Roche).
5. 4-6-Diamidino-2-phenylindole (DAPI) (2 μg/mL) (Sigma). A stock solution was prepared at 100 μg/mL in sterile water and stored at –20°C. The working solution was prepared at 2 μg/mL by diluting the stock solution in McIlvaine's buffer, pH 7.0 (82 mL of 200 m*M* Na_2HPO_4 and 18 mL of 100 m*M* citric acid).
6. Antifade solution (Citifluor).

3. Methods

The methods outlined below describe the four main steps of the FISH technique.

3.1. Mitotic Metaphase Chromosome Preparation

3.1.1. Plant Material

Chromosome preparations were made from the root tips of germinating seedlings as these give a good accumulation of metaphase chromosomes. The method of chromosome preparation follows the squashing protocol of Schwarzacher and Leitch ***(17)*** with some modifications.

1. Germinate the seeds on moist filter paper in Petri dishes at 25°C for 24 h, then transfer to 4°C for 24 h and transfer back to 25°C again for 24–30 h according to species (*see* **Note 1**). At this stage the root tips should be approx 1 cm long.
2. Cut the root tips from each seed (*see* **Note 2**). Using forceps, place root tips in 10-mL vials of aerated, icy, distilled water. Pack the vials in ice and store at 4°C for 24 h (*see* **Note 3**).
3. Remove the root tips from the ice water and immediately place in fixative solution overnight at room temperature. Thereafter, store the fixed material at 4°C (*see* **Note 4**).

3.1.2. Chromosome Preparation

High-quality chromosome preparations are required for *in situ* hybridization. The presence of cytoplasm and other cellular debris will reduce the quality of results.

1. Transfer root tips from the fixative solution into clean vials using forceps.
2. Remove the fixative by washing the root tips in 1X enzyme buffer (4 m*M* citric acid and 6 m*M* sodium citrate) for 3 × 5 min (*see* **Note 5**).
3. Digest the root tips in 1% (w/v) cellulase (Sigma) (*see* **Note 6**) and 20% (v/v) pectinase (Sigma) in 1X enzyme buffer for 90–100 min at 37°C.
4. Following digestion, remove the enzyme solution from the vial with a pipette and replace it with 1X enzyme buffer. Incubate the digested material in 1X enzyme buffer at room temperature for at least 15 min before preparing the chromosome spread.
5. Prepare microscope slides by cleaning with 100% ethanol. Chromosome spreads are made under a dissecting microscope.

6. Take enough root tip material for one preparation (usually a single root tip).
7. To the root tip, add a one or two drops of 45% acetic acid to disperse cell cytoplasm and after 2–3 min, using a fine syringe needle and working under a dissecting microscope, remove the root cap and as much outer material as possible to leave the central meristematic tissue on the slide (*see* **Note 7**).
8. Gently spread the meristematic tissue around a small area on the slide and cover the material with a clean cover slip. Firmly press down on the coverslip applying even pressure (*see* **Note 8**).
9. Place the slides with the chromosome spreads onto a metal tray placed onto a bed of dry ice for at least 10 min or until the slides turn opaque, then flick the coverslip off using a razor blade (*see* **Note 9**).
10. Allow the slides to air-dry and observe the chromosome spreads using phase-contrast microscopy. Good preparations should appear with high contrast. Any residual cytoplasm around chromosomes can be seen as a gray shadow. Only slides with at least 10 good chromosome spreads should be used for *in situ* hybridization. Whenever possible cytological preparations should be used immediately. However, slides can be stored at 4°C in a dry environment for several weeks. If stored at –20°C they may be kept for several years.

3.2. Probe Preparation

The method used for probe preparation is described by Salvo-Garrido et al. *(7)*. Essentially the method involves using, as a probe, digested fragments from the plasmid used for transformation. This approach led to significant improvements in the detection of low copy number transgenes over methods using intact plasmids as probes. The method described yields 2 × 50 µL probes.

3.2.1. Plasmid Digestion

1. Prior to plasmid digestion, check the plasmid to be used as a probe by running it on a 1% agarose gel in 1X TAE buffer. The gel should show clean plasmid bands of the expected size with no contamination (*see* **Note 10**). Determine the amount of plasmid needed to result in 1 µg per probe. Note that the probe sample goes through two purification steps that each reduce yield. Starting with approx 4 µg of plasmid gives good results.
2. Digest the plasmid with appropriate enzymes that will release fragments of 1–3 kb. Use either one or two different enzymes to give the required fragments and carry out the digestion at an appropriate temperature for the enzymes chosen in a total volume of 50 µL for approx 3 h. Use an enzyme buffer appropriate to the enzymes chosen (*see* **Note 11**).
3. Following digestion, run 2 µL of the digestion products on a 1% agarose gel in 1X TAE. Check that the expected digestion products are obtained.

3.2.2. Purification of the Digested Plasmid

Purify the digested plasmid using a QIAquick PCR Purification Kit. Follow the manufacturer's instructions.

3.2.3. Labeling of the Probe

Plasmid fragments are labeled by nick translation using DNase 1–DNA polymerase activity to incorporate labeled dUTP together with unlabeled dATP, dCTP, and dGTP.

1. Mix the following components in an Eppendorf tube:
   a. 10 µL of 10X nick translation buffer.
   b. 10 µL of unlabeled nucleotide mix.
   c. 2 µL of 0.1M DTT.
   d. 50 µL of digested plasmid.
   e. 10 µL DNA polymerase–DNase I.
   f. Either 5 µL of 1 m*M* biotin-11-dUTP
   or 2 µL of 1 m*M* dioxigenin-11-dUTP mixed with dTTP. (1 m*M* in 100 m*M* Tris-HCl, pH 7.5) to a concentration of 0.35 m*M* dioxigenin-11-dUTP, 0.65 m*M* dTTP.
   Sterile water to give a total volume of 100 µL.
2. Mix the contents of the tube gently and centrifuge briefly in an ultracentrifuge to bring the contents to the bottom of the tube.
3. Incubate at 15°C for 90 min (*see* **Note 12**).

3.2.4. Probe Purification

It is necessary to purify the probe to remove any unincorporated nucleotides.

1. Purify the probe using a QIAquick Nucleotide Removal Kit. As the total volume of the nick translation mix is 100 µL it is necessary to split this into 2 × 50 µL so as not to overload the QIAquick columns. Follow the manufacturer's instructions. After purification, 50 µL is eluted from each column. Each 50-µL aliquot should contain at least the minimum of 1 µg required for each probe.

3.2.5. Preparation of Marker Probes

A range of marker probes may be used to aid identification of the individual chromosomes so that the transgenes can be allocated to specific chromosomes and chromosome arms. For example, in barley, the probes pTa71 (18S–5.8S–26S rDNA) and pTa794 (5S rDNA), which hybridize to ribosomal DNA, give distinct and specific chromosomal patterns that unambiguously identify all seven individual chromosomes ***(18)***. These barley markers will be used as examples to describe the preparation of labeled marker probes.

1. pTa 71 is linearizsed with *Eco*RV and can be labelled by nick translation using the same protocol described above for the plasmid. Alternatively, pTa 794 can be labeled by PCR, and the protocol for PCR labeling of this marker is given below as an example.
2. Add the following to an Eppendorf tube keeping all components on ice:
   a. 5 µL 10X PCR buffer.
   b. 1.5 µL dNTP stock.

c. 1.5 µL 1 m*M* biotin-11-dUTP **or** digoxigenin-11-dUTP.
d. 1.5 µL 0.2 m*M* M13 reverse primer.
e. 1.5 µL 0.2 m*M* M13 single stranded primer.
f. 1 µL miniprep DNA of pTA794 diluted 1: 100 in 1X TE.
g. 0.5 µL *Taq* DNA polymerase.
h. 37.5 µL water = 50 µL total volume.

3. Carry out the PCR reaction using the following conditions:

   a. 94°C 5 min.
   b. 94°C, 30 s.
   c. 56°C, 30 s
   d. 72°C, 90 s.
   e. To **step b** for 30 cycles.
   f. 72°C, 5 min.

   Thereafter, store at –20°C or keep on ice when in use.

3.2.6. Checking the Incorporation of the Label in the Probe

1. Soak a small square of Hybond N+ membrane in buffer 1 in a sterile Petri dish for 5 min then blot dry between filter paper
2. In pencil, draw one circle on the membrane for each probe you wish to check. Load 1 µL of probe inside the circle and allow 10 min to air-dry.
3. Incubate the membrane in buffer1 for 1 min, then buffer 2 for 30 min at room temperature.
4. Incubate membrane for 30 min at 37°C in a 1:500 dilution of anti-biotin-AP Fab fragment in buffer 1 for the detection of biotin or a 1:1000 dilution of anti digoxigenin-AP Fab fragment in buffer 1 for the detection of digoxigenin.
5. Wash membrane in buffer 1, 3 × 5 min, then transfer the membrane to buffer 3 for 2 min.
6. Incubate membrane in the detection solution for 10 min in the dark.
7. Wash the membrane in water and air-dry.
8. Dark spots inside the circles on the membrane indicate that the labeled nucleotide has been incorporated into the probe.

3.3. Hybridization

3.3.1. Pretreatment

Pretreatment is required to reduce nonspecific hybridization of probe to non-target nucleic acid and to reduce nonspecific interactions with proteins that may bind to the probe.

1. Place slides in a humid chamber at 37°C (*see* **Note 13**).
2. To each slide add 200 µL of 100 µg/mL of RNase A in 2X SSC to the area containing the chromosome squash and cover with a plastic cover slip (*see* **Note 14**).
3. Place the chromosomes preparations back in the humid chamber in a 37°C oven for 1 h.

4. After RNase treatment, wash for 5 min with 2X SSC at 37°C and at 120 rpm in a shaking 37°C water bath to remove the plastic coverslip.
5. Add 200 µL of 25 µg/µL pepsin in 0.01 *M* HCl to the squash area on each slide, cover with a plastic cover slip, and incubate for 10 min at 37°C.
6. Wash the slides in water for 2 min to stop the reaction and to float off the cover slips, then wash slides in 2X SSC for 3 × 5 min (*see* **Note 15**).

3.3.2. Prehybridization Fixation

Incubate slides in 4% freshly depolymerized paraformaldehyde in water for 10 min at room temperature, then wash for 3 × 5 min in 2X SSC. This step is to stabilise the chromosomes prior to dehydration (*see* **Note 16**).

3.3.3. Dehydration

Dehydrate the slides successively in 70, 90, then 100% ethanol for 3 min each and allow to air-dry. Do not leave them for too long to dry or they will rehydrate. Dehydration prevents dilution of the probe.

3.3.4. Hybridization

1. Prepare the probe mix for hybridization by adding the following to an Eppendorf tube:
   a. 20 µL 100% formamide.
   b. 8 µL 50% dextran sulfate.
   c. 4 µL 20X SSC.
   d. 0.5 µL 10% SDS.
   e. 2 µL sheared salmon sperm DNA (5 µg/µL).
   f. 1 µL probe (1 µg/µL).
   g. 4.5 µL water = 40 µL total volume (this is the quantity needed for one slide).

   When making up the hybridization mix, first add the water to the dextran sulfate to dilute it so that it is easier to handle.
2. Denature the hybridization mix at 70°C for 15 min, pulse spin to bring the contents to the bottom of the tube, then transfer to ice immediately.
3. Add 40 µL of the hybridization mix to the squash area on each slide as quickly as possible, cover with a plastic coverslip, and place in a thermocycler (Omnislide) (*see* **Note 17**). Use the following conditions in the thermal cycler:
   a. 78°C, 10 min.
   b. 50°C, 1 min.
   c. 45°C, 90 s.
   d. 40°C, 2 min.
   e. 38°C, 5 min.
   f. 37°C, for 16 h overnight.

3.3.5. Posthybridization Washes

Posthybridization washes are carried out to remove any nonspecific or weakly bound probe.

1. Float cover slips away by washing in 2X SSC, in a 37°C water bath, shaking at 120 rpm.
2. Remove the SSC and wash in a stringent wash solution (20% [v/v] formamide in 0.1X SSC) for 2 × 5 min at 37–42°C and 120 rpm.
3. Wash slides with 0.1X SSC for 3 × 5 min in a 37°C waterbath with shaking at 120 rpm and then allow to cool to room temperature (*see* **Note 18**).

3.4. Probe Detection

1. Places slides in 4X SSC, 0.2% Tween-20 for 5 min. Add 200 µL of 5% BSA in 4X SSC, 0.2% Tween-20 to each slide and incubate the slides for 5 min at room temperature covered with a plastic cover slip.
2. Biotin-labeled probes can be detected with extra-avidin conjugated to Cy3 and digoxigenin labeled probes can be detected with anti-digoxigenin conjugated to fluorescein. Using the same cover slip, incubate the slides for 1 h at 37°C with the respective antibody and/or avidin at a concentration of 2 µg/mL in a total volume of 50 µL 5% BSA in 4X SSC, 0.2% Tween-20. As the antibody is light and temperature sensitive, prepare the mix in a foil-covered Eppendorf and keep on ice. To add the antibody, leave the plastic cover slip on the slide and pour away any excess 5% BSA in 4X SSC 0.2% Tween-20. Then using forceps, carefully lift the corner of the cover slip and add the 50 µL of antibody and 5% BSA in 4X SSC, 0.2% Tween-20.
3. After 1 h, float away the coverslips by washing the slides in 4X SSC 0.2% Tween-20 for 5 min at 120 rpm, then wash for a further 2 × 5 min in 4X SSC 0.2% Tween-20 at 120 rpm and allow to cool.

3.4.1. Counterstaining and Mountants

1. Add 100 µL DAPI (2 µg/mL) (*see* **Note 19**) to the squash area on each slide and incubate for 10 min at room temperature in the dark to visualize the chromosomes (*see* **Note 20**).
2. Rinse the slides in 4X SSC, 0.2% Tween-20, add a drop of antifade solution, and apply a cover slip. To seal the cover slip, paint the edges of the cover slip with nail varnish. Store slides at 4°C in a dry environment. Slides can be viewed immediately but can also be kept at 4°C for viewing later. Slides will keep for several months.

3.4.2. Microscopic Visualisation

1. Examine slides using a suitable microscope (e.g., Nikon Microphot-SA) as per the manufacturer's instructions. Choose a suitable filter set for the fluorochrome being visualized. An example of using *in situ* hybridization to identify transgene insertion sites in barley is shown in **Fig. 1**. Biotin-labeled pTa794 was used as a marker probe, which identified a subtelomeric region on the long arm of chromosome 4H. Digoxigenin-labeled fragments of the inserted plasmid were used as a transgene probe. In this way, the transgene insertion site was localized to a subtelomeric region of the short arm of chromosome 4H.

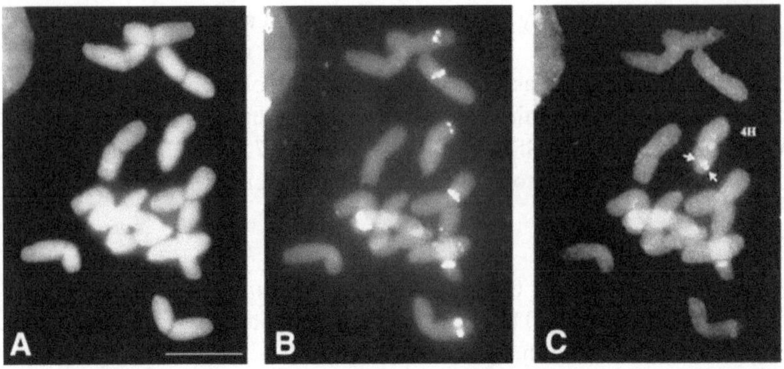

Fig. 1. FISH to determine transgene position in barley. **(A)** DAPI-stained chromosomes. **(B)** *In situ* hybridization with digoxigenin-labeled probe pTa794 detected by green fluorescence of anti-digoxigenin conjugated to fluorescein. **(C)** *In situ* hybridization with biotin labeled probe made from the plasmid used for transformation detected by red fluorescence of extra-avidin conjugated to Cy3. Bar on DAPI-stained cell = 17 µm.

2. Photographic images can be captured on Fuji 400 ISO (ASA) 35-mm film or similar film using an automatic camera on a photomicrographic attachment (*see* **Note 21**).

4. Notes

1. Germination should be carried out using sterile distilled water and in sterile Petri dishes. If necessary, the seed can be surface sterilized by rinsing with 70% ethanol, washing two times in sterile distilled water, rinsing for 1 min in sodium hypochlorite solution (6% [w/v] available chlorine) followed by at least three washes in sterile distilled water to prevent contamination of seeds.
2. Take care not to touch the end of the root tips, as this may cause cell damage.
3. It is really important that the water is cold enough to contain ice crystals. Shake the water vigorously to aerate it before putting in the root tips. A colchicine treatment is an alternative to the ice water treatment ***(17)***.
4. Make up the fixative solution fresh each time. It is also recommended to use fresh ethanol as an opened solution may absorb moisture from the air, which dilutes the ethanol.
5. Root tips can easily be transferred from vial to vial, without damaging the root cap, by using a Pasteur pipet.
6. Adjust the digestion time and concentration of the enzyme to suit the material. It is often necessary to increase the concentration of cellulase to 2%. The longer the material has been fixed, the longer digestion is required.
7. If the root cap is difficult to remove, this indicates that digestion should have been carried out for a little longer.

8. It is important to make sure that squashes are flat. Good indications of an unflat squash are air bubbles under the cover slip. Try to tap out any air bubbles using a syringe needle before pressing down on the cover slip.
9. Before placing the slide on dry ice, etch the surface of the slide, around the edge of the cover slip using a diamond pencil. This will ensure that, once the cover slip has been removed, one will still know the location of the squash on the slide. The metal tray should be placed on a bed of level dry ice, so that the surface freezes evenly. To help prevent cytological preparation loss, is it very important to flick off the cover slip while the slide if still frozen, that is, opaque. Despite this measure, material can be lost from the slide when removing the cover slip. However, if root tips have been incubated in icy water to allow accumulation of metaphase chromosomes then there should be plenty of cells left on the slide with chromosomes at the correct stage of division so if a few are lost it should not be a problem.
10. Use sterile water in the preparation of buffers and solutions to ensure that they are free from contamination.
11. Good laboratory practice is important, enzymes should be kept on ice and frozen buffers and sample stocks should be thawed at room temperature and thereafter be kept on ice.
12. It is very important that the temperature does not go above 16°C as this reduces the efficiency of nick translation. Therefore, it is better to incubate the mix at 15°C in a PCR machine rather than in a water bath to achieve better temperature control.
13. Prior to hybridization pretreatments, place slides in a humid chamber at 37°C. The humid chamber can be a metal tray lined with filter paper that is soaked in sterile distilled water.
14. Plastic cover slips can be cut from plastic autoclave bags (e.g., Guest Medical).
15. Be gentle during the washing steps, as violent washing will remove cells from the slide.
16. Take care to dispose of the waste paraformaldehyde solution correctly.
17. The denaturation step is the critical step and the other steps in the thermocycler just allow a slow cool down to 37°C. Therefore, the use of a thermocycler is not essential for denaturation. If a thermocycler is used add only sterile water to the thermocycler humid chamber. The presence of dextran sulphate in the hybridization solution makes the solution viscose and, therefore, difficult to spread over the squash area without disturbing the material on the slide. To overcome this, pipet the hybridization mixture onto the slide in 8×5 μL aliquots and when applying a plastic cover slip, ensure the aliquots of hybridization mixture join so that no air bubbles remain over the squash area.
18. Preincubate all solutions at 37°C before use. Note that waste formamide solution must be disposed of correctly.
19. **Caution:** Care is needed in handling DAPI, as it is carcinogenic.
20. These steps should be carried out as quickly as possible as the fluorescence of the slide will be lost by exposure to light.
21. FISH signals can fade quickly so that it is often only possible to take one or two pictures of any particular cell.

Acknowledgments

This work was supported by the Food Standards Agency and the Biotechnology and Biological Sciences Research Council of the UK. H. Salvo-Garrido acknowledges the support of the National Institute of Agriculture Research (INIA) Chile and S. Travella support from the European Commission (FAIR program).

References

1. Gall, J. and Pardue, M. L. (1969) Formation and detection of RNA–DNA hybrid molecules in cytological preparations. *Proc. Natl. Acad. Sci. USA* **63,** 378–383.
2. Ten Hoopen, R. , Robbins, T. P., Fransz, P. F., et al. (1996) Localization of T-DNA insertions in petunia by fluorescence *in situ* hybridization: physical evidence for suppression of recombination. *Plant Cell* **8,** 823–830.
3. Iglesias, V. A., Moscone, E. A., Papp, I., et al. (1997) Molecular and cytogenetic analyses of stably and unstably expressed transgene loci in tobacco. *Plant Cell* **9,** 1251–1264.
4. Dong, J., Kharb, P., Teng, W., and Hall, T. C. (2001) Characterisation of rice transformed via an *Agrobacterium*-mediated inflorescence approach. *Mol. Breed.* **7,** 187–194.
5. Snowdon, R. J., Böttinger, P., Pickardt, T., Köhler, W., and Friedt, W. (2001) Physical localisation of transgenes on *Vicia faba* chromosomes. *Chromosome Res.* **9,** 607–610.
6. Pedersen, C., Zimny, J., Becker, D., Jähne-Gartner, A., and Lörz, H. (1997) Localization of introduced genes on the chromosomes of transgenic barley, wheat and triticale by fluorescence *in situ* hybridisation. *Theor. Appl. Genet.* **94,** 749–757.
7. Salvo-Garrido, H., Travella, S., Schwarzacher, T., Harwood, W. A., and Snape, J. W. (2001) An efficient method for the physical mapping of transgenes in barley using *in situ* hybridisation. *Genome* **44,** 104–110.
8. Choi, H. W., Lemaux, P. G., and Cho, M.–J. (2002) Use of fluorescence *in situ* hybridisation for gross mapping of transgenes and screening for homozygous plants in transgenic barley (*Hordeum vulgare* L.). *Theor. Appl. Genet.* **106,** 92–100.
9. Abranches, R., Santos, A. P., Wegel, E., et al. (2000) Widely separated multiple transgene integration sites in wheat chromosomes are brought together at interphase. *Plant J.* **24,** 713–723.
10. Bourdon, V., Ladbrooke, Z., Wickham, A., Lonsdale, D., and Harwood, W. A. (2002) Homozygous transgenic wheat plants with increased luciferase activity do not maintain their high level of expression in the next generation. *Plant Sci.* **163,** 297–305.
11. Leggett, J. M., Perret, S. J., Harper, J., and Morris, P. (2000) Chromosomal localisation of co-transformed transgenes in hexaploid cultivated oat *Avena sativa* L. using fluorescence *in situ* hybridisation. *Heredity* **84,** 46–53.

12. Svitashev, S. K., Ananiev, E., Pawlowski, W. P., and Somers, D. A. (2000) Association of transgene integration sites with chromosomal rearrangements in hexaploid oat. *Theor. Appl. Genet.* **100,** 872–880.
13. Svitashev, S. K. and Somers, D. A. (2001) Genomic interspersions determine the size and complexity of transgenic loci in transgenic plants produced by microprojectile bombardment. *Genome* **44,** 691–697.
14. Svitashev, S. K. and Somers, D. A. (2002) Characterisation of transgene loci in plants using FISH: A picture is worth a thousand words. *Plant Cell Tiss. Org. Cult.* **69,** 205–214.
15. Kohli, A., Twyman, R. M., Abranches, R., Wegel, E., Stoger, E., and Christou, P. (2003) Transgene integration, organization and interaction in plants. *Plant Mol. Biol.* **52,** 247–258.
16. Schwarzacher, T. and Heslop-Harrison, P. (2000) *Practical* In Situ *Hybridisation*, BIOS, Oxford, UK.
17. Schwarzacher, T. and Leitch, A. R. (1994) Enzymatic treatment of plant material to spread chromosomes for *in situ* hybridization, in *Protocols for Nucleic Acid Analysis by Non-Radioactive Probes* (Isaac, P. G., ed.), Humana Press, Totowa, NJ.
18. Leitch, I. J. and Heslop-Harrison, J. S. (1993) Physical mapping of four sites of 5S rDNA sequences and one site of the alpha-amylase-2 gene in barley (*Hordeum vulgare*). *Genome* **36,** 517–523.

23

Amplification of Genomic Sequences Flanking T-DNA Insertions by Thermal Asymmetric Interlaced Polymerase Chain Reaction

Yao-Guang Liu, Yuanlin Chen, and Qunyu Zhang

Abstract

Thermal asymmetric interlaced polymerase chain reaction (TAIL-PCR) is an efficient tool for the recovery of DNA fragments adjacent to known sequences. A protocol is presented for the amplification of genomic sequences flanking DNA (T-DNA or transposon) insertions using the TAIL-PCR method. The amplified products are suitable as templates for direct sequencing, or for cloning in vectors. Examples are given in use of the protocol for T-DNA tagging in rice using the pCAMBIA binary vectors.

Key Words: DNA tagging; genomic flanking sequence; TAIL-PCR.

1. Introduction

DNA tagging by T-DNA and transposon insertions has become an important approach for study of functional genomics in plants. With this approach large numbers of DNA-insertion lines and important mutations have been created in *Arabidopsis* and rice. To identify the genes tagged by DNA insertions, it is necessary to recover genomic sequences flanking the insertion tags. However, the tagged gene sequences cannot be obtained simply by regular specific polymerase chain reaction (PCR) procedures because the genomic flanking sequences are unknown. So far several PCR-based methods such as inverse PCR *(1,2)* and thermal asymmetric interlaced (TAIL) PCR *(3–5)* have been developed for amplification of unknown DNA fragments flanked by known sequences. With the advantages of simplicity and high efficiency, TAIL-PCR has been widely used for molecular biology studies. This chapter presents a

From: *Methods in Molecular Biology, vol. 286: Transgenic Plants: Methods and Protocols*
Edited by: L. Peña © Humana Press Inc., Totowa, NJ

```
                    TR1
5'gtaatgcatgacgttatttatgagatgggtttttatgattagagtcc

   TR2
cgcaattatacatttaatacgcgatagaaaacaaaatatagcgcgcaaa

    TR3                   TR4
ctaggataaattatcgcgcgcggtgtcatctatgttactagatcgggaa

                       Right border
ttaaactatcagtgtttgacaggatatattggcgggtaaacctaag-3'


          TL1                            TL2
5'attagggttcctatagggtttcgctcatgtgttgagcatataagaaa

cccttagtatgtatttgtatttgtaaaatacttctatcaataaaatttc

       TL3                                 TL4
taattcctaaaaccaaaatccagtactaaaatccagatccccgaatta

attcggcgttaattcagtacattaaaaacgtccgcaatgtgttattaag

                          Left border
ttgtctaagcgtcaatttgtttacaccacaatatatcctgccacca-3'
```

Fig. 1. Specific primers used for TAIL-PCR of insertion flanking sequences. The T-DNA is from the pCAMBIA binary vectors. Primers TR1/TL1, TR2/TL2, TR3/TL3, and TR4/TL4 are corresponding to the P1, P2, P3, and P4 shown in **Fig. 2**, respectively.

detailed protocol for the TAIL-PCR with the examples of amplification of rice genomic sequences tagged by T-DNA using the pCAMBIA binary vectors. Specific products isolated by TAIL-PCR can be used as templates for direct sequencing or for cloning.

2. Materials

1. Genomic DNAs prepared by conventional methods (working DNA samples are diluted with 5 m*M* Tris-HCl, pH 8.0, or 1/4X TE buffer in concentrations of 30–50 ng/mL).
2. Specific primers (*see* **Fig. 1**): TR1 and TL1 (1.5 μ*M*); TR2 and TL2 (2.0 μ*M*); TR3 and TL3 (2.5 μ*M*), and AD primers (20 μ*M*). The primers are diluted with 5 m*M* Tris-HCl, pH 8.0
3. *Ex-Taq* DNA polymerase (Takara, Japan) (*see* **Note 1**).
4. 10X *Ex-Taq* buffer ($MgCl_2$ free).
5. 25 m*M* $MgCl_2$.
6. dNTPs mixture (2.5 m*M* each of dATP, dCTP, dGTP and dTTP).
7. Low melting agarose.
8. β-Agarase.
9. Gel purification kit.
10. Sequencing kit.

3. Methods

TAIL-PCR uses three nested specific primers in consecutive reactions together with an arbitrary degenerate (AD) primer having a lower T_m (melting temperature), so that the relative amplification efficiencies of specific and non-specific products can be thermally controlled. In the primary reaction, one low stringency PCR cycle is conducted to create one or more annealing sites for the AD primer in the targeted sequence. Specific product is then preferentially amplified over nonspecific ones by swapping of two high-stringency PCR cycles with one reduced-stringency PCR cycle (**Fig. 2**). This is based on the principle that in the high-stringency PCR cycles with high annealing temperature only the specific primer having a higher T_m can efficiently anneal to target molecules, whereas the AD primer is much less efficient for annealing because of its lower T_m. The seminested PCR amplifications help to achieve higher specificity. By two rounds of TAIL-PCR reactions, specific products that are primed at one end by specific primers and the other end by AD primer are amplified to levels visible on agarose gel.

3.1. Primer Design

1. Specific primers: Three specific primers are set on the region near the T-DNA right or left border (*see* **Note 2**). A fourth specific primer is used for the sequencing. The first and second specific primers are designed to have 22–26 nt with T_m of 62–65°C as calculated by the formula: $T_m = 69.3 + 0.41(G/C)\% - 650/L$ (L = primer length). The third specific primer can be shorter (18–20 nt) because normal PCR cycling is used for the tertiary reaction. **Figure 2** shows the locations of the specific primers for T-DNA of the pCAMBIA binary vectors. A distance of 50–70 bp is set between the first and second specific primers to facilitate confirmation of the product specificity by the differential shift on agarose gel.
2. AD primers. AD primers are designed to have 15–16 nt with average T_m of approx 45°C. Degenerate bases are introduced in the primers with 64–256 times of degeneracy.

AD1: 5'-NTCGA(G/C)T(A/T)T(G/C)G(A/T)GTT-3'
AD2-1: 5'-NGACGA(G/C)(A/T)GANA(A/T)GAA-3' (*see* **Note 3**)
AD2-2: 5'-NGACGA(G/C)(A/T)GANA(A/T)GTT-3'
AD2-3: 5'-NGACGA(G/C)(A/T)GANA(A/T)GAC-3'
AD2-4: 5'-NGACGA(G/C)(A/T)GANA(A/T)CAA-3'
AD2-5: 5'-NGACGA(G/C)(A/T)GANA(A/T)CTT-3'
AD3: 5'-NGTA(A/T)AA(G/C)GTNT(G/C)CAA-3'

3.2. PCR Reactions

3.2.1. Primary Reaction

1. Prepare primary reaction mixture, each reaction (25 μL) consisting of:
   a. 2.5 μL of 10X PCR buffer.

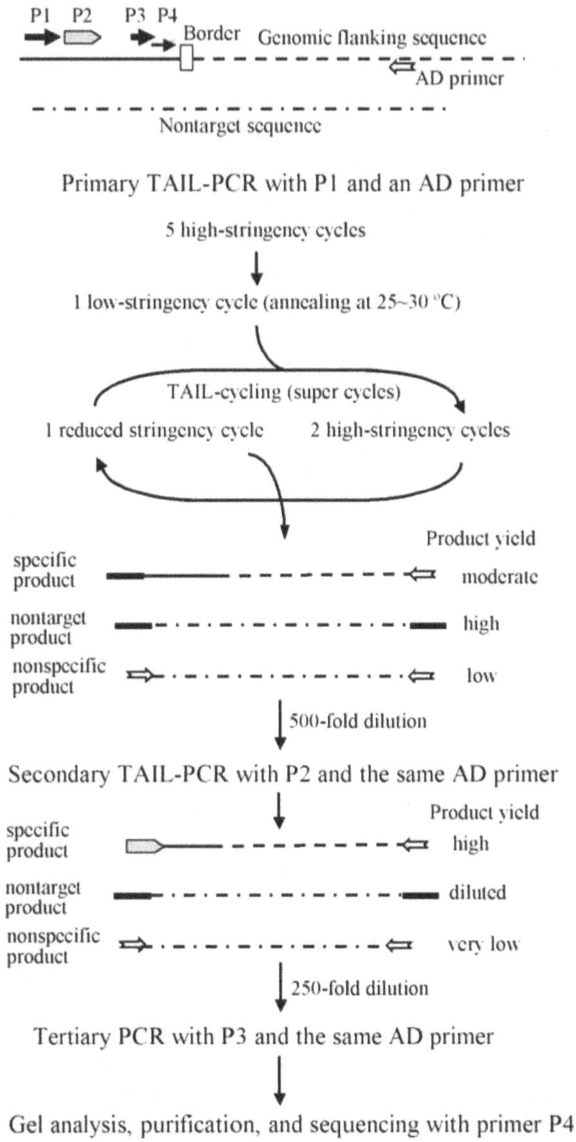

Fig. 2. Schematic diagram of TAIL-PCR procedure. In high-stringency cycles, the high-temperature annealing favors the specific primer having higher T_m for priming, resulting in linear amplification for target molecules and no amplification for nonspecific ones. By interspersing reduced-stringency cycles to allow AD primer for priming, double-stranded molecules can be formed, and the preferential amplification of target molecules becomes logarithmic. In the secondary and tertiary PCR, the nontarget product primed by P1 at both ends fails to be reamplified owing to the lack of the P1 primer.

Table 1
Cycling Conditions Used for TAIL-PCR

Reaction	File no.	Cycle no.	Thermal condition
Primary	1	1	92°C (3 min), 95°C (1 min)
	2	5	94°C (30 s), 65°C (1 min), 72°C (2.5 min)
	3	1	94°C (30 s), 25–30°C (3 min), ramping to 72°C over 2 min, 72°C (2.5 min)
	4	14	94°C (15 s), 65°C (1 min), 72°C (2.5 min), 94°C (15 s), 65°C (1 min), 72°C (2.5 min), 94°C (15 s), 44°C (1 min), 72°C (2.5 min)
	5	1	72°C (5 min)
Secondary	6	11–12	94°C (15 s), 65°C (1 min), 72°C (2.5 min), 94°C (15 s), 65°C (1 min), 72°C (2.5min), 94°C (15 s), 45°C (1 min), 72°C (2.5 min)
	5	1	72°C (5 min)
Tertiary	7	12–14	94°C (40 s), 45°C (1 min), 72°C (2.5 min)
	5	1	72°C (5 min)

Note: The files of each reaction are linked automatically.

   b. 2.0 µL of 25 m*M* $MgCl_2$ (final concentration 2.0 m*M*; *see* **Note 4**).
   c. 2.0 µL of dNTPs mixture.
   d. 2.5 µL of specific primer TR1 or TL1 (3.75 pmol, final concentration 0.15 µ*M*, *see* **Note 5**).
   e. 2.5 µL of any of the AD primers (50 pmol, final concentration 2 µ*M*)
   f. 0.75–0.8 U of *Ex-Taq* DNA polymerase.
   g. dH_2O to 25 µL.
2. To each reaction add 1 µL (30–50 ng) of genomic DNA.
3. Perform primary amplification with thermal conditions as summarized in **Table 1**.
4. Run 10 mL of the product on 1.0% agarose gel (*see* **Note 6**).

3.2.2. Secondary Reaction

1. Prepare secondary reaction mixture, each reaction (25 µL) consisting of:
   a. 2.5 µL of 10X PCR buffer.
   b. 2.0 µL of 25 m*M* $MgCl_2$.
   c. 2.0 µL of dNTPs mixture.
   d. 2.5 µL of specific primer TR2 or TL2 (5 pmol).
   e. 2 µL of the same AD primer (40 pmol).
   f. 0.7 U of *Ex-Taq* DNA polymerase.
   g. dH_2O to 25 mL.

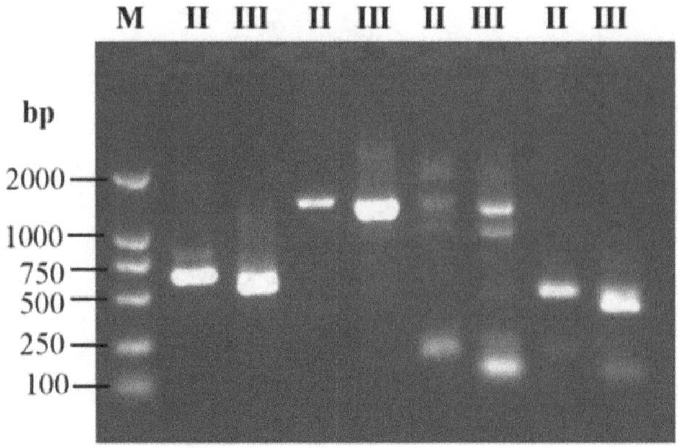

Fig. 3. Agarose gel analysis of TAIL-PCR products from T-DNA tagging rice lines. *Ex-Taq* was used for the PCR reactions. Each set of two lanes contains products from secondary (II) and tertiary (III) reactions. The product specificity is confirmed by the size shift between lanes II and III. Multiple product bands observed in some samples may be nested fragments derived from single insertion by annealing of the AD primer at more than one site along the target sequence molecules, or different sequences if there are multiple insertions. *Lane* M, Molecular weight marker.

2. Dilute 1 μL of the primary PCR product with 20 μL of H_2O and add 1 μL of the diluted DNA to each reaction.
3. Perform secondary amplification with thermal conditions as summarized in **Table 1**.

3.2.3. Tertiary Reaction

1. Prepare tertiary PCR mixtures, each reaction (25 μL) consisting of (*see* **Note 7**):
   a. 2.5 μL of 10X PCR buffer.
   b. 2.0 μL of 25 m*M* $MgCl_2$.
   c. 2.0 μL of dNTPs mixture.
   d. 2.5 μL of specific primer TR3 or TL3 (6.25 pmol).
   e. 1.5 μL of the same AD primer (30 pmol).
   f. 0.6 U of *Taq* DNA polymerase.
   g. dH_2O to 25 mL.
2. Dilute 1 μL of the secondary PCR product with 10 μL of H_2O and add 1 μL of the diluted DNA to each reaction.
3. Perform tertiary amplification using thermal conditions as summarized in **Table 1**.

3.2.4. Agarose Gel Analysis

The secondary and tertiary products (10 μL) were run on 1.0 % agarose gel (**Fig. 3**). The specificity of the products is confirmed by the expected size differences between the secondary and tertiary products (*see* **Note 8**).

3.2.5. Scaled-Up Tertiary Amplification

If specific products are detected, the tertiary amplification is repeated with larger scale of 50 µL per reaction.

3.3. Agarose Gel Purification

3.3.1. Purification by Low-Melting Agarose Gel

1. Run the tertiary PCR products on 0.8% low-melting agarose gel containing 0.3 µg/mL of ethidium bromide.
2. Recover the gel bands containing target DNA fragments and wash twice for 20 min with 5 vol of 1X β-agarase buffer (10 m*M* Tris-HCl, pH 6.5, 1 m*M* EDTA).
3. Melt the gel at 70°C for 5 min, and digest the gel at 40°C for 60–90 min with 0.1–0.2 U of β-agarase/100 mL gel (*see* **Note 9**).
4. Extract the digested gel with an equal volume of equilibrated phenol (without chloroform).
5. Extract the DNA with equal or 2/3 volumes of chloroform (optional).
6. Add 0.1 volume of 3 *M* sodium acetate and 2 vol of ethanol to precipitate the DNA.
7. Resuspend the DNA in 10–15 µL of 5 m*M* Tris-HCl, pH 8.0 (without EDTA), and run 1–2 µL of the DNA on agarose gel with DNA of known amounts (e.g., 10, 20, 40 ng) to check the concentrations of the recovered DNA.

3.3.2. Purification Using a Gel Purification Kit

Run the tertiary PCR products on 0.8% agarose gel containing 0.3 µg/mL ethidium bromide, recover the gel bands containing target DNA fragments, and purify the DNA using a gel purification kit.

3.3.3. Purification by Direct Ethanol-Precipitation

If the tertiary PCR products show single bands, the products can be purified by phenol–chloroform extraction and ethanol precipitation.

3.4. Sequencing

The purified TAIL-PCR products can be used for direct sequencing using the fourth specific primer (*see* **Note 10**), or cloned into a TA-cloning vector.

4. Notes

1. *Ex-Taq* has higher performance and can amplify larger fragments. However, other *Taq* DNA polymerases such as Ampli-*Taq* also can be used for TAIL-PCR. The amount of the *Taq* enzyme for the primary reaction is about 10–20% more than normal PCR.
2. If the successful rate of TAIL-PCR is low in combination with any AD primers, try to design another specific primer for the primary reaction.
3. Because the AD2-1 primer usually gives higher successful rate, AD2-2, AD2-3, AD2-4, and AD2-5 are derived from AD2-1 by modifying the bases at the 3'-end.

4. When other *Taq* polymerases are used, the final concentration of $MgCl_2$ in the primary reaction is 2 m*M*, and those in the secondary and tertiary reactions are 1.5 m*M*.
5. The concentration of primer TR1 or TL1 should not be higher than 0.2 μ*M*.
6. The visible products of the primary PCR are nontarget products primed at both ends by TR1 or TL1, which are created from nonspecific priming and amplified with highest efficiency. The target ones primed at one end by TR1 or TL1 and at another end by AD are still at low levels of yield and nonvisible on agarose gel. Agarose gel analysis of primary products is usually unnecessary for routine amplification.
7. Control reactions with AD primer or specific primers only also can be set, but they are usually unnecessary.
8. If nonspecific products primed at both ends by AD primer are observed, which show the same fragment sizes between the secondary and tertiary reactions, try to decrease the concentration of the AD primer.
9. To reduce the cost, the amount of β-agarase for the gel digestion is decreased and the gel may not be digested completely. However, the partially digested gel can be completely removed by phenol extraction. Although low-melting agarose without digestion with β-agarase also can be removed by repeated phenol extraction, the recovery rate of DNA is relatively low.
10. To reduce the cost for sequencing, about one fourth of a standard reaction of the kit is recommended. For example, use 2 μL of ABI-Big-Dye Sequence Mix and about 30–50 ng of template DNA in 8-μL reaction volume for the sequencing.

References

1. Ochman, H., Gerber, A. S., and Hartl, D. L. (1988) Genetic applications of an inverse polymerase chain reaction. *Genetics* **120,** 621–623.
2. Triglia, T., Peterson, M. G., and Kemp, D. J. (1988) A procedure for in vitro amplification of DNA segments that lie outside the boundaries of known sequences. *Nucl. Acids Res.* **16,** 8186.
3. Liu, Y.-G. and Whittier, R. F. (1995) Thermal asymmetric interlaced PCR: Automatable amplification and sequencing of insert end fragments from P1 and YAC clones for chromosome walking. *Genomics* **25,** 674–681.
4. Liu, Y.-G., Mitsukawa, N., Oosumi, T., and Whittier, R. F. (1995) Efficient isolation and mapping of *Arabidopsis thaliana* T-DNA insert junctions by thermal asymmetric interlaced PCR. *Plant J.* **8,** 457–463.
5. Liu, Y.-G. and Huang, N. (1998) Efficient amplification of insert end sequences from bacterial chromosome clones by thermal asymmetric interlaced PCR. *Plant Mol. Biol. Rep.* **16,** 175–181.

VI

Risk Assessment

24

Agrobacterium Persistence in Plant Tissues After Transformation

Jaime Cubero and María M. López

Summary

Agrobacterium spp. are routinely used in plant transformation to introduce genes of interest in valuable economic species. However, several agrobacteria species are also plant pathogens with ability to survive in different environments including the inner part of the plants. To avoid the release of genetic modified bacteria a successful plant transformation protocol must include the total elimination of agrobacteria by the use of antibiotics. Because sometimes these antibiotics failed in removing the bacteria entirely, confirmation of agrobacteria absence after plant transformation and regeneration is required. Different methodologies can be used for this purpose: isolation techniques followed by identification are used if detection of viable and culturable bacteria is necessary and techniques based on the polymerase chain reaction can be used to detect agrobacteria independently of their physiological state. Here we present several protocols to detect *Agrobacterium* in tissues of transformed plants as well as methods to identify the strains isolated. These identification methods can help to elucidate if they are the engineered bacteria used in the transformation process or just part of the natural endophytic microbiota.

Key Words: Biovar; detection, enrichment; genetically modified organism; isolation; polymerase chain reaction.

1. Introduction

The ability of *Agrobacterium* to transfer a fragment of plasmid DNA into plant cells is routinely used for plant transformation, and some of the strains of this bacterium have become important biotechnological tools *(1)*.

Agrobacterium is a phytopatogenic Gram-negative bacterium able to infect a large number of host plants naturally *(2)* as well as to survive in different envi-

From: *Methods in Molecular Biology, vol. 286: Transgenic Plants: Methods and Protocols*
Edited by: L. Peña © Humana Press Inc., Totowa, NJ

ronments that includes not only plant tissues but the soil, water, and even humans *(3–6)*. Tumors resulted after transformation, represent theoretically the optimal environment for agrobacteria to survive and multiply because of the production of specific substances called opines that are essential for bacterial division and plasmid transference *(7)*. However, it is also possible to find *Agrobacterium* outside the tumors after its translocation within the plant, and furthermore pathogenic agrobacteria have been detected inside nonsymptomatic plants without tumors. Internal migration of *Agrobacterium* has been described in several hosts; primarily in grapevine but also in some herbaceous plants as well as in rose and recently in fruit trees *(8–20)*. In grapevine *Agrobacterium* is considered a systemic organism able to move internally along the plants *(13)*; in fact, protocols to disinfect plant material from pathogenic agrobacteria are used before plant multiplication for sanitation *(21)*. In rose, *Agrobacterium* has been also observed as systemic and able to induce tumors outside the first site of bacterial entrance *(18)*. This phenomenon has been also recently reported in fruit trees where distribution of the bacteria along cherry, apricot, peach × almond hybrids, and walnut has been revealed in symptom and symptomless plants *(19,20)*. Moreover, persistence of the strain K84, a nonpathogenic strain used in biological control, has been observed in the stem of peach × almond plants *(19)*. Furthermore, nonpathogenic agrobacteria are frequently found as endophytic bacteria and have been discovered even in *Brassica napus* seeds, indicating the potential risk of bacterial spread via adult plants by unexpected dissemination means *(22)*. Taken together, these data demonstrate the ability of *Agrobacterium* to persist in low populations in different organs of the plants, to migrate inside them, and to survive in widely diverse environments. In addition, presence of engineered *Agrobacterium tumefaciens* has been confirmed in agroinfected plants of several species including tomato, avocado, grapefruit *(23)*, and in ex vitro transgenic tobacco plants after regeneration *(24)*. Experiments performed in our laboratory suggest the persistence of genetic modified agrobacteria even in nonnatural host plants such as citrus. In these assays, a strain of *Agrobacterium* used for transformation was recovered after regeneration in 45–65% of citrus transgenic explants (Cubero, unpublished results). Consequently there is a possibility to detect low populations of the agrobacteria used for transformation in the regenerated plants, and this should be a matter of concern.

Plant transformation protocols should include the elimination of the bacteria used in the transformation process *(25)*. A successful removal of bacterial cells is required to avoid the risk of releasing a genetically modified organism (GMO) to the environment and/or the possible transfer of artificial plasmids to wild-type strains via conjugation and their subsequent disemination in nature.

Usually elimination of *Agrobacterium* is achieved by the addition of antibiotics to the culture media *(25)*. Nevertheless, sometimes the elimination of bacterial cells could not be absolute because the inappropriate selection or relative low efficiency of the substances used. Sometimes the bacteria can inactivate the antibiotics; in other cases the antibiotics could have a bacteriostatic and no bactericidal effect inhibiting the bacterial growth in a reversible mode, instead of killing the bacterial cells. In addition, some effective antibiotics cannot be used because may inhibit plant regeneration *(25–27)*.

The different methodologies proposed for *Agrobacterium* detection and identification in engineered plants can be divided into two main groups: methods based on bacterial isolation that require the existence of viable and culturable bacteria, and methods designed to detect the bacteria in whatever condition bacterial cell might be. Advantages of isolation methods are the simplicity as well as the capacity of detecting only culturable cells, which guarantees the detection solely of living cells. To improve the sensitivity of these techniques and overcome the drawback of the low sensitivity of the regular isolation techniques, it is possible to include an enrichment step in selective medium previous to plating *(19,20)*. Methods autonomous of the bacterial fitness are mainly based in the use of polymerase chain reaction (PCR). They are considered very sensitive, but their main disadvantage is that the bacteria can be detected even if they are already dead. In the case of transgenic plants this is irrelevant because dead cells do not imply any ecological risk. However, these techniques have the advantage over the isolation methods that they are able to detect bacteria when they are still alive but not able to multiply in regular culture media. This condition has been defined as viable but nonculturable (VBNC) state and has been described for several bacterial genera including *Agrobacterium (28)*. VBNC is supposed to be a starvation status for bacteria to avoid unfavorable environmental circumstances while waiting on optimal situation for growth and, consequently, it is a reversible condition. In fact, recovering from VBNC has been described in some bacterial species *(29)*. As the induction of the VBNC is frequent under stress conditions *(30,31)*, it would be possible that the agrobacteria could adopt this state in the transformed tissue.

Combination of isolation and molecular techniques is the best way to accomplish the sensitive and reliable detection of *Agrobacterium* to guarantee the production and multiplication of plants free of engineered bacteria. To reduce the controversy regarding the liberation of GMOs to the environment *(32)* is essential to ensure the elimination of *Agrobacterium* used in the transformation process. An excellent description of material and methods used for detection and characterization of pathogenic agrobacteria is provided in Moore et al. *(33)*. Some of the methods used for detection of this bacterial pathogen can be applied for detection and characterization of the engineered agrobacteria.

2. Material

2.1. Detection and Identification of Agrobacterium by Isolation

1. PYGA medium (peptone yeast extract glycerol agar): 5 g of bactopeptone, 3 g of yeast extract, 10 mL of glycerol, 20 g of agar, 1 L of H_2O. Dissolve ingredients and sterilize the solution at 121°C for 20 min.
2. Medium Schroth ***(34)*** for biovar 1 of *Agrobacterium* (now *A. tumefaciens*): First part: 10 g of mannitol, 4 g of $NaNO_3$, 1.2 g of calcium propionate, 2 g of $MgCl_2 \cdot 6H_2O$, 100 mg of $MgSO_4 \cdot 7H_2O$, 75 mg of $MgCO_3$, 75 mg of $NaHCO_3$,, 20 g of agar, 1 L of H_2O. Second part: 275 mg of berberine, 250 mg of cycloheximide, 100 mg of Na_2SeO_3, 100 mg of penicillin G, 1 mg of tyrothricin, 100 mL of H_2O. Dissolve ingredients of the first part and adjust the pH to 7.0–7.2 before adding the agar. Sterilize the solution at 121°C for 20 min. Cool to approx 50°C, and then add filter-sterilized second part. For broth preparation, do not include agar.
3. Medium 1A for biovar 1 of *Agrobacterium* (now *A. tumefaciens*) ***(35)***: First part: 3.04 g of L-(–)-arabitol, 160 mg of NH_4NO_3, 540 mg of KH_2PO_4, 290 mg of sodium taurocholate, 250 mg of $MgSO_4 \cdot 7H_2O$, 15 g of agar, 2 mL of crystal violet, 0.1% (w/v) aqueous, 1 L of H_2O. Second part: 1 mL of cycloheximide (2% solution), 6.6 mL of Na_2SeO_3 (1% solution). Dissolve ingredients of the first part, adjust the pH to 7.0–7.2, and sterilize the solution at 121°C for 15 min. Cool to approx 50°C, then filter sterilize the second part. For broth preparation, do not include agar.
4. Medium New and Kerr for biovar 2 of *Agrobacterium* (now *A. rhizogenes*) ***(36)***: First part: 5 g of erythritol, 2.5 g of $NaNO_3$, 100 mg of KH_2PO_4, 200 mg of $CaCl_2$, 200 mg of $MgSO_4 \cdot 7H_2O$, 10 mL of biotine 0.02% (w/v) aqueous, 15 g of agar, 1 L of H_2O. Second part: 250 mg of cycloheximide, 10 mg of Na_2SeO_3, 10 mg of bacitracin, 1 mg tyrothricin, 100 mL of H_2O. Dissolve ingredients of the first part and adjust the pH to 7.0–7.2 before adding the agar. Sterilize the solution at 121°C for 15 min. Cool to about 50°C, and then add filter-sterilized second part. For broth preparation, do not include agar.
5. Medium 2E also for biovar 2 of *Agrobacterium* (now *A. rhizogenes*) ***(35)***: First part: 160 mg of NH_4NO_3, 3.05 g of erythritol, 540 mg of KH_2PO_4, 1.04 g of K_2HPO_4, 250 mg of $MgSO_4 \cdot 7H_2O$, 290 mg of sodium taurocholate, 1 mL of yeast extract, 1% (w/v) aqueous, 5 mL of malachite green 0.1% (w/v), 15 g of agar, 1 L of H_2O. Second part: 20 mg of cycloheximide, 6.6 mL of Na_2SeO_3 (1% solution). Dissolve ingredients of the first part, adjust the pH to 7.0–7.2, and sterilize the solution at 121°C for 15 min. Cool to about 50°C, then add filter-sterilized second part. For broth preparation, do not include agar.
6. Roy and Sasser modified medium for biovar 3 of *Agrobacterium* (now *A. vitis*) ***(37)***: First part: 4 g of adonitol, 1 g of H_3BO_3, 140 mg of yeast extract, 200 mg of $MgSO_4 \cdot 7H_2O$, 700 mg of KH_2PO_4, 900 mg of K_2HPO_4, 200 mg of NaCl, 20 g of agar, 900 mL of H_2O. Second part: 80 mg of triphenyltetrazolium chloride, 250 mg of cycloheximide, 100 mL of H_2O. Dissolve ingredients of the first part and adjust the pH to 7.0–7.2 before adding the agar. Sterilize the solution at 121°C for 20 min. Cool to approx 50°C, and then add filter-sterilized second part. For broth preparation, do not include agar.

2.2. Tests to Determine Agrobacterium Biovar

1. The 3-ketolactose test *(38)* consists of a basal Bernaert's medium and Benedict's reagent: Basal medium: 10 g of lactose, 1 g of yeast extract, 20 g of agar, 1 L of H_2O. Benedict's reagent:
   Solution A: 173 g of sodium citrate, 100 g of Na_2CO_3, 600 mL of H_2O.
   Solution B: 17.3 g of $Cu_2(SO)_4$, 150 mL of H_2O.
   Solutions A and B must be prepared separately and mixed under constant stirring. Basal medium should be sterilized at 121°C for 20 min. Benedict's reagent is added at the time of reading the test. A yellow color becomes visible if 3-ketolactose is present after 10–15 min.
2. Simmons citrate utilization *(39)*: 5 g NaCl, 200 mg $MgSO_4 \cdot 7H_2O$, 1 g of $NH_4H_2PO_4$, 1 g of K_2HPO_4, 2 g of sodium citrate, 40 mL of 0.2% bromothymol blue (w/v) in 20 g of agar, 1 L of H_2O. Adjust the pH to 6.8–7 before adding the agar and sterilize at 120°C for 20 min. The medium turns blue if citrate is utilized.
3. Ferric ammonium citrate test *(40)*: 10 g of ferric ammonium citrate, 500 mg of $MgSO_4 \cdot 7H_2O$, 500 mg of K_2HPO_4, 200 mg of $CaCl_2$, 1 L of H_2O. Adjust to pH 7.0 and sterilize at 120°C for 20 min. A brown pellicle becomes visible at the surface in positive samples.
4. Malonic acid test *(41)*: 2 g of $(NH_4)_2SO_4$, 400 mg of KH_2PO_4, 600 mg of K_2HPO_4, 200 mg of NaCl, 1 L of H_2O, 100 mg of yeast extract, 3 g of malonic acid sodium salt. Adjust to pH 7.0 and sterilize at 120°C for 20 min. The medium turns blue if alkali is produced.
5. Acid and alkali production from sucrose, melezitose, L-tartaric, and mucid acid:
   The same basal medium is used in the four tests: Basal medium *(42)*: 1 g of $NH_4H_2PO_4$, 200 mg of KCl, 200 mg of $MgSO_4 \cdot 7H_2O$, 75 mL of bromothymol blue 0.2% (w/v), 1 L of H_2O. Stock solutions are prepared for each sugar or acid: sucrose 10% (w/v) aqueous, melezitose 10 % (w/v) aqueous, L-tartaric acid 1% (w/v) aqueous, mucid acid 1% (w/v) aqueous. Adjust the basal medium to pH 7.0 and sterilize at 121°C for 20 min. Add the sugars or acids from 10 or 1% filter-sterilized stock solutions to adjust the basal medium to 1% sucrose or melezitose and 0.1% L-tartaric or mucid acids. The medium turns yellow or blue if acid or alkali are produced, respectively (*see* **Note 2**).

2.3. Detection and Identification of Agrobacterium by PCR

1. DNA extraction buffer: 200 m*M* Tris-HCl, pH 7.5, 25 m*M* EDTA, 0.5% (w/v) sodium dodecyl sulfate (SDS), 2% (w/v) polyvinylpyrrolidone (PVP) 10,000, 250 m*M* NaCl. Dissolve ingredients in 1 L and filter-sterilize (*see* **Note 3**).
2. Primers for *Agrobacterium* amplification: Primers based on sequences from genes transferred to the plant cell are desirable to identify a purified bacterial culture as the engineered agrobacteria used for transformation. These primers can be designed for example based in the *gus* or *gfp* genes (if those have been used as transformation markers) as well as based in the gene/s of interest introduced into the plant.

   However, when trying to detect *Agrobacterium* in plant material is convenient to use primers that amplify parts of the genome present in the plasmid but not transferred to the plant tissue after the transformation. Some of these primer sets, based in virulence genes within pTi are shown in **Table 1**.

3. Methods

3.1. *Agrobacterium Isolation*

1. Wash portions of the plant to be analyzed in soapy water, flame superficially, and deposit in a sterile plastic Petri dish.
2. Remove epidermis aseptically and comminute small pieces of the sample in sterile distilled water.
3. After 15–30 min maceration, plate comminuted tissues on the semiselective media described for the different agrobacterial biovar and on PYGA medium. Culture medium must be selected according to the biovar of the strain used for transformation. Previous verification of the growth of the strain used for transformation in the semiselective medium is advised.
4. Streak 50–100 µL of the macerated suspensions on the plates. After 3–5 d of incubation at 25–27°C, select and culture colonies for further analysis. If an enrichment step is desired, add 0.5 mL of the macerated suspension to 5 mL of the appropriate semiselective medium broth according to the biovar of the engineered strain used.
5. After 48–72 h of incubation at 25–27°C, plate 50–100 µL of the enrichment solutions on the same semiselective medium used in the liquid growth step.

3.2. *Biochemical Characterization of Bacterial Colonies*

1. Colony morphology varies among different strains as well as according to the culture media used. Nevertheless, *Agrobacterium* is generally cream-white, circular, convex and glossy; in Schroth medium, bacteria typically show a yellow/orange feature, and in New and Kerr and Roy and Sasser media, they often have a dark red center with white edges.
2. Select characteristic *Agrobacterium* colonies from the media described above and purify by subculturing twice in the PYGA medium from bacterial suspensions prepared in sterile distilled water. Cultures can be maintained at –80°C in sterile 30% (v/v) water glycerol solution.
3. Biovar determination can be used as a first approach to determine the origin of the strain isolated. All the tests may be performed in microtiter plates ***(46)***. Dispense 150 µL media in horizontal rows on sterile culture microplate and add 15 µL of a bacterial suspension of 10^8 colony-forming units (cfu)/mL of each strain to be analyzed to each vertical row using a multichannel pipet. Include in all the tests a reference strain of each biovar. In the cases of citrate utilization, use solid medium, and add strains by puncture from a 48-h plate culture. 3-Ketolactose tests can be performed in solid or liquid media, add two drops of Benedict's reagent after 48 h of incubation, and read results in <20 min. After 48 to 72 h of *Agrobacterium* incubation at 25–27°C the majority of other tests can be read and analyzed according to **Table 2** (*see* **Note 2**).

Table 1
Primers Used to Identify *Agrobacterium* Strains Based on Virulence pTi Sequences

Primers	Characteristics
VirB11+21 5' TGC CGC ATG GCG CGT TGT AG 3' virG15 5' GAA CGT GTT TCA ACG GTT CA 3'	This set of primers amplifies a fragment of 246 bp within the intercistronic region between virulence genes *virB* and *virG* in nopaline plasmid type strains ***(43)***
VCF5' ATC ATT TGT AGC GAC T 3' VCR 5' AGC TCA AAC CTG CTT C 3'	This set of primers amplifies a fragment of 730 bp within *virC* operon and they were reported universal for *Agrobacterium* carrying pTi plasmid ***(44)***
Primer A 5' ATG CCC GAT CGA GCT CAA GT 3' Primer C 5' TCG TCT GGC TGA CTT TCG TCA TAA 3' Primer E' 5' CCT GAC CCA AAC ATC TCG GCT GCC CA 3'	Primers designed to amplify fragment of 224 bp (A–C') and 338 bp (A–E') in the *virD2* endonuclease domain and reported to be universal for all agrobacteria carrying the Ti plasmid ***(45)***

Table 2
Test Selected to Determine *Agrobacterium* Biovars and Expected Results

	Biovar 1	Biovar 2	Biovar 3	Positive result
3-Ketolactose production	+	–	V	yellow precipitate
Citrate utilization	V–	+	+	turn to blue
Ferric ammonium citrate	+	–	–	brown precipitate
Alkali from malonate	–	+	+	turn to blue
Acid from:				
Sucrose	+	–	V	turn to yellow
Melezitose	+	–	–	turn to yellow
Alkali from:				
L-Tartaric acid	–	+	+	turn to blue
Mucic acid	–	+	–	turn to blue

V, variable; V–, most of the strains are negative for this test. (Modified from Moore et al. *[33]*.)

4. Usually, 3-ketolactose production is the only test necessary to determine whether the isolated strain is biovar 1, the main biovar of strains used in plant transformation. This test should be complemented by the use of PCR for a positive identification (*see* **Subheading 3.3.**).

3.3. Identification of Agrobacterium *by PCR*

1. PCR can be used to identify strains isolated. Generally, no complicate DNA extraction protocols are required to amplify from water suspensions of *Agrobacterium*. Bacterial suspensions of 10^7–10^8 cfu/mL are heated for 10 min at 95°C and centrifuged 2 min at 13,000*g*. A 5-µL sample is enough to perform PCR amplifications of resuspended pellets.
2. If a more effective DNA extraction protocol is required, follow the protocol described below from plant material after precipitation of the bacterial cells by centrifugation.
3. Primers to amplify specifically genes present in the plasmid used for transformation can be used in these PCR amplifications. Those indicated in **Table 1** can also be used.

3.4. Detection of Agrobacterium *From Plant Material*

DNA extraction from plant material suspected to contain *Agrobacterium* can be performed according to the protocol described in Cubero et al. *(47)* for detection of *Agrobacterium* in tumors.

1. Centrifuge 1 mL of the suspensions prepared as above for *Agrobacterium* isolation centrifuged at 13,000*g* for 5 min.

2. Discard supernatant, resuspend pellet in the DNA extraction buffer described in material **Subheading 2.3.**, and shake for 30 min at room temperature.
3. Then, centrifuge at 2,000*g* for 2 min to clarify the suspension and remove most of the plant material.
4. Heat 300 µL of the supernatant fluids for 5 min at 93°C.
5. Centrifuge samples at 12,000*g* for 1 min.
6. Collect 200 µL of the supernatant fluid and add 200 µL isopropanol; mix the solution gently, and allow to stand at least 1 h at room temperature for DNA precipitation.
7. Next, centrifuge the mixture at 12,000*g* for 15 min, remove the supernatant fluid, dry the precipitate under vacuum or room temperature, resupend in 100 µL of ultrapure water, and use 5 µL of this DNA solution in PCR reactions.
8. Include one negative control without plant material for every DNA extraction and, when possible, perform a negative control of the same plant material analyzed as a sample.
9. To improve the sensitivity, a previous enrichment step in a semiselective medium for *Agrobacterium* can be used. Perform the enrichment procedure as described in **Subheading 3.1.** Centrifuge 1 mL of the medium enriched at 13,000*g* and resuspend the pellet in DNA extraction buffer and shake as described earlier. Follow the protocol as described earlier for plant material without enrichment (*see* **Note 4**).

4. Notes

1. All the media described can be stored for months at 4°C. Second parts of selective media are usually difficult to dissolve. They can be heated at 50°C to facilitate the solution of the components. Their second parts can be prepared as a 100-mL or 1-L stock and stored for several months at 4°C for further use.
2. All the culture medium stock solutions for biovar determination can be stored at 4°C. As described in **Subheading 3.**, section, the entire test can be performed in microtiter plates ***(46)***. The majority of results can be read after 48 and 72 h, although some additional time is required for slow growing strains. Sterile culture microplates must be used and they must be carefully sealed with Parafilm to avoid external contamination after addition of bacteria. The test for 3-ketolactose production is best performed in separate microplates or strips to avoid contamination of the other tests when adding the Bernaert's reagent after 48 h.
3. DNA extraction buffer must be stored after filter sterilization at 4°C. It may be heated at 50°C before it is used to dissolve all the components because usually the SDS precipitates at low temperatures. DNA precipitation can be improved by the use of Pellet Paint coprecipitant (Novagen, Darmstadt, Germany) ***(48)*** as described in other bacterial models ***(49)***.
4. The major limitation of PCR detection in plant material is related to the frequent presence of inhibitor compounds, especially in woody plant material, which can interfere in the amplification reaction and result in a false-negative detection ***(50,51)***. To recognize false negatives resulting from inhibition of PCR, different approaches of internal controls can be used ***(49,51,52)***. Each internal control must

be designed based on the sequence of the primers used, therefore different constructions are required. Coamplification of the internal control and target sequence ensures the attainment of at least one PCR product in every amplification reaction if the DNA extracted is of high enough quality to be amplified ***(51)***.

References

1. Hooykaas, P. J. J. and Schilperoort, R. A. (1992) *Agrobacterium* and plant genetic-engineering. *Plant. Mol. Biol.* **19,** 15–38.
2. De Cleene, M. and De Ley, J. (1976) The host range of crown gall. *Bot. Rev.* **42,** 389–466.
3. Agrios, G. N. (1988) Bacterial galls, in *Plant Pathology*, 3rd Ed., Academic Press, San Diego, CA, pp. 558–565.
4. Moore, L. W. and Candfield, M. (1996) Biology of *Agrobacterium* and management of crown gall disease, in *Principles and Practice of Managing Soilborne Plant Pathogens* (Robert, H., ed.), APS Press, St. Paul, MN.
5. Southern, P. M. (1996) Bacteremia due to *Agrobacterium tumefaciens* (*radiobacter*). Report of infection in a pregnant woman and her stillborn fetus. *Diag. Microbiol. Infect. Dis.* **24,** 43–45.
6. Yu, W. L., Wang, D. Y., and Lin, C. W. (1997) *Agrobacterium radiobacter* bacteremia in a patient with chronic obstructive pulmonary disease. *J. Formos. Med. Assoc.* **96,** 664–666.
7. Tempe, J., Guyon, P., Tepfer, D., and Petit, A. (1979) The role of opines in the ecology of Ti plasmids of *Agrobacterium*, in *Plasmids of Medical, Environmental, and Commercial Importance* (Timmis, K. N. and Puhler, A., eds.), Elsevier/North Holland Biomedical Press, Amsterdam, The Netherlands, p. 353.
8. Riker, A. J. (1923) Some relations of the crown gall organism to its tissue. *J. Agric. Res.* **25,** 119–132.
9. Hill, J. B. (1928) The migration of *Bacterium tumefaciens* in the tissue of tomato plants. *Phytopathology* **18,** 553–564.
10. Suit, R. F. and Eardley, E. A. (1935) Secondary tumor formation on herbaceous hosts induced by *Pseudomonas tumefaciens*. *Sci. Agric.* **15,** 345–357.
11. Stapp, C., Muller, H., Dame, F., and Pfeil, E. (1938) Derpflanzen Krebs und sein Erregen *P. tumefaciens* VIII. Mitt unterssuchungen uber die Moglicheit einer wirksamen Bekampfung an Kernobstgeholzen. *Zentrabl. Bakteriol. II* **99,** 210–276.
12. Braun, A. C. (1941) Development of secondary tumors and tumor strands in the crown gall of sunflowers. *Phytopathology* **31,** 135–149.
13. Lehoczky, J. (1968) Spread of *Agrobacterium tumefaciens* in the vessels of the grapevine, after natural infection. *Phytopath. Z.* **63,** 239–246.
14. Lehoczky, J. (1971) Further evidences concerning the systemic spreading of *Agrobacterium tumefaciens* in vascular system of grapevine. *Vitis* **10,** 215–221.
15. El Khalifa, M. D., El Nur, E. E., Lippincott, B. B., and Lippincott, J. A. (1973) Crown gall on castor bean leaves. *J. Exp. Bot.* **24,** 1117–1129.
16. Miller, H. N. (1975) Leaf, stem, crown and root galls induced in chrysanthemum by *Agrobacterium tumefaciens*. *Phytopathology* **65,** 805–811.

17. Bouzar, H., Chilton, W. S., Nesme X., et al. (1995) A new *Agrobacterium* strain isolated from aerial tumours on *Ficus benjamina* L. *Appl. Environ. Microbiol.* **61,** 65–73.
18. Marti, R., Cubero, J., Daza, A., et al. (1999) Evidence of migration and endophytic presence of *Agrobacterium tumefaciens* in rose plants. *Eur. J. Plant Pathol.* **105,** 39–50.
19. Cubero, J. (1998) Detección y movimiento de *Agrobacterium tumefaciens* en distintas especies vegetales. PhD Thesis, Universidad de Valencia, Spain.
20. Cubero, J., Lastra, B., Salcedo, C. I., Piquer, J., and López, M. M. (2004) Detection of *Agrobacterium* translocation inside plants of different species. In preparation.
21. Burr T. J., Ophel, K., Katz, B. H., and Kerr, A. (1989) Effect of hot water treatment on systemic *Agrobacterium tumefaciens* biovar 3 in dormant grape cuttings. *Plant Dis.* **73,** 242–245.
22. Weller, S. A., Simpkins, S. A., Stead, D. E., Kurdziel, A., Hird, H., and Weekes, R. J. (2002) Identification of *Agrobacterium* spp. present within *Brassica napus* seed by Taqman PCR—implications for GM screening procedures. *Arch. Microbiol.* **178,** 338–343.
23. Mogilner, N., Zutra D., Gafny, R., and Bar-Joseph, M. (1993) The persistence of engineered *Agrobacterium tumefaciens* in agroinfected plants. *Mol. Plant Microbe Interact.* **5,** 673–675.
24. Matzk, A., Sinclair, M., and Sciemann, J. (1996) Localization of persisting agrobacteria in transgenic tobacco plants. *Mol. Plant Microbe Interact.* **9,** 373–381.
25. Tang, H. and Krczal, R. G. (2000) An evaluation of antibiotics for the elimination of *Agrobacterium tumefaciens* from walnut somatic embryos and for the effect on the proliferation of somatic embryos and regeneration of transgenic plants. *Plant Cell Rep.* **19,** 881–887.
26. Hammerschlag, F. A., Zimmerman, R. H., Yadava, U. L., Hunsucher, S., and Gercheva, P. (1995) An evaluation of antibiotics for elimination of *Agrobacterium tumefaciens* from apple leaf explants in vitro and for effect of regeneration. *Hort. Sci.* **30,** 876.
27. Shackelford, N. J. and Chlan, C. A. (1996) Identification of antibiotics that are effective in eliminating *Agrobacterium tumefaciens. Plant. Mol. Biol. Rep.* **14,** 50–57.
28. Manahan, S. H. and Steck, T. R. (1997) The viable but nonculturable state in *Agrobacterium tumefaciens* and *Rhizobium meliloti. FEMS Microbiol. Ecol.* **22,** 29–37.
29. Whitesides, M. D. and Oliver, J. D. (1997) Resuscitation of *Vibrio vulnificus* from the viable but nonculturable state. *Appl. Environ. Microbiol.* **63,** 1002–1005.
30. McDougald, D., Rice, S. A., Weichart, D., and Kjelleberg, S. (1998) Nonculturability: adaptation or debilitation? *FEMS Microbiol. Ecol.* **25,** 1–9.
31. Alexander, E., Pham, D., and Steck, T. R. (1999) The viable-but-nonculturable condition is induced by copper in *Agrobacterium tumefaciens* and *Rhizobium leguminosarum. Appl. Environ. Microbiol.* **65,** 3754–3756.

32. Stewart, C. N., Richards, H. A., and Halfhill, M. D. (2000) Transgenic plants and biosafety: science, misconceptions and public perceptions. *BioTechniques* **29,** 832–843.
33. Moore, L. W., Bouzar, H., and Burr, T. (2001) *Agrobacterium*, in *Laboratory Guide for Identification of Plant Pathogenic Bacteria*. (Schaad, N. W., Jones, J. B., and Chun, W., eds.), APS Press, St. Paul, MN, pp. 17–34.
34. Schroth, M. N., Thompson, J. P., and Hildebrand, D. C. (1965) Isolation of *A. tumefaciens–A. radiobacter* group from the soil. *Phytopathology* **55,** 645–647.
35. Brisbane, P. G. and Kerr, A. (1983) Selective media for three biovars of *Agrobacterium*. *J. Appl. Bacteriol.* **54,** 425–431.
36. New, P. B. and Kerr, A. (1971) A selective medium for *Agrobacterium radiobacter* biotype 2. *J. Appl. Bacteriol.* **34,** 233–236.
37. Roy, M. A. and Sasser, M. (1983). A medium selective for *Agrobacterium tumefaciens* biotype 3 (abstr.). *Phytopathology* **73,** 810.
38. Bernaerts, M. J. and De Ley, J. (1963) A biochemical test for crown gall bacteria. *Nature* **197,** 406–407.
39. Simmons, J. S. (1926) A culture medium for differentiating organisms of typhoid-colon aerogenes groups and for isolation of certain fungi. *J. Infect. Dis.* **39,** 209–214.
40. Hendrickson, A. A., Baldwin, I. L., and Riker, J. (1934) Studies on certain physiological characters of *Phytomonas tumefaciens*, *Phytomonas rhizogenes*, and *Bacillus radiobacter*. II. *J. Bacteriol.* **28,** 597–618.
41. Kerr, A. and Panagopoulos, C. G. (1997) Biotypes of *Agrobacterium radiobacter* var. *tumefaciens* and their biological control. *Phytopath. Z.* **90,** 172–179.
42. Ayers, S. H., Rupp, P., and Johnson, W. T. (1919) A study of the alkali-forming bacteria in milk. United States Department of Agriculture Bulletin 782.
43. Nesme, X., Leclerc, M. C., and Bardin, R. (1989) PCR detection of an original endosymbiont: the Ti plasmid of *Agrobacterium tumefaciens*, in *Endocytobiology IV*. (Nardon, P., Gianinazzi-Peason, V., Greines, A. M., Margulis, L. and Smith, D. C., eds.). Institute National de Recherche Agronomique, Paris, pp. 47–50.
44. Sawada, H., Ieki, H., and Matsuda, I. (1995) PCR detection of Ti and Ri plasmids from phytopathogenic *Agrobacterium* strains. *Appl. Environ. Microbiol.* **61,** 828–831.
45. Haas, J. H., Moore, L. W., Ream, W., and Manulis, S. (1995) Universal PCR primers for detection of phytopathogenic *Agrobacterium* strains. *Appl. Environ. Microbiol.* **61,** 2879–2884.
46. Cubero, J. and López, M. M. (2001) An efficient microtiter system to determine *Agrobacterium* biovar. *Eur. J. Plant Pathol.* **107,** 757–760.
47. Cubero, J., Martínez, M. C., Llop, P., and López, M. M. (1999) A simple and efficient PCR method for the detection of *Agrobacterium tumefaciens* in plant tumors. *J. Appl. Microbiol.* **86,** 591–602.
48. Taggart, E. M., Byngton, C. L., Hillyard, D. R., Robinson, J. E., and Carrol, K. C. (1999) Enhancement of the amplicor enterovisus PCR test with a coprecipitant. *J. Clin. Microbiol.* **36,** 3408–3409.

49. Cubero, J., Graham, J. H., and Gottwald, T. R. (2001) Quantitative PCR method for diagnosis of citrus bacterial canker. *Appl. Environ. Microbiol.* **67,** 2849–2852.
50. Wilson, I. G. (1997) Inhibition and facilitation of nucleic acid amplification. *Appl. Environ. Microbiol.* **63,** 3741–3751.
51. Cubero J., van der Wolf, J., van Beckhoven, J., and Lopez, M. M. (2002) An internal control for the diagnosis of crown gall by PCR. *J. Microbiol. Methods* **51,** 387–392.
52. Sachadyn, P. and Kur, J. (1998) The construction and use of an internal control. *Mol. Cell. Probes* **12,** 259–262.

25

Transgene Dispersal Through Pollen

Laura C. Hudson, Matthew D. Halfhill, and C. Neal Stewart, Jr.

Summary

Techniques used for the transfer of novel genes into host plant genomes have created new possibilities for crop improvement. The implementation of transgenic crop species into agriculture has introduced the possibility of transgene escape into the environment via pollen dispersal. Although the movement of pollen is a critical step in transgene escape, there is currently no system to monitor transgenic pollen movement under field conditions. The development of an effective in vivo monitoring system suitable for use under field conditions is needed for research and commercial purposes so potential risks can be quantified and evaluated. This chapter describes the development of a model system using green fluorescent protein (GFP) expression in pollen as a marker to monitor pollen distribution patterns. A pollen specific promoter was used to express the GFP gene in tobacco (*Nicotiana tabacum* L.). GFP was visualized in pollen and growing pollen tubes using fluorescent microscopy. Furthermore, the goal of this research was to compare the dynamics of pollen movement with that of gene flow by using another method of whole plant expression of GFP (*see* Chapter 15) to estimate out-crossing frequencies by progeny analysis. Pollen movement and gene flow were quantified under field conditions. Pollen traps were collected and screened for presence of GFP-tagged pollen using fluorescence microscopy. Progeny from wild type plants were screened with a hand held ultraviolet light for detection of the GFP phenotype.

Key Words: Gene flow; green fluorescent protein; *Nicotiana tabacum*; out-crossing; pollen flow; transgenic.

1. Introduction

Over the past decade, the use of molecular techniques in plant breeding has led to the widespread use of transgenic crops in agriculture. These technological advances present new opportunities for developing plants that are resistant to pests and diseases, better able to withstand stressful environments, and have

From: *Methods in Molecular Biology, vol. 286: Transgenic Plants: Methods and Protocols*
Edited by: L. Peña © Humana Press Inc., Totowa, NJ

the capacity to produce better quality food products. As with many technical advances in agriculture and biotechnology, concerns are raised about the potential consequences of these developments to the environment.

One of the principal concerns of genetically modified crops is the likelihood and possible consequence of the introduced transgenes being transferred through pollen dispersal to wild relatives or nontransgenic crops. For pollen-mediated gene flow to occur among plant populations, dispersal of pollen to a different population must occur with successful fertilization of an ovule. Therefore, a complete description of gene flow in plants must include an assessment of the relative importance of pollen as the agent of gene flow. Currently, there are few systems for the direct monitoring of pollen movement under field conditions. Previous attempts to measure gene flow have evolved around the analyses of genetic markers ***(1)***. For instance, population genetic structure gathered from isozyme surveys that can be fit to data models of population differentiation have been used ***(2)***. Other research approaches have concentrated exclusively on gene flow by using paternity exclusion analysis ***(3–6)*** or microsatellite markers ***(7)***. These systems have limitations because they are species-specific, requiring the use of expensive assays that cannot yield results in real time or in the field. More recently, visual markers such as GFP have been proposed for use, using whole plant expression to monitor gene flow under agricultural conditions ***(8–10)***. This method has been used successfully to assess out-crossing events in canola (*Brassica napus*) under field conditions ***(11)***.

A direct method could be the use of GFP-tagged pollen to monitor pollen movement under field conditions. This system would allow the quantification of pollen flow directly from a group of individuals in the field and would determine the distance and directional patterns of pollen dispersal within a plant population. GFP expression in plant pollen will not only enable the tracking of pollen movement but also can be used to differentiate between pollen from individual plants of the same species. GFP-tagged pollen could also be used to assess multiple pollination mechanisms. Because GFP can be expressed in pollen under the control of a pollen specific promoter, a system to monitor and detect pollen distribution and gene flow patterns can be developed on a large scale, thus revealing answers to many questions involving ramifications of the introgression of transgenic crop species into the environment and to evaluate the adequacy of current isolation distances for the prevention of outcrossing.

In current research, we used the pollen-specific LAT59 promoter to express GFP in pollen grains of tobacco (*N. tabacum*), an easily engineered model plant ***(12)***. The tomato LAT59 promoter ***(13)*** is a pollen-active promoter that is preferentially expressed in the anthers and pollen of tomatoes ***(13)***. The LAT59 promoter controls the expression of a gene that shows similarity to two regions found to have conserved sequences between all *Erwinia* pectate lyases. LAT59 has a 61 and 54% similarity to regions I and II, respectively, which have been described for seven pectate lyases of *Erwinia chrysanthemi* and *E. carotovora* ***(14)***. To compare pollen movement in the field with actual out-crossing events, we describe methods to determine pollen-mediated gene flow of transgenic tobacco to nontransgenic tobacco at various distances from a source population using whole plant expression of GFP. Gene flow was determined by screening progeny from wild type recipient plants growing at various distances from the source population.

This chapter describes experimental methods on the construction of a transformation vector using a pollen specific promoter to express GFP in pollen grains. It also discusses the instrumentation and methods used to visualize GFP in pollen and pollen tubes. We describe an experimental field design, which can be used to track pollen movement under field conditions with pollen-tagged tobacco or similar plants. Finally, we describe the use of whole plant expression of GFP to detect gene flow under field conditions.

2. Materials

2.1. Plasmid Construction

1. Plasmid construct pBINmGFP5-ER ***(15)*** containing the CaMV35s promoter and an *npt*II kanamycin resistance cassette.
2. Plasmid construct containing the LAT59 promoter ***(13)***.
3. Restriction enzymes *Hin*dIII and *Bam*HI (Promega, Madison, WI).
4. T4 DNA ligase (Promega).

2.2. Plant Transformation

1. Surface sterilized seeds (20% bleach and 0.001% Tween-20 solution for 8 min) from *Nicotiana tabacum* cv "Xanthi."
2. Murashige and Skoog (MS) basal media ***(16)*** is used for seed germination.
3. All plant medium use 0.2% Gelrite gellan gum as a solidifying agent, and all agents are autoclaved prior to media being poured into plates with the exception of antibiotics.
4. *Agrobacterium tumefaciens* strain GV 3850 containing the pBINDC1 expression vector ***(12)*** with the LAT59 pollen specific promoter controlling the mGFP5-ER gene along with an *npt*II cassette for kanamycin selection.
5. DBI medium containing 1 mg/L of indoleacetic acid for tobacco shoot organogenesis (leaf as explant source).
6. MSO medium for rooting.
7. Antibiotics kanamycin (Sigma, St. Louis, MO) and timentin (GlaxoSmithKline, Philadelphia, PA).
8. 100-mm Petri dishes and GA 7 Magenta boxes for tissue culture.
9. Laminar flow hood.

2.3. Fluorescence Microscopy

1. An epifluorescence microscope (Olympus Reflected Fluorescence system BX51) under blue light using a fluorescein isothiocyanate (FITC) filter set was used to visualize GFP expression in pollen.
2. Pollen was photographed on microscope slides using a digital camera (Olympus Q color 3 with Q Capture software).
3. BK medium: 10% sucrose, 100 mg/L of boric acid, 300 mg/L of calcium nitrate, 200 mg/L of magnesium sulfate, 100 mg/L of potassium nitrate *(17)* for pollen tube germination.

2.4. Plant Material

1. GFP pollen specific tobacco plants containing the pBINDC1 plasmid (PGFP).
2. Tobacco plants containing the pBINmGFP5-ER plasmid and expressing GFP throughout the entire plant (WPGFP).

2.5. Field Experiment

1. The experimental field design was based on Saeglitz et al. *(18)* and consisted of a central donor plot split into four quadrants.
2. Two quadrants of the center donor plot contained PGFP tobacco plants.
3. The two remaining two quadrants contained WPGFP tobacco plants.
4. Each of the four quadrants contained six rows with eight tobacco plants per row, giving a total of 192 transgenic tobacco plants located within the center donor plot.
5. A pollen trap was placed in eight different directions (N, S, E, W, NW, SW, NE, and SE) at distances of 5, 10, 15, 20, and 25 m from the center donor plot as well as inside each quadrant within the center plot to measure wind dispersed pollen flow.
6. Replicate pollen traps were constructed from double-sided sticky adhesive tape on glass microscope slides.
7. Slides were covered with petroleum jelly and attached to vertical wooden stakes with collection heights of 50 and 100 cm from the soil surface.
8. Pollen dispersion was measured from the onset of anthesis.
9. Two wild-type recipient tobacco plants cv "Xanthi" were placed in a spatial grid around the center donor plot in eight directions at distances of 10–100 m from the center plot.
10. 10 × 15 seeds germination paper (Anchor Paper Co., St. Paul, MN).
11. 0.2 g/L of calcium sulfate ($CaSO_4$).
12. A hand-held long wave ultraviolet light, (model B-100AP 100 W: 365 nm, UVP, Upland, CA).

3. Methods

3.1. Plasmid Construction

The CaMV 35S promoter cassette from the *Agrobacterium tumefaciens* expression vector pBINmGFP5-ER (courtesy of J. Haseloff) was excised by a

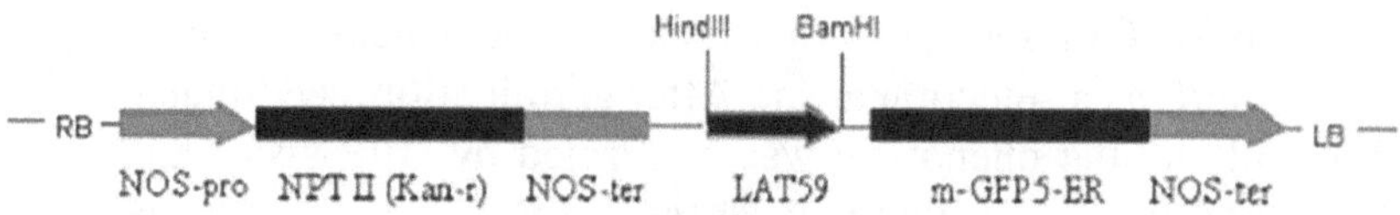

Fig. 1. Construction of pBINDC1 plasmid. The LAT59 pollen-active promoter was subcloned into the *Agrobacterium tumefaciens* expression vector pBINmGFP5-ER in the place of the CaMV35s promoter to create pBINDC1 as shown above.

*Hin*dIII and *Bam*HI restriction digest. The LAT59 promoter (courtesy of S. McCormick) was ligated into the vector to replace CaMV35S promoter. The plasmid was renamed pBINDC1 (**Fig. 1**) *(12)*. This vector contained an *npt*II cassette (kanamycin resistance) that was under the control of the nopaline synthase promoter and terminator.

3.2. Plant Transformation

Nicotiana tabacum cv "Xanthi" was transformed with pBINDC1 *(12)* using the *Agrobacterium*-mediated leaf-disk transformation method *(19)*. Transgenic plants were selected on MS media *(15)* containing kanamycin (200 mg/L) and timentin (400 mg/L). Shoots arising from leaf discs were rooted on agar solidified MSO medium *(20)*. After the plantlets formed roots, they were transferred to soil and grown to maturity under growth chamber conditions.

3.3. Fluorescence Microscopy

For observing GFP in pollen, freshly dehisced pollen grains were removed from anthers and placed on microscope slides. No staining or cover slide was necessary (*see* **Note 1**). GFP-tagged pollen was viewed under blue light conditions using an epifluorescent microscope (*see* **Note 2**). To observe GFP expression in growing pollen tubes, pollen grains were removed from anthers and placed in BK media (*see* **Note 3**). After 2 h, 30 μL of the BK pollen mixture was placed on microscope slide (*see* **Note 4**). Pollen tubes were observed at 100× magnification under a microscope (*see* **Note 5**). A 16 ms exposure time was used when photographing pollen under white light conditions and 2.75-s exposure under blue light conditions (*see* **Note 6**).

3.4. Plant Material

The field design of this experiment incorporated two types of transgenic tobacco. Tobacco plants expressing GFP throughout the entire plant (WPGFP) contained the *mgfp5-er* transgene, driven by the CaMV 35S constitutive promoter. WPGFP tobacco plants were used to measure gene flow in the field.

Homozygous WPGFP seeds were germinated on MS medium containing kanamycin 200 mg/L as a selection agent. After germination, seedlings were transferred to soil and the phenotype was confirmed by GFP visualization with a handheld, long-wave UV light (*see* **Notes 7–14**). Plantlets were placed in the greenhouse until transferred to the field sites.

GFP pollen-specific tobacco plants (PGFP) expressed the *mgfp5-er* transgene, driven by the LAT59 pollen specific promoter. PGFP tobacco plants expressed the GFP protein exclusively within pollen grains and were used to measure pollen movement in the field. Homozygous PGFP seeds (T_2) were germinated on MS media with 200 mg/L of kanamycin. After germination, plantlets were placed in soil and grown in greenhouse conditions until planted at the field sites.

3.5. Field Design

Pollen flow was measured with pollen traps to sample pollen distribution at specified distances (*see* **Notes 15–17**). Pollen slides were collected at 24-, 48-, and 72-h periods after pollen shed from the donor tobacco population within the center plot. The presence of GFP-tagged pollen was assessed by screening the slides collected from the field site. Slides were screened using an epifluorescent microscope with blue light at 100×–400× magnification without staining. Gene flow was measured by analyzing progeny from the wild-type recipient plants for the GFP phenotype. Seed capsules were harvested from the receptor plants, which surround the donor plot at various distances and directions. The progeny from these seeds were screened using either the germination paper method, or the soil germination method. Using the germination paper method seeds were germinated in a dark incubator at 27°C on filter paper soaked in a 0.2 g/L calcium chloride solution. In the soil germination method, seeds were germinated in soil and grown under greenhouse conditions. After 3 wk gene flow was quantified by progeny analysis of seedlings from recipient plants and plants expressing GFP in the pollen (PGFP) within the donor plot. Seedlings were screened for the GFP phenotype with a handheld UV light. Outcrossing frequencies were calculated from the summed progeny at each coordinate and represent the average outcrossing frequency per plant.

4. Notes

1. When viewing GFP-tagged pollen under the microscope, it is not necessary to use a stain, such as aniline blue. Spread pollen evenly over the slide and do use a not cover slip. When the pollen grains become crowded on a slide, a cover slip tends to mash the pollen and makes it difficult to see each grain clearly. We found that a magnification of 400× was most effective for viewing GFP in pollen; however, it is visible at lower magnifications (i.e., 40×–100×).

2. When using an epifluorescence microscope, it is necessary to turn on the light source approx 30 min prior to viewing the specimen to allow proper warm-up of the burner. In general, when using epifluorescence microscopes, leaving the light source on for at least 30 min will prolong bulb life.
3. Pollen tube germination requires the use of a pollen tube germination media. Several variations of BK media exist so it is important to review the current literature to choose a variant of BK media that is optimal for the plant species being used. These variations have been modified to be more effective for pollen tube germination in specific plant species. Freshly dehisced pollen and fresh germination media must be used when germinating pollen tubes for best results. Depending on the species, binucleate pollen grains will germinate and tubes will grow in excess of 5 h. Many plant species with trinucleate pollen will germinate and grow but will have less longevity.
4. Pollen grains viewed under dry conditions will have a different shape (oblong) from that of hydrated pollen (round) in an aqueous solution. This is important when screening for pollen on pollen traps, as petroleum jelly will hydrate the pollen grains.
5. GFP can be visualized in the pollen tubes during any time of growth. No cover slip or stain is required. We found that ×100 magnification was best for visualizing GFP in pollen tubes.
6. When photographing GFP in pollen grains, exposure time is crucial. Wild-type tobacco pollen grains have slight autofluorescence under blue light, which might be confused with the GFP phenotype in photographs when using different exposure times. We found 16 ms to be the optimal exposure time for photographing pollen under white light conditions. The optimal exposure time for photographing GFP-tagged pollen under blue light was 2.75 s.
7. We have used two methods to screen large numbers of seedlings for the GFP phenotype: sowing seeds on germination paper and soil.
8. One strength of the germination paper method includes the ability to rapidly screen thousands of seedlings in a relatively small space by a single researcher. This method has been efficient for seeds produced under ideal conditions, that is, clean and healthy seeds. This method is also especially effective for large seeded plant species, such as many from the genus *Brassica*.
9. The germination paper method also has shown some weaknesses when the seeds are dirty (as is often the case with field collected material) or produced from plants grown in suboptimal conditions. In these cases, the germination paper can grow a large amount of contamination from dirty seed, which interferes with seedling health and the ability to accurately score the GFP phenotype. Also, when the seeds are from a sick parental plant, the seedlings are often of poor health, and grow poorly on the germination paper. Plant health is important when screening for GFP, and suboptimal seeds and seedlings will reduce the ability to accurately score the presence or absence of GFP.
10. The soil germination method is good for small seeded plants that require a period of growth before the GFP status can be determined. In the case of tobacco and

Arabidopsis thaliana, seedlings from these plants require several weeks of growth before they are large enough to accurately screen for GFP. In these cases, sowing the seeds on soil under greenhouse conditions is an efficient method to produce material suitable for GFP screening.

11. The soil germination method also has some difficulties, including sowing seeds at proper densities and the space required for large numbers of plants. With regard to sowing density, a balance must be reached between the numbers of seedlings in each container compared with the ability to accurately screen each plant for GFP. If the seedlings are at extreme densities, they will crowd each other and it will be difficult to see each plant to score the GFP phenotype. If the density is too low, the greenhouse space will become a limiting factor.
12. We have found that screening a large number of plants on soil requires several researchers. In our case, we found that it was most efficient to have people dedicated to UV screening in a dark environment and others dedicated to bringing and removing plant containers to be screened.
13. Multiple UV lights may also be employed to increase the accuracy of scoring low expressing GFP individuals. From our experience, the power and number of UV lamps can be increased to help discern between plants that exhibit slight differences in fluorescence.
14. Overall, one of the most important factors in the ability to screen GFP is overall plant health. Plants grown in suboptimal conditions are very difficult to screen for GFP.
15. Wild-type plants placed at coordinates around the center plot of transgenic plants must be germinated and planted in the field at the same time as those in the center plot to ensure coinciding flowering times. Planting large numbers of wild-type plants at the coordinate locations increases the amount of seeds that can be collected and screened. This increases the chance of detecting a rare out-crossing event. However, increasing the number at each coordinate could also limit the ability to detect an outcrossing event because cross-pollination will be occurring between wild-type plants at each coordinate. It is important to balance the number of plants at each location to maximize the amount of seeds that can be collected without decreasing chances of outcrossing between the transgenic and wild-type plants in field plot.
16. To use GFP-tagged pollen to effectively monitor pollen movement, it is our suggestion to use a plant species that is known to outcross under field conditions. Homozygous plants must be used in the field experiments. In homozygous plants, 100% of the pollen will express GFP maximizing the ability to see pollen movement. We used tobacco as a model plant, with designs toward employing the system for monitoring canola pollen.
17. Many types of pollen traps exist that could be used to track pollen movement. To maximize the chance of seeing pollen movement in the field, pollen traps need to be appropriate heights depending on the plant species being used. Placing traps around the center plot at a high density will ensure maximum efficiency.

References

1. Slatkin, M. (1985) Gene flow in natural populations. *Annu. Rev. Ecol. Syst.* **16,** 393–430.
2. Slatkin, M. and Barton, N. H. (1989) A comparison of three indirect methods for estimating average levels of gene flow. *Evolution* **43,** 1349–1368.
3. Ellstrand, N. C., Delvin, B., and Marshall, D. L. (1989) Gene flow by pollen into small populations: data from experimental and natural strands of wild radish. *Proc. Natl. Acad. Sci. USA* **86,** 9044–9047.
4. Ellstrand, N. C. (1992) Gene flow among seed plant populations. *New Forests* **6,** 241–256.
5. Friedman, S. T. and Adams, D. L. (1985) Estimation of gene flow into two seed orchards of loblolly pine (*Pinus taeda* L.). *Theor. Appl. Genet.* **69,** 609–615.
6. Adams, W. T. and Birkes, D. S. (1990) Estimating mating patterns in for trees populations, in *Biochemical Markers in the Population Genetics of Forest Trees,* (Hattemer, H. H. and Fineschi, S. eds.), S.P.B. Academic Publishing, The Hague, The Netherlands, pp. 157–172.
7. Dow, B. D. and Ashley, M. V. (1998) High levels of gene flow in bur oak revealed by paternity analysis using microsatellites. *J. Hered.* **89,** 62–70.
8. Stewart, C. N. Jr. (1996) Monitoring transgenic plants using in vivo markers. *Nat. Biotech. 14,* 682.
9. Leffel, S., Mabon, S. A., and Stewart, C. N. Jr. (1997) Application of green fluorescent protein in plants. *BioTechniques* **23,** 912–918.
10. Harper, B. K., Mabon, S. A., Leffel, S. M., et al. (1999) Green fluorescent protein as a marker for expression of a second gene in transgenic plants. *Nat. Biotech.* **17,** 1125–1129.
11. Halfhill, M. D., Warwick, S. I., Raymer, P. L., Millwood, R. J., and Weissinger, A. K. (2003) Gene flow from transgenic oilseed rape and crop x weed hybrids under field conditions. *Environ. Bio. Res.*, in press.
12. Hudson, L. C., Chamberlain, D., and Stewart, C. N., Jr. (2001) GFP-tagged pollen to monitor pollen flow of transgenic plants. *Mol. Ecol. Notes* **1,** 321–324.
13. Twell, D., Yamaguchi J., Wing, R. A., Ushiba, J., and McCormick, S. (1991) Promoter analysis of genes that are coordinately expressed during pollen development reveals pollen-specific enhancer sequences and shared regulatory elements. *Genes Develop.* **5,** 496–507.
14. Wing, R. A., Yamaguchi J., Larabell S. K., Ursin, V. M., and McCormick S. (1989) Molecular and genetic characterization of two pollen expressed genes that have sequence similarity to pectate lyases of the plant pathogen *Erwinia. Plant Mol. Biol.* **14,** 17–28.
15. Haseloff, J., Siemering, K. R., Prasher, D. C., and Hodge, S. (1997) Removal of a cryptic intron and subcellular localization of green fluorescent protein are required to mark transgenic *Arabidopsis* plants brightly. *Proc. Natl. Acad. Sci. USA* **94,** 2122–2127.
16. Murashige, T. and Skoog, F. (1962) A revised medium for rapid growth and bioassays with tobacco tissue cultures. *Physiol. Plant* **15,** 473–497.

17. Brewbaker, J. L. and Kwack, B..H. (1963) The essential role of calcium ion in pollen germination and tube growth. *Am. J. Bot.* **50,** 589–865.
18. Saeglitz C., Pohl, M., and Bartsch, D. (2000) Monitoring gene escape from transgenic sugar beet using cytoplasmic male sterile bait plants. *Mol. Ecol.* **9,** 2035–2040.
19. Horsch, R. B., Fry, J. E., Hoffman, N. L., Eichholts, D., Rogers, S. G., and Fraley R. T. (1985) A simple method for transferring genes into plants. *Science* **227,** 1229–1231.
20. McCormick, S., Niedermeyer, J., Fry, J., Barnanson, A., Horsch, R., and Fraley, R. (1986) Leaf disc transformation of cultivated tomato (*L. esculentum*) using *Agrobacterium tumefaciens. Plant Cell Rep.* **5,** 81–84.

VII

Transgenic Crops

26

Transgenic Crops

The Current and Next Generations

Jim M. Dunwell

Summary

This chapter describes the present status and future prospects for transgenic (genetically modified) crops. It concentrates on the most recent data obtained from patent databases and field trial applications, as well as the usual scientific literature. By these means, it is possible to obtain a useful perspective into future commercial products and international trends. The various research areas are subdivided on the basis of those associated with input (agronomic) traits and those concerned with output (e.g., food quality) characteristics. Among the former group are new methods of improving stress resistance, and among the latter are many examples of producing pharmaceutical compounds in plants.

Key Words: *Agrobacterium*; antibody; genetically modified; pharmaceutical; transformation; vaccine.

1. Introduction

This review concentrates on recent advances in the production and use of transgenic crops and should be read in conjunction with previous reviews *(1–5)*. It does not consider issues such as the possible impact of genetically modified (GM) crops on the environment *(6)* or the safety of GM food *(7)*, as these subjects have been adequately reviewed elsewhere *(8)*.

During the preparation of this chapter, as well as consulting the usual sources of research publications, extensive use has been made of the freely available patent databases in the United States (http://www.uspto.gov/patft/index.html), Europe (http://ep.espacenet.com/), World International Patent Organization (http://pctgazette.wipo.int/) and other international sites (e.g., http://www.surfip.gov.sg/sip/site/sip_home.htm). In addition, this review includes data from

From: *Methods in Molecular Biology, vol. 286: Transgenic Plants: Methods and Protocols*
Edited by: L. Peña © Humana Press Inc., Totowa, NJ

various sites providing information on the field tests of GM crops in the United States (http://www.nbiap.vt.edu/cfdocs/fieldtests1.cfm) and elsewhere (http://www.nbiap.vt.edu/cfdocs/globalfieldtests.cfm) (http://gmoinfo.jrc.it), as these often provide useful perspectives into future commercial products and international trends.

The following subheadings first summarizes the state of those GM crops already being commercialized and those in the development pipeline ***(9,10)***. It then describes some of the recent research and patent publications detailing advances in the underlying technology for GM crop production and some of the novel approaches to modifying both agronomic (input) ***(3)*** and product quality (output) ***(4)*** traits.

2. Present Status of GM Crops

Table 1 provides data on the global areas of GM crops grown over the last 2 yr. The estimated global area for 2003 was 67.7 million hectares or 167.2 million acres, grown by 7 million farmers in 18 countries. The increase in area between 2002 and 2003 was 15%, equivalent to 9 million hectares or 22 million acres. Since the introduction of GM crops in 1996, the average growth in area has been more than 10% per year, an overall increase of 40-fold; this ranks as one of the highest adoption rates for any crop technology. Of the global area, almost all (99%) is found in six principal countries: the United States, Argentina, Canada, Brazil, China, and South Africa. The latter two countries have had the highest year-on-year growth, with 33% increases in their total areas of GM crops. Specifically, China showed a 40% increase in its area of cotton expressing the *Bacillus thuringiensis* (Bt) protein with this area comprising 58% of the total national cotton area of 48 million hectares. An increasing number are grown in developing countries, accounting for more than one-third (increase from 27%) of the global crop area. Two countries, Brazil and Philippines grew approved transgenic crops for the first time in 2003. Of the global area of transgenic crops in 2003, herbicide tolerant soybean occupied 61%, accounting for 41.4 million hectares, with Bt corn second at 9.1 million hectares. The other six crops, planted on 5% or less of the global transgenic crop area, include herbicide-tolerant canola, herbicide-tolerant cotton, Bt cotton, Bt/herbicide-tolerant cotton, herbicide-tolerant corn, and Bt/herbicide-tolerant corn. There is an increasing trend towards stacked genes (combinations of herbicide and insect resistance).

An additional method by which to access the impact of GM crop is to estimate their having increased from $4.0 billion in 2002. These values are based on the sale price of GM seed plus any technology fees that apply.

Within Europe, the most comprehensive recent survey is probably that prepared by the European Science and Technology Observatory ***(10)***. Fourteen crops

Table 1
Areas of GM Crops Grown Worldwide in 2002 and 2003

Country	Area 2002 (million hectares)	Area 2003 (million hectares)
United States	39.0	42.8
Argentina	13.5	13.9
Canada	3.5	4.4
Brazil	?	3.0
China	2.1	2.8
South Africa	0.3	0.4
Australia	0.1	0.1
India	<0.1	0.1
Romania	<0.1	<0.1
Spain	<0.1	<0.1
Uruguay	<0.1	<0.1
Mexico	<0.1	<0.1
Bulgaria	<0.1	<0.1
Indonesia	<0.1	<0.1
Colombia	<0.1	<0.1
Honduras	<0.1	<0.1
Germany	<0.1	<0.1
Total	58.7	67.7

Adapted from Website: http://www.isaaa.org/.

have been approved for commercialisation to date; these are maize (4), oilseed rape (4), carnation (3), chicory (1), soybean (1), and tobacco (1). Although no further authorizations have been granted since October 1998, 13 applications were pending approval under the old Directive 90/220/EEC and 19 applications have been submitted under the new Directive 2001/18/EC.

Summaries of field trial data for the United States and Europe are provided in **Figs. 1** and **2**, respectively. The overall total for the United States is now over 9300 (from 227 institions) since 1987, with an average annual total of about 1000. In the United States, two companies (Monsanto and DuPont/Pioneer) account for about 50% of all field trials, with almost two-thirds of the trials on maize, potato, and soybean. Analysis of the data for 2003 shows that about 32% of these trials involved herbicide tolerance, 22% product quality traits (starch, sugar, proteins, etc.) (**Table 2**), 19% insect resistance, and 12% agronomic traits, with most of the remainder covering resistance to fungi, viruses, bacteria, and nematodes.

In the European Union, by contrast, the number peaked at 250 in 1998 and has declined by 80% since then, because of the 1999 decision to block any

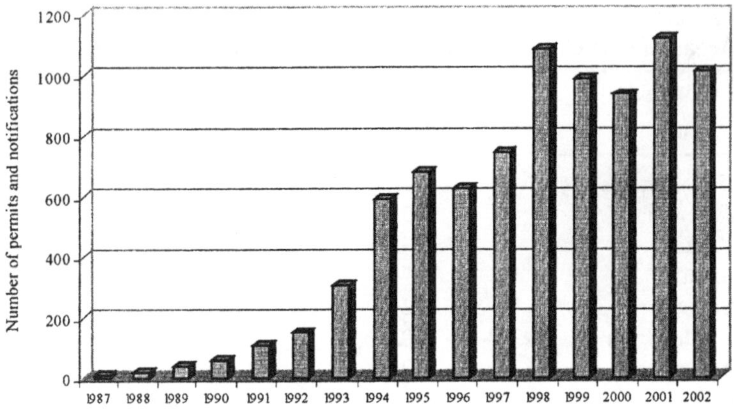

Fig.1. Total number of US field trial permits and notifications approved by year. (Adapted from Website: http://www.nbiap.vt.edu/cfdocs/fieldtests1.cfm)

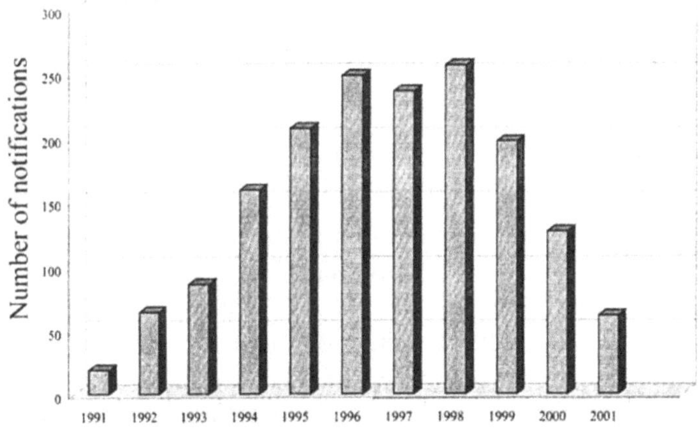

Fig. 2. Annual number of EU field trial notifications. (Adapted from **ref. *10*.**)

European Union new commercial release of GMOs and the general opinion of the European Union public. Further details of the present position within the United Kingdom can be obtained from the DEFRA websites on GM crop farm scale evaluations (http://www.defra.gov.uk/environment/gm/fse/index.htm), the Royal Society pages on GM plants (http://www.royalsoc.ac.uk/gmplants/),

Table 2
Selected Examples of Recent US Field Trial Applications for Product Quality Traits

Number	Crop	Applicant	Gene	Trait
04-083-09	Barley	Washington State	Lysozyme	Novel protein
04-029-02	Barley	Washington State	Glucanase	Heat stable enzyme
04-021-01	Corn	Pioneer	Amino polyolamine oxidase	Fumonisin degradat.
04-020-02	Alfalfa	Forage Genet. Inc	CBI[a]	Altered lignin
04-033-15	Wheat	ARS	Glutelin	Storage protein
03-345-05	Tomato	BHN Research	Sucrose phosphate	Altered sugar synthetase
03-288-26	Fescue	Noble Foundat.	Cinn. alc. dehydrog.	Decreased lignin
03-268-07	Soybean	Univ. Nebraska	δ-6 desaturase	Altered oil
03-253-05	Sweetgum	ArborGen	CBI	Decreased lignin
03-203-09	Eucalyptus	ArborGen	CBI	Decreased lignin
03-153-01	Pineapple	Univ. Hawaii	ACC oxidase	Fruit ripening
03-136-01	Paspalum	Univ. Florida	*O*-methyl transfer.	Decreased lignin
03-091-08	Tobacco	Vector Tobacco	Aquaporin Quino. phos. trans.	Reduced nicotine
03-090-11	Lettuce	Harris Moran	Trans. int. fact.	Reduced senescence
03-073-07	Apple	Cameron Nursery	?	Brown spot resistance Polyphenol oxidase
03-057-02	Potato	Univ. Idaho	CBI	Reduced bruising
03-052-07	Petunia	Scotts	CBI	Altered color
03-035-10	Tobacco	Virginia Tech.	Gulono-lactone oxidase	Increased vitamin C
03-034-05	Wheat	Montana State	Purindoline	Better breadmaking

[a] Denotes confidential business information.
(Adapted from Website: http://www.nbiap.vt.edu/cfdocs/fieldtests1.cfm)

the Friends of the Earth report on farm scale evaluations (http://www.foe.co.uk/resource/reports/science_smokescreen.pdf), and the Food Standards Agency public forum on GM food (http://www.food.gov.uk/gmdebate). Medical Association (http://www.bma.org.uk/GMFoods) and on the ethics of GM crops in developing countries from the Nuffield Council on Bioethics (http://www.nuffieldbioethics.org/filelibrary/pdf/gm_crops_paper_final.pdf)

3. Regeneration and Transformation Techniques

The most commonly used transformation technologies are those involving either particle bombardment or *Agrobacterium* ***(11)*** and these are not considered in detail. A noticeable recent trend is the development of efficient *Agrobacterium*-mediated methods for cereals and other crops previously considered recalcitrant. Among the associated recent improvements in underlying regeneration technology for important crops is that of wheat, in which it is claimed that the addition of copper at a concentration range from 50 to 300 μ*M* and a growth hormone at a concentration from 0.05 to 10 mg/L gives greatly improved results ***(12)***. Such culture medium allows reliable high-frequency regeneration from a range of current elite wheat germplasm leading to up to 12-fold increases in the production of fertile transgenic plants.

The latest proposed method for soybean transformation ***(13)*** comprises *Agrobacterium* treatment of an explant (particularly after pretreatment with high doses of cytokinin), transferring embryonic axes explants of the mature seeds incubated on wet filter papers in the presence of at least one phenol compound, to induce *vir* genes, and incubation in the dark at 20–25°C for at least 24 h. After incubation, the explants are transferred to a medium to develop shoots from explants, control *Agrobacterium* growth, and after shoot elongation, separated shoots—with or without roots—are either transferred to soil or treated with at least 1 mg/L Indole butyric acid (IBA) before transplantation.

Other recent improvements include novel methods for *Eucalyptus* ***(14)*** and *Tagetes* ***(15)***, the latter involving *Agrobacterium*-mediated transformation of cell cultures.

For certain less amenable crop species (e.g., maize) only particular genotypes are easy to transform. In these cases, it may be preferable to transform a hybrid between the target genotype and a transformation competent line of the same species or of another closely related species. The gene of interest can then be introgressed into the genetic line from which the original recipient parent was derived or into other genetic lines ***(16)***.

With the increasing regulatory pressure to avoid antibiotic selectable marker systems there has been a search for alternatives, the latest being a development of positive selection methods that involve conferring on cells the ability to metabolize certain compounds, preferably arabitol, ribitol, raffinose, sucrose, mannitol, or combinations of these compounds ***(17)***. Another related advance developed in potato is a strategy that relies on the transformation of tissue explants or cells with a *Agrobacterium* strain and selection of transformed cells or shoots after polymerase chain reaction (PCR) analysis ***(18)***. For example, incubation of explants with *Agrobacterium* strain AGL0 resulted in transformed shoots at an efficiency of 1–5%, depending on the genotype used. Because this system does not require genetic segregation or site-specific DNA deletion systems to remove marker genes, it may provide a reliable and efficient tool for generating transgenic plants for commercial use, especially in vegetatively propagated species such as potato and cassava.

Another variation *(19)* designed to avoid the possibility of transfer of non-transfer (T)-DNA sequences is one in which the construct includes non-T-DNA sequence comprising a lethal gene. Selection then allows the identification of plant cells that contain the T-DNA sequence and not the lethal gene.

A final alternative *(20)* is provided by a system that enhances the selection of transgenic plants having two T-DNA molecules integrated into the genome at different physical and genetic loci. The constructs comprise novel arrangements of T-DNA molecules containing genes of interest, positive selectable marker genes, and conditional lethal genes. In this system, first the transgenic plant and independent transgene loci are identified; subsequently, the selectable marker genes and introduced genes of interest can be segregated in the progeny.

As well as the "traditional" process of introducing genes into GM crops, there is increasing interest in the selected inactivation of endogenous genes or targeted integration *(21)*. For example, one such method *(22)* involves the use of a combination of target sites for two site-specific recombinases and expression of a chimeric recombinase with dual target site specificity.

As well as the research described in **Subheadings 4.** and **5.** concerning the introduction of specific coding regions, there are also interesting developments in the design of specific novel regulatory constructs (e.g., those comprising zinc finger transcription factors *[23,24]*, and also in the identification of novel promoter sequences such as those targeting expression to the seed *(25)* or caryopsis *(26)*, and those that can be regulated by the application of exogenous chemicals *(27)*.

Chloroplast transformation *(28–32)* has also become a method attracting both academic and commercial interest (http://chlorogen.com) in recent years, partly because of the ability of these organelles to accumulate introduced proteins at very high yield but also because of the theoretical ecological advantages of reduced transfer of the transgene via pollen dispersal. Another associated technology concerns a method of modifying mitochondrial-encoded traits in plants by physically inserting this organelle by spraying a filtrate under pressure onto target cells *(33)*.

4. Input Traits

The priority for most plant breeders is maximising yield, and improvements in this character has been claimed in several recent patents involving the plastid expression of cyanobacterial fructose-1,6-bisphosphatase/sedoheptulose-1,7-bisphosphatase *(34)* or other enzymes *(35)*, and the use of the e2f transcription factor *(36)*. General approaches have been reviewed previously *(3)*.

4.1. Agronomic Traits

The first transgenic products were those that contained tolerance to various herbicides, and these make up the majority of the GM acreage (*see* **Subheading 2.**). However, there is still further extension and fine-tuning of these technologies either to improve efficacy or extend the technology to other species ***(37)***. One example of this latter type is the project ***(38)*** designed to improve resistance to broomrape (*Orobanche* spp.), parasitic weeds that are major constraints to vegetable crop production in the Mediterranean basin and in localized areas in India, China, and the United States. Transgenic target-site herbicide resistance (e.g., to acetolactate synthase [ALS] inhibitors) allows for movement of unmetabolized herbicide through the crop to the photosynthate sink in the parasite, as well as through the soil. The authors report the successful engineering of a mutant *ALS* gene into carrot, allowing control of broomrape by imazapyr, an imidazolinone *ALS* inhibitor.

Apart from the more common strategies for herbicide tolerance, another novel approach ***(39)*** consists of introducing genes encoding the enzymes of the complete mevalonate pathway.

Regarding insect resistance, there is also continuing development of strategies based on Bt. For example, there is recent claim ***(40)*** for a novel Bt δ-endotoxin, designated CryET29, that exhibits insecticidal activity against siphonapteran insects, including larvae of the cat flea (*Ctenocephalides felis*), as well as against coleopteran insects, including the southern corn rootworm (*Diabrotica undecimpunctata*), Western corn rootworm (*D. virgifera*), Colorado potato beetle (*Leptinotarsa decemlineata*), Japanese beetle (*Popillia japonica*), and red flour beetle (*Tribolium castaneum*). Such research needs to be assessed against a background suggesting that long-term regional pest suppression after deployment of Bt crops (e.g., Bt cotton resistant to the major pest, pink bollworm *Pectinophora gossypiella*) may contribute to reducing the overall need for insecticide sprays ***(41)***.

4.2. Abiotic Stress

There are many, varied transgenic approaches to improving the resistance of a crop to abiotic stress. Among the recent suggestions are the use of a novel heat-shock protein with high homology to chloroplast elongation factor EF-Tu ***(42)***; protein phosphatase stress-related polypeptides ***(43)***; an *Arabidopsis* Na^+/H^+ exchanger polypeptide that allows crops to survive in soil with high salt levels ***(44)***; and a farnesyl transferase that improves the tolerance of plants to environmental stresses and senescence ***(45)*** (expression of inhibitors of these enzymes enhance drought tolerance, improve resistance to senescence and modify growth habit).

In this context, the promoter of the *Wcs120* gene, which encodes a highly abundant protein induced during cold acclimation of wheat, has been proposed to drive the expression of genes needed for low-temperature tolerance in sensitive species ***(46)***.

4.3. Sterility and Flowering

A recently described example of modifying these characteristics involves manipulation of the first example of a phytochrome-regulated transcription factor; this protein, designated CCA1, binds to the promoter region of a chlorophyll binding protein gene of *Arabidopsis* ***(47)***. When CCA1 is overexpressed, the normal circadian rhythms of the plant are disrupted, and the transgenic plants take a significantly longer time to reach flowering, even in the presence of day length conditions that normally induce flowering. Thus, this method may represent a valuable means of extending vegetative growth and delaying flowering.

In a similar example, a late-flowering transgenic radish has been produced by the expression of an antisense GIGANTEA (*GI*) gene fragment ***(48)***. This study provides evidence that downregulation of the *GI* gene by cosuppression could delay bolting in a cold-sensitive long-day (LD) plant. Production of late-flowering germplasm of radish may allow this important crop to be cultivated over an extended period and also provide further food to the famine countries of Southeast Asia. Modification of flowering time or flowering period is also reported as a consequence of altering the levels of a specific transcription factor ***(49)***.

Induction of floral sterility is claimed to result from expression of selected floral homeotic genes from poplar- ***(50)*** or biotin-binding compounds ***(51)***, and female sterility has been induced by expression of a deacetylase gene under the control of a tissue-specific promoter that is specifically active in the female organs; these organs are killed by treatment with *N*-acetyl-PTC or *B*-acetyl-phosphinothricin ***(52)***.

Other examples of altered phenotype include methods for inducing dwarfism ***(53)*** and reduced senescence ***(54)*** linked to expression of deoxyhypusine synthase.

5. Output Traits

These have been recently reviewed ***(4,5)***, and selected examples of field trials of such transgenics are included in **Table 2**.

5.1. Seed Quality

Apart from general attempts to modify seed size ***(55)***, seed quality can also be modified. For example, chlorophyll reduction in the seed of *Brassica* can be achieved by downregulating its synthesis. To achieve this, expression of an antisense glutamate 1-semialdehyde aminotransferase gene (*gsa*), directed by a *Brassica* napin promoter, was targeted specifically to the embryo of the developing seed of oilseed rape. These transgenic lines have provided useful materials for the development of a low chlorophyll seed variety of this crop ***(56)***.

Another novel method of improving processing grain in crops such as corn and soybeans involves utilizing thioredoxin and/or thioredoxin reductase to enhance extractability and recovery of starch and protein ***(57)***. Other methods of reducing fungal toxin content of seed to improve palatability have also been developed ***(58,59)***.

5.2. Protein

The nutritive value of storage organs for human and/or animal consumption was achieved by transferring a gene that encodes a sulfur-rich protein, such as sunflower seed albumin (SSA) containing 16% methionine and 8% cysteine, placed under the control of a promoter that confers storage organ-specific expression ***(60)***. It was discovered that, in addition to the expected changes in sulfur-rich protein content of seeds, the overall composition of the seed was altered unexpectedly so as to produce a dramatic improvement in many different, unrelated nutritive parameters. In particular, the process produced an increase in the total protein content (e.g., in rice, peas and chickpeas), altered fiber composition (lupins and peas), modified oil content and composition (lupins), altered starch content (peas), and a decrease in the content of endogenous antinutritional factors (peas and chickpeas).

Two related studies describe different approaches to improving the amounts of free amino acids in plants. In the first, Kisaka ***(61)*** described the introduction of a glutamate dehydrogenase gene as a means of increasing the amount of at least one of the following glutamic acid, asparagine, aspartic acid, serine, threonine, alanine, and histidine. This process is also claimed to increase potato yield.

In a related study on maize ***(62)***, a prolamin box binding factor peptide, or a subunit thereof, was used to increase the amount of a preselected amino acid, such as lysine and/or methionine, in the seed. This beneficial alteration occurred without substantially altering the total protein content of the seed, which might be deleterious to other agronomic characteristics of the transgenic plant.

Several plant proteins, particularly those from seeds ***(63)***, have proven or possible allergenic potential, and various transgenic approaches have been made to reduce their amount ***(64)***. Among these is one study ***(65)*** on hypoallergenic soybean with reduced amounts of the major soybean allergen, the vacuolar protein known as P34, as well as other allergens. Soybean protein products made from these hypoallergenic soybeans should be substantially free of P34, as well as a series of other minor allergens such as various glycinins and conglycinins.

5.3. Carbohydrate

One proposed method of producing modified starch relates to use of a rice gene encoding a plastid protein referred to as an R1 protein ***(66)***. It is probable that this protein exists in the plastids in a form bound to the starch granules as well as in a soluble form, and that this protein is involved in starch phosphorylation. Similar cases include those claiming the use of reserve polysaccharide biosynthetic enzymes, such as glycogen biosynthetic enzymes, glycogen synthase, and/or ADP-glucose pyrophosphorylase ***(67)***; pea plastidial phosphoglucomutase for altering the sucrose and starch content of plants ***(68)***; modified sucrose binding proteins ***(69)*** with enhanced sucrose uptake activity ***(70)***; and a method of decreasing the oil content of seeds by expression of ADP-glucose pyrophosphorylase ***(71)***.

Among studies on specific sugars rather than total carbohydrate is one involving palatinose (isomaltulose, 6-*O*-α-D-glucopyranosyl-D-fructose), a structural isomer of sucrose with very similar physicochemical properties. Owing to its noncariogenicity and low calorific value, it is an ideal sugar substitute for use in food production. Usually, palatinose is produced on an industrial scale from sucrose by an enzymatic rearrangement using immobilized bacterial cells, but production in potato has now been achieved by use of a chimeric sucrose isomerase gene from *Erwinia rhapontici* under control of a tuber-specific promoter ***(72)***. Despite the soluble carbohydrates being altered within the tubers, growth of transgenic plants was indistinguishable from wild-type plants.

5.4. Oils and Fatty Acids

Modification of these compounds has been reviewed recently ***(73)***. cDNAs for a wide variety of unusual fatty acid biosynthetic enzymes have been identified, particularly through the use of expressed sequence tags. Amongs the GM crops undergoing field tests in the United States in the last 2 yr are plants expressing Δ-6 desaturase (from *Borago officinalis*), Δ-12 saturase (from soybean), palmitoyl thioesterase (from soybean), lysophosphatidate acyltransferase (from *Saccahromyces cerevisiae*), and stearoyl ACP desaturase (from *Rattus norvegicus*). Among the most comprehensive generic applications is one from Pioneer covering GM soybean-expressing genes (not described) from *Euphorbia lagascae*, *Isochrysis galabana*, *Mortierella alpina*, *Parthenium argentatum*, *Saprolegnia diclina*, *Schizochytrium aggregatum*, *Thraustochytrium aureum*, and *Veronia galamensis*.

One specialized example involves exploitation of long chain polyunsaturated fatty acids (LCPUFAs) such as arachidonic acid (ARA, 20:4*n*-6) and docosahexaenoic acid (DHA, 22:6*n*-3) ***(74)***. These compounds are highly concentrated in the phospholipid bilayer of biologically active brain and retinal neural membranes and are important in phototransduction (retina) and neuronal function (brain). They are present in large quantities in human milk, and it has been recommended that plant-derived versions could be added to infant food.

5.5. Other Food Components

Isoflavones are compounds with claimed health benefits, and there is much interest in modifying the expression of genes encoding enzymes in their biosynthetic pathway; these enzymes include chalcone isomerase, isoflavone reductase, and vestitone reductase *(75)*.

Similarly, several eukaryotic genes encoding ε-cyclase, isopentenyl pyrophosphate isomerase, and β-carotene hydroxylase have been used as a means of increasing the production of novel and rare carotenoids *(76)*. Among these is zeaxanthin, an important dietary carotenoid, although its abundance in food is low. To provide a better supply of this compound in a staple crop, two different potato varieties were genetically modified *(77)* by transformation with sense and antisense constructs encoding zeaxanthin epoxidase; zeaxanthin conversion to violaxanthin was inhibited. Both approaches (antisense and cosuppression) yielded potato tubers with higher levels of zeaxanthin (between 4- and 130-fold).

Among the various efforts to modify food quality is one involving flavour improvement by increased expression of a lipoxygenase gene to produce transgenic grapes with optimal levels of the enzyme *(78)*.

Other recent examples of plants with modified vitamins include the field testing (US field trial APHIS no. 03-035-10n) (**Table 3**) of a tobacco line expressing L-gulono-γ-lactone oxidase from *Rattus norvegicus* and containing increased amounts of vitamin C. Similarly, a barley gene encoding 4-hydroxyphenylpyruvate dioxygenase was overexpressed in tobacco plants under control of the 35S promoter with the aim of enhancing the vitamin E content *(79)*. Seeds from transgenic lines had up to twofold enhanced levels of this vitamin without any change in the ratio of γ-tocopherol and γ-tocotrienol.

A particularly specialized example of this type relates to the use of transgenic plants for the expression of vitamin B_{12} (cobalamin) binding proteins *(80)*. Such recombinant proteins can be used in analytical tests and in the treatment of vitamin B_{12} deficiency.

5.6. Industrial Products

1. Lignin. Expression of several of the enzymes that comprise the biosynthetic pathway for lignin have been altered in efforts to modify lignin, either as a means of improving paper-making quality or digestibility for animals. A recent example of the latter approach is provided by a study on alfalfa in which reduction in caffeic acid 3-*O*-methyltransferase and caffeoyl CoA 3-*O*-methyltransferase led to a dramatic decrease in lignin content (approx 20%) and modest increase in cellulose (approx 10%) *(81)*. These compositional changes potentially allow enhanced use of alfalfa as a major forage crop by increasing the digestibility of its stem fraction.

Table 3
Predicted Areas of GM Crops in the European Union (EU) up to 2013

Crop/trait	Commercially available in EU	% EU area planted 2008	% EU area planted 2013
Insect-resistant maize	2005–2007	10	25–30
Herbicide-tolererant maize	2005–2007	10	35–45
Herbicide-tolererant oilseed rape	2006–2008	0–5	20–30
Herbicide-tolererant sugar beet	2006–2008	5–10	40–50
Insect-resistant cotton	2006–2008	5–10	40–50
Herbicide-tolererant cotton	2006–2008	5–10	40–50
Herbicide-tolererant wheat	2008–2011	0	15–25
Herbicide-tolererant soybeans	2007–2009	0–10	30-40
Herbicide-tolererant rice	2007–2009	0–5	30–40
Nematode- and fungus-resistant potatoes	2010–2012	0	5–10
Fungus-resistant oilseed rape	2010–2012	0	5–10

(Adapted from **ref.** ***9***.)

2. Plastics. Polyhydroxyalkanoates (PHAs) and polyhydroxybutyrates (PHBs), polyesters of hydroxyacids naturally synthesized in bacteria as a carbon reserve, have properties of biodegradable thermoplastics and elastomers and their synthesis in crop plants is seen as an attractive system for the sustained production of large amounts of polymers at low cost ***(82)***. Various PHAs and PHBs having different physical properties have now been synthesized in the cytoplasm, plastid, or peroxisome of numerous transgenic plants, including *Arabidopsis*, rape, corn ***(83)***, and hairy roots of sugar beet ***(84)***. This latter study is the first example of plastidic PHB production in roots of a carbohydrate-storing crop plant.
3. Cellulose. Modifying the amount of this compound in plants is now feasible following an understanding of its biosynthetic pathway ***(85)***.

5.7. Enzymes

A bacterial thermostable cellulase, the endo-1,4-β-D-glucanase E1 from *Acidothermus cellulolyticus*, has been expressed in chloroplasts, and an active enzyme recovered both in vitro and in vivo ***(86)***.

5.8. Plant-Based Pharmaceuticals

Much investment has been committed recently to this subject; a wide range of different production systems have been proposed and are at various stages of commercial development. These include expression in chloroplasts (http://www.chlorogen.com/), oil bodies (http://www.sembiosys.ca/), and those using specific species for production, for example, maize, rice, and barley (http://www.ventriabio.com/). This latter system is being used for production of lacto-ferrin, lysozyme, α_1-antitrypsin, fibrinogen, and thioredoxin.

Other companies are specializing in noncrop species such as the water plant *Lemna* (http://www.biolex.com/), being tested in a joint venture with Bayer for the production of human plasminogen, or the moss *Physcomitrella* (http://www.greenovation. com/). Other specialized products include antibodies (http://www.epicyte.com/, http://www.planetbiotechnology.com/) such as that active against one of the bacteria responsible for dental caries. Information on other commercial programs can be obtained from the following sites: http://www.prairieplant.com/intro.htm, http://www.lhsc.on.ca/plantigen/, http://www.geneart.de/, http://www. mpt.monsanto.com/, http://www.chromatininc.com/, http://www.medicago. com/, http://www.meristem-therapeutics.com/, http://www.ceres-inc.com/index. html, and http://www.prodigene.com/. Among the recently announced projects from this latter company are a Phase I clinical trial in cooperation with the National Institutes of Health's (NIH) National Institute of Allergy and Infectious Diseases (NIAID), studying the safety and immunogenicity of an oral vaccine against traveller's diarrhea (a condition caused by enterotoxigenic *E. coli*); and an agreement with Sigma Aldrich to manufacture and distribute a plant-derived recombinant trypsin.

Other less developed projects in this area include one on the humanizing of plant cDNAs ***(87)***. Another potentially significant advance is the production of enzymatically active recombinant human and animal lysosomal enzymes in plants that has been accomplished by construction and expression of recombinant constructs encoding human glucocerebrosidase and α-galactosidase sequences ***(88)***. Various novel extraction technologies have also been developed for extraction of enzymes produced in this manner ***(89)***.

In a recent regulatory change, a problem with "contamination" by corn expressing pharmaceutical proteins has led to stricter isolation distances for such transgenics growing in the field (http://www.usda.gov/news/releases/2003/03/aphisfactsheet030603.pdf).

5.9. Vaccines

The prospects for plant-derived vaccines ***(89)***, sometimes called "edible vaccines," have been well reviewed ***(90,91)***. Recent examples include the production of transgenic carrots expressing an immunodominant antigen, hemagglutinin (H) glycoprotein, of the measles virus ***(92)***.

6. The Future

It has been predicted that the global area and the number of farmers planting GM crops will continue to grow in 2004 in the six principal countries already growing GM crops (**Table 1**) (http://www.isaaa.org/). Among the other 12 countries growing such crops, India, is expected to increase its Bt cotton significantly and one or more new countries will also grow GM crops for the first time. The three most populous countries in Asia, namely China, India, and Indonesia, with 2.5 billion people, are all now growing GM crops commercially.

Looking further into the future, detailed predictions of the prospects for commercialization in Europe have recently been published *(9)* and are summarized in **Table 3**. The key elements from this study are first that it is likely to be another 2–3 yr before GM seed is widely available to European Union producers of maize and possibly 3–4 yr for other crops such as oilseed rape and sugar beet; GM wheat is unlikely to be available until 2008–2010.

Second, in a 5 yr time period the extent of GM crops is likely to be limited to no more than 10% of cultivation in some crops such as maize. This largely reflects the time required to complete the various regulatory approvals and to introgress introduced genes into leading varieties, and the continued existence of anti-GM sentiment among some consumers.

Thirdly, in a 10 yr time period, it is predicted that the acreage of GM crops will largely reflect the degree to which specific pests and weeds (which are targeted by GM traits) are considered to be a problem for farmers. Thus, takeup of insect resistant and herbicide tolerant crops such as oilseed rape, maize, and sugarbeet will be concentrated in regions where pests and weeds are perceived to be causing significant crop losses and/or conventional control methods have been found to be of limited effectiveness (or are more expensive than the GM alternative). However, in 2013, areas of GM wheat and potatoes will probably be more limited than for other crops, mainly because the traits available will be fairly new to the market. The potential impact of biotechnology in improving pest management in Europe is also the subject of a recent detailed case study analysis (*see* Website: http://www.ncfap.org/reports/Europe/ExecutiveSummary December.pdf).

Whether these predictions are fulfilled will depend not on any limit to available novel genotypes, but rather on the uncertain combination of economic pressures, regularoty systems (*see* Website: http://www.pewagbiotech.org/research/regulation/RegulationExecSum.pdf), biosafety assessments like those that operated under the Cartagena Protocol that came into force in September 2003 (*see* Website: http://www.biodiv.org/doc/publications/bs-brochure-02-en.pdf), and public perception. This is well-demonstrated in the United Kingdom where, despite an extensive debate on the future of GM crops (*see* Website: http://www.gmsciencedebate.org.uk/), an assessment of various environmental impacts (the Farm Scale Evaluations, *see* Website summary at http://www.defra.gov.uk/environment/gm/fse/results/fse-summary.pdf), and conditional permission from the government, commercial planting has recently (April 2004) been postponed.

References

1. Dunwell, J. M. (1998) Novel food products from genetically modified crop plants: methods and future prospects. *Int. J. Food Sci. Tech.* **33,** 205–213.
2. Dunwell, J. M. (1999) Transgenic crop plants: the next generation or an example of 2020 vision. *Ann. Bot.* **84,** 269–277.
3. Dunwell, J. M. (2000) Transgenic approaches to crop improvement. *J. Exp. Bot.* **51,** 487–496.
4. Dunwell, J. M. (2001) Modifying quality traits: future prospects, in *BCPC Conference "Weeds 2001,"* British Crop Protection Council, Farnham. pp. 421–430.
5. Dunwell, J. M. (2002) Future prospects for transgenic crops. *Phytochem. Rev.* **1,** 1–12.
6. Conner, A. J., Glare, T. R., and Nap, J-P. (2003) The release of genetically modified crops into the environment. Part II. Overview of ecological risk assessment. *Plant J.* **33,** 19–46.
7. Cockburn, A. (2002) Assuring the safety of genetically modified (GM) foods: the importance of an holistic, integrative approach. *J. Biotechnol.* **98,** 79–106.
8. Ford, B. J., ed. (2003) GM Crops: *The Scientists Speak, Rothay House*, Cambridge, UK.
9. Brooks, G. and Barfoot, P. (2003) GM *Crops in Europe—Planning for the End of the Moratorium*, PG Economics Limited, Dorset, UK.
10. Lheureux, K., Libeau-Dulos, M., Nilsagård, H., et al. (2003) *Review of GMOs Under Research and Development and in the Pipeline in Europe*, European Science and Technology Observatory.
11. Gelvin, S. B. (2003) *Agrobacterium*-mediated plant transformation: the biology behind the "Gene-Jockeying" tool. *Microbiol. Mol. Biol. Rev.* **67,** 16–37.
12. Lazzeri, P. A. and Rasco-Gaunt, S. (2003) Transformation and regeneration of wheat using increased copper levels. US patent application no. 20030018991.
13. Dias, K. M. (2003) Transformation of soybeans. US patent application no. 20030046733.
14. Matsunaga, E., Sugita, K., and Ebinuma, H. (2003) Method for introducing genes into plants having improved transformation efficiency. US patent application no. 20030033639.
15. Kunze, I., Herbers, K., and Heim, U. (2003) Production of transgenic plants of the Tagetes species. US patent application no. 20030033638.
16. Ranch, J. P., Marsh, W. A., Meyer, T. E., et al. (2003) Methods of transforming plants and identifying parental origin of a chromosome in those plants. US patent application no. 20030046724.
17. Parrott, W., LaFayette, P., and Kane, P. (2003) Arabitol or ribitol as positive selectable markers. US patent application no. 20030041352.
18. De Vetten, N., Wolters, A. M., Raemakers, K., et al. (2003) A transformation method for obtaining marker-free plants of a cross-pollinating and vegetatively propagated crop. *Nat. Biotechnol.* **21,** 439–442.
19. Gutterson, N. and Hanson, W. G. (2003) Compositions and methods for improved plant transformation. US patent no. 6,521,458.

20. Barton, K., Crow, L., Gilbertson, L., et al. (2003) Methods for enhancing segregation of transgenes in plants and compositions thereof. Patent application no. WO03/003816.
21. Baszczynski, C. L., Bowen, B. A., Duesing, J. H., Peterson, D. J., Tagliani, L. A., and Zhu, T. (2003) Targeted manipulation of genes in plants. US patent no. 6,528,700.
22. Baszczynski, C. L., Lyznik, L. A., Gordon-Kamm, W. J., Guan, X., Rao, A. G., and Tagliani, L. A. (2003) Method for the integration of foreign DNA into eukaryotic genomes. US patent no. 6,541,231.
23. Barbas, C. F., Stege, J. T., Guan, X., and Dalmia, B. (2003) Methods and compositions to modulate expression in plants. US patent application no. 20030037355.
24. Segal, D. J., Stege, J. T., and Barbas, C. F. (2003) Zinc fingers and a green thumb: manipulating gene expression in plants. *Curr. Opin. Plant Biol.* **6,** 163–168.
25. Linnestad, C., Lappegard, K. K., Abbitt, S. E., Martino-Catt, S. J., and Olsen, O-A. (2003) Seed-preferred promoters from end genes. US patent no. 6,528,704.
26. Sprunck, S., Kluth, A., Becker, D., Luetticke, S., and Loerz, H. (2003) Promoters for gene expression in caryopses of plants. US patent application no. 20030046731.
27. Padidam, M. (2003) Chemically regulated gene expression in plants. *Curr. Opin. Plant Biol.* **6,** 169–177.
28. Nehra, N. S., Schaaf, D. J., Sidorov, V., Stalker, D. M., and Ye, G. (2003) Plastid transformation of solanaceous plants. US patent no. 6,541,682.
29. Koop, H-U., Muhlbauer, S., Klaus, S., Eibl, C., Huang, F-C., and Golds, T. J. (2003) Gene expression in plastids based on replicating vectors. Patent application no. WO03/004658.
30. Staub, J. M. (2003) Expression of eukaryotic peptides in plant plastids. US patent application no. 20030033636.
31. Daniell. H. (2002) Molecular strategies for gene containment in transgenic crops. *Nat. Biotechnol.* **20,** 581–586.
32. Daniell, H. and Moar, W. (2003) Multiple gene expression for engineering novel pathways and hyperexpression of foreign proteins in plants. US patent application no. 20030041353.
33. Downer, C. P. (2003) Method of organelle transformation. US patent application no. 20030046725.
34. Yokota, A., and Shigeoka, S. (2003) Method for improving productivity of higher plants. US patent no. 6,528,705.
35. Kaplan, A., Lieman-Hurwitz, J., Rachmilevitch, S., Schatz, D., and Mittler, R. (2003) Plants characterized by enhanced growth and methods and nucleic acid constructs useful for generating same. US patent application no. 20030037356.
36. Beeckman, T., De Veylder, L., Inze, D., et al. (2003) A method to modify cell number, architecture and yield of plants by overexpressing the e2f transcription factor. Patent application no. WO03/025185.
37. Mannerloef, M., Tenning, P. P., and Steen, P. (2003) Transgenic sugar beet plant expressing cp4/epsps enzyme activity. US patent no. 6,531,649.

38. Aviv, D., Amsellem, Z., and Gressel, J. (2002) Transformation of carrots with mutant acetolactate synthase for *Orobanche* (broomrape) control. *Pest. Manag. Sci.* **58,** 1187–1193.
39. Hahn, F. M. and Kuehnle, A. R. (2003) Manipulation of genes of the mevalonate and isoprenoid pathways to create novel traits in transgenic organisms. US patent application no. 20030033626.
40. Rupar, M. J., Donovan, W. P., Tan, Y., and Slaney, A. C. (2003) *Bacillus thuringiensis* CryET29 compositions toxic to coleopteran insects and *Ctenocephalides* spp. US patent no. 6,537,756.
41. Carriere, Y., Ellers-Kirk, C., Sisterson, M., et al. (2003) Long-term regional suppression of pink bollworm by Bacillus thuringiensis cotton. *Proc. Natl. Acad. Sci. USA* **100,** 1519–1523.
42. Ristic, Z., Bhadula, S. K., Yang, G., Elthon, T. E., and Habben, J. E. (2003) Maize chloroplast protein synthesis elongation factors and methods of use for same. US patent application no. 20030044972.
43. Bohnert, H. J., Chen, R., Ishitani, M., Van Thielen, N., and Da Costa E Silva, O. (2003) Protein phosphatase stress-related polypeptides and methods of use in plants. Patent application no. WO03/020914.
44. Blumwald, E. and Apse, M. (2003) Increasing salt tolerance in plants by overexpression of vacuolar cation-proton antiporters. US patent application no. 20030046729.
45. McCourt, P., Ghassemian, M., Cutler, S., and Bonetta, D. (2003) Stress tolerance and delayed senescence in plants. US patent application no. 20030061636.
46. Sarhan, F. and Ouellet, F. (2003) Universal plant promoter inducing gene transcription in response to low-temperatures. US patent application no. 20030065162.
47. Tobin, E., Sun, L., and Wang, Z-Y. (2003) Phytochrome regulated transcription factor for control of higher plant development. US patent application no. 20030056247.
48. Curtis, I. S., Nam, H. G., Yun, J. Y., and Seo, K. H. (2002) Expression of an antisense GIGANTEA (GI) gene fragment in transgenic radish causes delayed bolting and flowering. *Transgen. Res.* **11,** 249–256.
49. Reuber, L. (2003) Methods for modifying flowering phenotypes. US patent application no. 20030041356.
50. Strauss, S. H., Rottmann, W., Brunner, A., and Sheppard, L. (2003) Floral homeotic genes for manipulation of flowering in poplar and other plant species. US patent application no. 20030033628.
51. Albertsen, M. and Huffman, G. (2002) Biotin-binding compounds for induction of sterility in plants. Patent application no. WO02/071834.
52. Bartsch, K. (2003) Process for producing female-sterile plants. US patent application no. 20030051274.
53. Liu, A., Van Winkle, J., Bovee-Picciano, S., and Mathews, H. (2003) Identification and characterization of a dwarf and late flowering 2 phenotype (DLF2) in *Arabidopsis*. US patent no. 6,534,695.

54. Thompson, J. E., Wang, T-W., and Lu, D. L. (2003) DNA encoding a plant deoxyhypusine synthase, a plant eukaryotic initiation factor 5A, transgenic plants and a method for controlling senescence programmed and cell death in plants. US patent no. 6,538,182.
55. Yanofsky, M. F., Martienssen, R., Ferrandiz, C., and Gu, Q. (2003) Method of decreasing seed size in a plant. US patent no. 6,541,683.
56. Tsang, E. W., Yang, J., Chang, Q., et al. (2003) Chlorophyll reduction in the seed of Brassica napus with a glutamate 1-semialdehyde aminotransferase antisense gene. *Plant Mol. Biol.* **51,** 191–201.
57. Lanahan, M. B., Desai, N. M., Gasdaska, P. Y., and Goff, S. A. (2003) Grain processing method and transgenic plants useful therein. US patent no. 6,531,648.
58. Duvick, J., Maddox, J., and Gilliam, J. (2003) Compositions and methods for fumonisin detoxification. US patent no. 6,538,177.
59. Chatterjee, R., Duvick, J. P., and English, J. (2003) AP1 amine oxidase variants. US patent application no. 20030056245.
60. Higgins, T. J., Tabe, L. M., and Schroeder, H. E. (2003) Method for altering storage organ composition. US patent application no. 20030066101.
61. Kisaka, H. and Kida, T. (2003) Method of constructing transgenic plant with improved amino acid composition and yield. Patent application no. WO03/000041.
62. Schmidt, R. J., Moose, S. P., and Vicente-Carbajosa, J. (2003) Transgenic maize comprising recombinant PBF genes. US patent application no. 20030051272.
63. Dunwell, J. M. (2003) Structure, function, and evolution of vicilin and legumin seed storage proteins, in *Biopolymers*, Vol. 8. *Polyamides and Complex Proteinaceous Materials II* (Steinbüchel, A., and Fahnestock, S. R., eds.), Wiley-VCH, Weinheim, Germany, pp. 223–253.
64. Kjaerulff, S. and Roggen, E. L. (2003) Transgenic plants. US patent application no. 20030041354.
65. Kinney, A. J. and Jung, R. (2003) Hypoallergenic transgenic soybeans. US patent application no. 20030041350.
66. Frohberg, C. (2003) Nucleic acid molecules from rice and their use for the production of modified starch. US patent no. 6,521,816.
67. Stalker, D. M., Shewmaker, C. K., and Oakes, J. V. (2003) Glycogen biosynthetic enzymes in plants. US patent no. 6,538,181.
68. Lacey, C. N., Hughes, S. G., Harrison, C. J., Wang, T. L., and Hedley, C. L. (2003) Method for increasing sucrose content of plants. US patent no. 6,538,180.
69. Dunwell, J. M., Khuri, S., and Gane, P.J. (2000) Microbial relatives of the seed storage proteins of higher plants: conservation of structure and diversification of function during evolution of the cupin superfamily. *Microbiol. Mol. Biol. Rev.* **64,** 153–179.
70. Grimes, H. D. and Chao, W. S. (2003) Sucrose-binding proteins. US patent application no. 20030061634.
71. Barry, G. F., Kishore, G. M., and Stark, D. M. (2003) Enhanced starch biosynthesis in seeds. US patent no. 6,538,179.

72. Bornke, F., Hajirezaei, M., and Sonnewald, U. (2002) Potato tubers as bioreactors for palatinose production. *J. Biotechnol.* **96,** 119–124.
73. Jaworski, J. and Cahoon, E. B. (2003) Industrial oils from transgenic plants. *Curr. Opin. Plant Biol.* **6,** 178–184.
74. Feussner, I., Hornung, E., Pernstich, C., Korfei, M., and Kindl, H. (2003) Method for producing arachidonic acid in transgenic organisms. Patent application no. WO03/012092.
75. Fader, G. M. (2003) cDNA sequences from plants that encode activities associated with isoflavone biosynthesis. US patent no. 6,521,433.
76. Cunningham, F. X. and Sun, Z. (2003) Methods of increasing or decreasing carotenoids and other isoprenoids using IPP isomerase. US patent no. 6,524,811.
77. Romer, S., Lubeck, J., Kauder, F., Steiger, S., Adomat, C., and Sandmann, G. (2002) Genetic engineering of a zeaxanthin-rich potato by antisense inactivation and co-suppression of carotenoid epoxidation. *Metab. Eng.* **4,** 263–272.
78. Descenzo, R. A. and Irelan, N. A. (2003) Lipoxygenase genes from *Vitis vinifera* and polypeptides encoded thereby are provided. US patent application no. 20030033627.
79. Falk, J., Andersen, G., Kernebeck, B., and Krupinska, K. (2003) Constitutive overexpression of barley 4-hydroxyphenylpyruvate dioxygenase in tobacco results in elevation of the vitamin E content in seeds but not in leaves. *FEBS Lett.* **540,** 35–40.
80. Berglund, L. E., Petersen, T. E., Fedosov, S. N., Nexo, E., Laursen, N. B., and Jensen, E. O. (2003) Transgenic plants expressing cobalamin binding proteins. Patent application no. WO03/006661.
81. Marita, J. M., Ralph, J., Hatfield, R. D., Guo, D., Chen, F., and Dixon, R. A. (2003) Structural and compositional modifications in lignin of transgenic alfalfa down-regulated in caffeic acid 3-*O*-methyltransferase and caffeoyl coenzyme A 3-*O*-methyltransferase. *Phytochemistry* **62,** 53–65.
82. Poirier, Y. (2002) Polyhydroxyalknoate synthesis in plants as a tool for biotechnology and basic studies of lipid metabolism. *Prog. Lipid. Res.* **41,** 131–155.
83. Peoples, O. P. and Sinskey, A. J. (2003) Polyhydroxybutyrate polymerase. US patent no. 6,528,706.
84. Menzel, G., Harloff, H. J., and Jung, C. (2003) Expression of bacterial poly(3-hydroxybutyrate) synthesis genes in hairy roots of sugar beet (*Beta vulgaris* L.). *Appl. Microbiol. Biotechnol.* **60,** 571–576.
85. Saxena, I. M., Lin, F. C., and Brown, R. M. (2003) Recombinant cellulose synthase. US patent no. 6,541,238.
86. Jin, R., Richter, S., Zhong, R., and Lamppa, G. K. (2003) Expression and import of an active cellulase from a thermophilic bacterium into the chloroplast both *in vitro* and *in vivo*. *Plant Mol. Biol.* **51,** 493–507.
87. Kumagai, M. H., Della-Cioppa, G. R., Erwin, R. L., and McGee, D. R. (2003) Method of humanizing plant cDNAs by transfecting a nucleic acid sequence of a non-plant donor into a host plant in an anti-sense orientation. US patent application no. 20030064392.

88. Garger, S. J., Turpen, T. H., and Kumagai, M. H. (2002) Production of lysosomal enzymes in plants by transient expression. Patent application no. WO02/008404.
89. Bratcher, B., Garger, S. J., Holtz, B. R., and McCulloch, M. J. (2003) Flexible processing apparatus for isolating and purifying viruses, soluble proteins and peptides from plant sources. WO patent application no. 03/028432.
90. Lam, D. M. K., Zeng, F., and Leung, F. C. (2003) Novel vaccine compositions and methods of vaccine preparation for veterinary and human diseases. Patent application no. WO03/013598.
91. Mason, H. S., Warzecha, H., Mor, T., and Arntzen, C. J. (2002) Edible plant vaccines: applications for prophylactic and therapeutic molecular medicine. *Trends Mol. Med.* **8,** 324–329.
92. Marquet-Blouin, E., Bouche, F. B., Steinmetz, A., and Muller, C. P. (2003) Neutralizing immunogenicity of transgenic carrot (*Daucus carota* L.)-derived measles virus hemagglutinin. *Plant Mol. Biol.* **51,** 459–469.

27

Comparative Development and Impact of Transgenic Papayas in Hawaii, Jamaica, and Venezuela

Gustavo Fermín, Paula Tennant, Carol Gonsalves, David Lee, and Dennis Gonsalves

Summary

We present data concerning the creation of transgenic papayas resistant to *Papaya ringspot virus* (PRSV) and their adoption by three different countries: the United States (e.g., Hawaii), Jamaica, and Venezuela. Although the three sets of transgenic papayas showed effective resistance to PRSV, the adoption rate in each country has varied from full utilization in Hawaii to aggressive testing but delay in deregulating of the product in Jamaica to rejection at an early stage in Venezuela. Factors that contributed to the rapid adoption in Hawaii include a timely development of the transgenic product, PRSV causing severe damage to the papaya industry, close collaboration between researchers and the industry, and the existence of procedures for deregulating a transgenic product. In Jamaica, the technology for developing the initial field-testing of the product progressed rather rapidly, but the process of deregulation has been slowed down owing to the lack of sustained governmental efforts to complete the regulatory procedures for transgenic crops. In Venezuela, the technology to develop and greenhouse test the transgenic papaya has moved abreast with the Jamaica project, but the field testing of the transgenic papaya within the country was stopped very early on by actions by people opposed to transgenic products. The three cases are discussed in an effort to provide information on factors, other than technology, that can influence the adoption of a transgenic product.

Key Words: Biotechnology adoption; *Papaya ringspot virus*; pathogen-derived resistance; technology transfer; transgenic papayas.

1. Introduction

Transgenic commodity crops with resistance to herbicide or lepidoptera insects have been widely adopted in North America by the United States and

From: *Methods in Molecular Biology, vol. 286: Transgenic Plants: Methods and Protocols*
Edited by: L. Peña © Humana Press Inc., Totowa, NJ

Canada, and in other countries as Argentina (*see* Chapter 1). However, the progress in commercialization of minor crops even within the United States has been rather slow. The transgenic papaya with resistance to *Papaya ringspot virus* (PRSV) is one of these few minor crops that have been commercialized in the United States (Hawaii). Interestingly, because papaya is widely grown in the tropical and subtropical lowlands of the world and is severely infected by PRSV, there is widespread interest in using PRSV-resistant transgenic papaya for these areas outside of the United States. Shortly after the Hawaiian transgenic papaya was developed and shown to be resistant, personnel from Jamaica and Venezuela launched efforts to develop transgenic papaya for their countries in concert with the Hawaii transgenic papaya effort ***(1)***. As a result, PRSV-resistant transgenic papayas have been developed for Jamaica and Venezuela ***(2,3)***.

This chapter covers the technical aspects of the transgenic papaya but focuses on describing the factors that affected the differential adoption of the transgenic papaya in Hawaii, Jamaica, and Venezuela. We feel that the approach in this chapter might provide insights that may help future efforts at transferring transgenic products to countries.

2. Transgenic Papaya Produced for Hawaii, Jamaica, and Venezuela

Papayas are fast-growing and high-yielding fruit trees (actually, the papaya "tree" is a gigantic herb) very popular in the lowlands of tropical and subtropical regions ***(4)***. Papayas are grown domestically or at industrial scale; 6 million Mt of papaya were produced worldwide in 2002 ***(5)***. However, papaya orchards can be severely attacked by PRSV, arguably the most important pathogen that attacks the plant.

Since the first demonstration that the *coat protein* (*CP*) gene of a plant virus inserted into a plant genome could protect transgenic plants from the homologous virus ***(6)***, many plants have been engineered for viral resistance using the concept of pathogen-derived resistance ***(7,8)***. PRSV ***(9,10)*** belongs to the Potyviridae family of plant viruses. Potyviruses have a sense polarity RNA genome ***(11)***, which is translated into a polyprotein that, on sequential processing by different virus-encoded proteases, yields all viral proteins including the CP ***(12–14)***.

CP-transgenic papayas resistant to PRSV were commercially released in Hawaii in 1998 and have helped to save the papaya industry ***(1)***. The transgenic cultivars, "SunUp" and "Rainbow," contain the *CP* gene of PRSV HA 5-1, a mild PRSV mutant strain from Hawaii ***(15)***. "Rainbow" is a hemizygous (*CP+/–*) transgenic papaya as it is an F1 hybrid that was obtained by crossing homozygous (CP+/CP+) "SunUp" and nontransgenic "Kapoho" ***(16)***. "SunUp" was obtained by particle bombardment ***(17)***, and it contains only one *CP* insert in homozygous state *(18,19)*. Resistance to PRSV in transgenic "Rainbow" and "SunUp" has been the target of numerous studies that can be summarized as follows: the CP protein can be detected in transgenic tissues, although at low levels; there is only one transgenic insertion in which all transgenes present, *uidA* for β-glucuronidase, *CP* for the PRSV coat protein, and *npt*II for neomycin phosphotransferase, have suffered no duplications or rearrangements *(19)*; nuclear run-on analysis has detected a high level of expression of the *CP* gene in the nucleus, but low levels in the cytoplasm *(18)*. The transgene messenger thus is degraded posttranscriptionally and resistance is RNA mediated. Because resistance is better achieved when the challenging virus shares a high degree of similarity to the transgene, resistance is said to be homology dependent. Resistance is also affected by gene dosage and the developmental stage at which the transgenic plants are challenged with the virus *(18)*.

Slight, but important differences exist between the *CP* transgene in "Rainbow" and the engineered *CP* transgene in the transgenic papayas from Jamaica *(2)* and Venezuela *(3)*. The engineered PRSV HA 5-1 CP transgene in "Rainbow" is fused to the nucleotides that code for the first 16 amino acids (aa) of the *Cucumber mosaic virus* (CMV) *CP* gene (**Fig. 1**). The transgene in the engineered Jamaican and Venezuelan papayas includes only the PRSV *CP* gene sequence four aa downstream of the Q/S cleavage site (**Fig. 1**). Also, the "Rainbow" transgene contains the 5' UTR of CMV, whereas the other two possess the translational enhancer of the "white leaf strain" of CMV. In all cases, the CaMV 35S promoter drives the transgenes; the terminator sequence in "Rainbow" is derived from the CaMV 35S DNA, whereas the other transgenes use the *nos* terminator. However, more remarkable is the fact that the transgene in "Rainbow" was isolated from a mild mutant of PRSV HA, while in all other cases the transgenes were isolated from aggressive geographical isolates of the virus.

Resistance in transgenic papayas transformed with the *CP* gene from Jamaica and Venezuela share with transgenic "Rainbow" the following features: transgene-derived resistance is RNA-mediated, homology-dependent and not related to the number of *CP* insertions. Yet the range of resistance in the Jamaican and Venezuelan transgenic papayas is wider as compared to "Rainbow" because resistance can be observed even with isolates showing less than 90% similarity to the transgenes *(2,3)*. However, "SunUp" shows a wide range of resistance to PRSV isolates, because of its homozygous *CP* state *(18)*. The untranslatable PRSV HA transgene is able to confer resistance to transformed papayas in the very same way the translatable transgene does *(20)*. However, the nontranslatable Jamaica transgene was not able to confer resistance to transformed papaya plants (*see* **Subheading 4.**); the corresponding version of the untranslatable Venezuelan *CP* transgene has not been used for transformation.

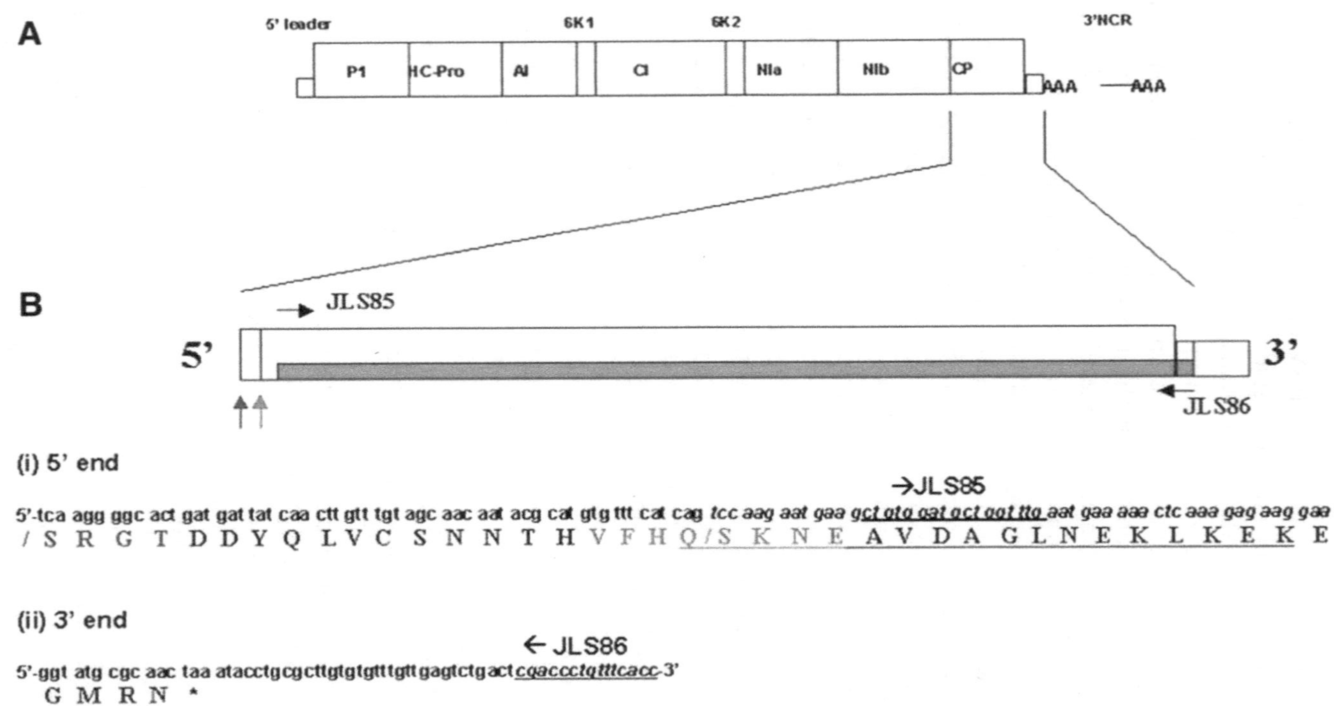

Fig. 1. (A) Genetic map of the *Papaya ringspot virus* genome showing all potential viral products after auto-proteolytic processing of the precursor polyprotein. **(B)** Schematic representation of the coat protein (*CP*) transgene with the leader sequence (white box) and the CP sequence as engineered in 'Rainbow' (dark box), with the corresponding location of primers JLS85 and JLS86 (horizontal arrows) used to engineer the PRSV *CP* gene from Jamaica and Venezuela (light box); (i) the 5' end *CP* coding region with its translated product showing the putative sites of proteolytic cleavage (/), and (ii) the 3' UTR that is followed by a terminator sequence.

3. Transgenic Papayas in Hawaii

The transgenic papaya story in Hawaii represents a case in which a transgenic product was introduced in a timely manner to stop further devastation of an industry by PRSV. The transgenic papayas developed for the papaya industry in Hawaii have been the subject of several recent reviews ***(20,21)***. This section focuses primarily on nontechnical factors that led to the development of the transgenic papaya in Hawaii, its subsequent adoption and impact on the industry, and challenges facing Hawaii's papaya industry.

3.1. The Rationale for Developing Control Measures for PRSV in Hawaii

PRSV was discovered in the mid-1940s on Oahu Island, where Hawaii's papaya industry was located ***(20)***. However, by the mid-1950s PRSV had caused severe damage to papaya grown on Oahu, pushing the industry to relocate to the Puna district of Hawaii Island in the late 1950s. Why relocate to Puna? In addition to the absence of PRSV, Puna had lots of land available for leasing at a reasonable price, had an abundance of sunshine and rain, and the well-drained lava-based soil was suited for papaya. Another very important reason was the grower-selected "Kapoho" variety, which had excellent flavor and shipping qualities and was well adapted to Puna. In fact, "Kapoho" generally does not grow well outside of the Puna area. By the 1970s, Puna was producing 95% of Hawaii's papaya crop. However, the potential threat of PRSV existed because, by the 1970s, PRSV was in papaya growing in backyards in the town of Hilo, which was only 19 miles from Puna. Recognizing the consequences to the industry if PRSV got established in Puna, the Hawaii Department of Agriculture (HDOA) formed a small task force that was responsible for identifying and roguing PRSV-infected trees in Hilo and nearby areas. Quarantine on movement of papaya seedlings into Puna was also put in place.

Another major factor that contributed to the success of the Hawaii papaya story is that efforts to develop control measures were started early. In fact, research to breed papaya with tolerance to PRSV began in the late 1970s. In 1979, scientists at Cornell University and the University of Hawaii started research on cross-protection to control PRSV ***(22)***. Although cross-protection did not reach the stage of large-scale commercial adoption in Hawaii, these research efforts made important contributions to the control of PRSV. For example, the now well-known PRSV HA isolate was purified and characterized ***(23)***, serology was used for rapid detection, the mild strain PRSV HA 5-1 was selected following nitrous acid treatment of the severe PRSV HA ***(15)***, and the *CP* gene of PRSV HA 5-1 was cloned and sequenced (*see* **Subheading 2.**).

3.2. Development of Transgenic Papaya for Hawaii

Research on the development of transgenic papaya was started in the mid-1980s. As with the cross-protection work, the research was supported by modest levels of special USDA grants that were aimed at helping agriculture in Hawaii and the Pacific Basin. The research team had a proper balance in expertise and a strong desire to develop a practical control measure for PRSV in Hawaii. The team consisted of Richard Manshardt, a horticulturist at the University of Hawaii; Maureen Fitch, a graduate student of Richard Manshardt, Jerry Slightom, a molecular biologist with whom D. Gonsalves had collaboration to develop virus resistance transgenic vegetables, and D. Gonsalves, a virologist and coauthor of this chapter.

After unsuccessful attempts to develop transgenic papaya by the transformation of papaya leaf disks with *Agrobacterium tumefaciens*, efforts to transform somatic embryogenic cultures via biolistics were started in fall of 1988. The yellow-flesh cultivar "Kapoho" and the red-fleshed cultivar "Sunset" were targeted for transformation. "Kapoho" was the dominant cultivar, accounting for more than 90% of the commercial papaya grown in Hawaii. Transformation efficiency was relatively low and resulted in less than 20 transgenic papaya lines. In 1991, clones of an R0 transgenic "Sunset" (designated 55-1) were resistant to inoculation by PRSV HA under greenhouse conditions ***(24)***. This line would be carried through to eventual commercialization.

Line 55-1 was a female and thus would need to be crossed with nontransgenic papaya to obtain seeds for further characterization, which would take a year. While the process of obtaining seeds under greenhouse conditions was being carried out, a permit to perform a field trial to test clones of line 55-1 was obtained from APHIS (Animal and Plant Health Inspection Service) in 1991. The plants were put in the field in June 1992. The value of this trial cannot be overestimated: it served to confirm the resistance of line 55-1 under field conditions, showed that line 55-1 was of suitable horticultural characteristics, and served to provide material for the deregulation process, and for developing the red-flesh cultivars "SunUp" and yellow-flesh cultivar "Rainbow" ***(25)***. "SunUp" is transgenic line 55-1 that is homozygous for the *CP* gene insert, and the yellow-flesh "Rainbow" is an F1 hybrid of "SunUp" and nontransgenic "Kapoho" ***(16)***.

The timely development of the transgenic virus-resistant papaya became evident when PRSV was identified in Puna in May 1992, just about the same time that the field trial was started on Oahu Island. The severity and speed of the virus spread in Hawaii has been reviewed ***(1)*** ; nevertheless, pictures of healthy papaya in Puna in 1992 (**Fig. 2A**) when the virus was discovered and a devastated field in 1994 (**Fig. 2B**) illustrate the rapid damage that was occurring to the industry in Puna. In brief, the Kapoho area of Puna was the first to become severely infected because it was close to Pahoa, the initial site of infection in

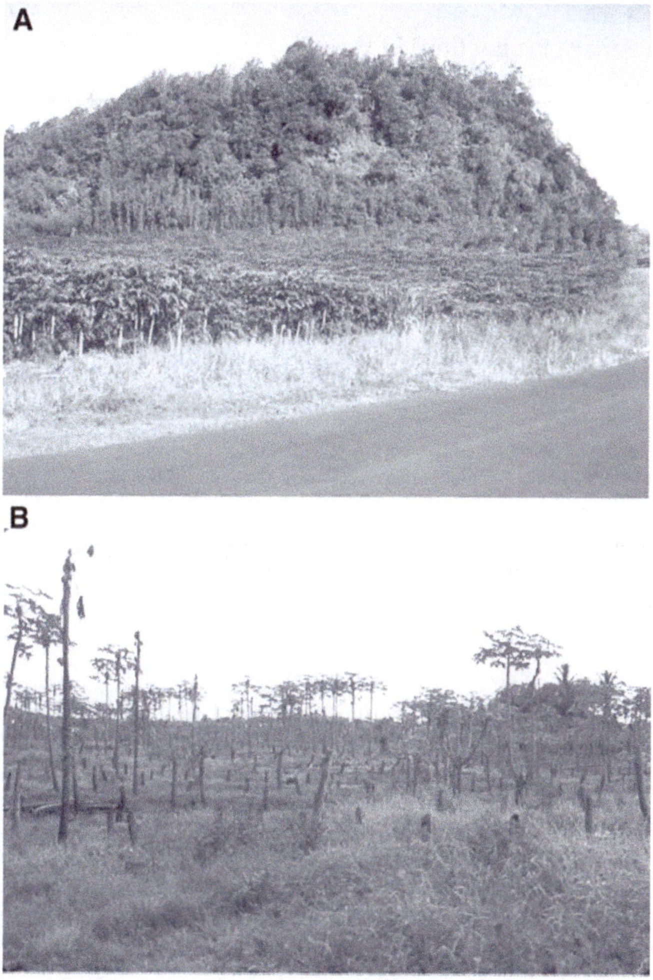

Fig. 2. (**A**) Papaya in Puna in 1992. (**B**) Severely infected papaya fields in Puna in 1994.

Puna. Roguing infected trees to suppress the spread of PRSV in Puna were begun immediately by HDOA, but the virus spread could not be contained. In late 1994, these efforts were abandoned, causing even faster spread of the virus because of abandonment of infected trees by growers. By 1997, PRSV was widespread in Puna, making it nearly impossible for growers to economically grow papaya in many areas of Puna. Yield of papaya in Puna dropped from 53 million pounds of fresh papaya in 1992 when PRSV was discovered in Puna to 26 million pounds in 1998 (**Table 1**). The industry was clearly in a crisis.

Table 1
Fresh Papaya Production[a] in Hawaii and in Puna District From 1992 to 2001

	Fresh Papaya Production in Hawaii		
Year	Total	Puna (× 1000 pounds)	
1992	55,800	53,010	PRSV enters puna
1993	58,200	55,290	
1994	56,200	55,525	
1995	41,900	39,215	
1996	37,800	34,195	
1997	35,700	27,810	
1998	35,600	26,750	Transgenic seeds released
1999	39,400	25,610	
2000	50,250	33,950	
2001	52,000	40,290	

[a] Data were compiled from USDA Statistical Reports of Papaya Grown in Hawaii (*see* Website: http://www.nass.usda.gov/hi/stats)

The growing of papaya in the Hamakua district of Hawaii Island and Oahu Island helped the state of Hawaii continue production but statewide yields of fresh papaya were only 35 million pounds in 1998 compared to 55 million pounds in 1992 (**Table 1**).

A field trial was started on a farm in the Kapoho area of Puna in October 1995 *(26)*. This trial, which was headed by Steve Ferreira of the University of Hawaii, was pivotal for several reasons. It allowed us to: (a) test the resistance of "SunUp" and "Rainbow" under severe virus pressure in Puna, (b) develop data on the horticultural characteristics of the cultivars, and (c) provide a demonstration site for growers and packers to observe the cultivars that they might produce and sell in the future. To accommodate these aims, the field trial consisted of replicated plots, and a large solid block of "Rainbow," which was surrounded by several rows of nontransgenic "Sunrise." The nontransgenic "Sunset" is a sib selection of "Sunrise." The large "Rainbow" block was to serve as a "simulated" commercial planting and the latter would provide data that could be analyzed statistically. Even though "Rainbow"-type papaya had not been tested in Puna, it was used in the simulated commercial plantings because Hawaiian growers overwhelmingly prefer the yellow-flesh papaya, as exemplified by "Kapoho."

Data from the field trial were collected for a period of 28 mo after planting *(26)*. "SunUp" and "Rainbow" were resistant under heavy disease pressure, the performance of "SunUp" and especially "Rainbow" were acceptable to growers and packers, and the comparative effect of growing virus resistant and susceptible papaya plants in an area with heavy and continual virus pressure were dramatic. For example, the annual yield of "Rainbow" in the solid block was about 2242 kg/ha of marketable fruit and yields were steady over the harvest period, and the susceptible "Sunrise" matrix annualized yields were 419 kg/ha initially and dropped to 56 kg/ha after 1.5 yr of harvest. Clearly, the data showed that "Rainbow" was an excellent substitute for the nontransgenic "Kapoho" as judged by its resistance, horticultural characteristics, and the opinions expressed by growers and packers.

3.3. Deregulation and Commercialization of Transgenic Papaya

The papaya crisis in Hawaii spurred the researchers and the industry to move aggressively toward deregulation and commercialization of the transgenic papaya. The details of the events have been described in a review *(1)*, and thus are not repeated here. Instead, some circumstances and efforts will be described to illustrate the interrelations of the researchers with the industry. The Papaya Administrative Committee (PAC) represented the organized section of the papaya industry. The PAC consisted of growers who voluntarily created a Marketing Order that was under the oversight of USDA. The PAC focused primarily on solving market problems such as quality regulations, promotion, and standardization of containers or packs. Funds for these projects came through fee assessment on papaya that was sold through the packinghouses.

The governmental agencies that oversee the regulatory aspects of transgenic crops in the United States are APHIS, Environmental Protection Agency (EPA), and the Food and Drug Administration (FDA). Petitions usually are developed and filed by private companies. Because private companies were not involved in the papaya project, the task of moving the papaya through the regulatory processes fell in the hands of the researchers who lacked experience in this subject. Thus, they simply learned by doing. These efforts were started in late 1995.

The key steps that led to the deregulation of the transgenic papayas in Hawaii *(1)* can be summarized as follows. APHIS, the US federal agency concerned with the potential risks of the papayas released on the environment, deregulated the transgenic plants in November 1996 after establishing that there were no risks associated with potential virus heteroencapsidation or plant weediness. Because the *CP* gene confers virus resistance, it is considered a pesticide and subjected to EPA regulations. The PAC provided the filing fees that accompanied the petition to the EPA. It was argued that PRSV-infected nontransgenic fruits contain detectable levels of CP and have been consumed with no reported ill effects. For instance, large numbers of fruit from trees deliberately infected with the attenuated strain PRSV HA 5-1 were consumed when the technique of cross-protection was being used for papaya *(22)* and the fruits eaten after they were infected. Data also showed that the transgenic fruits contained less CP than the infected nontransgenic fruits. In August 1997, the EPA granted an exemption from tolerance levels of CP in transgenic line 55-1. Finally, the FDA assessed an application with data and statements on the product's safety to human health. The information submitted included data on vitamin content, including vitamin C, presence of the *uidA* and *npt*II transgenes, levels of benzyl isothiocyanate (normally present in nontransgenic papayas) and more. FDA approval was granted in September 1997.

What remained was to obtain the necessary licenses for commercialization of the transgenic papaya *(1)*. The PAC took on the tasks to obtain the licenses for use of intellectual property rights that were used to develop the transgenic papaya and to produce and distribute seeds of the transgenic papaya. The necessary licenses were obtained in April 1998, and seeds were distributed to growers on May 1, 1998, almost 6 yr to the day after PRSV was discovered in Puna. Although seeds were distributed for free, prior to obtaining seeds the person had to register with the PAC, attend an educational session, and sign a material transfer and proprietary rights agreement. The latter states that the transgenic papaya seeds cannot be grown outside of Hawaii. Thus, the papaya industry was ready to start the reclamation of the Puna areas that were devastated by PRSV and to plant newly cleared areas (**Fig. 3A,B**).

3.4. Adoption of the Transgenic Papaya in Hawaii

The Hawaiian efforts also presented a rather unique opportunity to critically measure the adoption and impact of a transgenic product. The geographical area to be surveyed was rather small and the efforts were undertaken by Carol Gonsalves, who was intimately familiar with the activities involved in developing the transgenic papaya and understood the nature of the people in the papaya industry *(27)*.

Following the distribution of transgenic seeds on May 1, 1998, a survey was conducted to examine the rate of adoption by papaya farmers in the Puna area *(27)*. This study was necessary to ascertain the usefulness of the new virus-resistant papaya varieties for the farmers. It was an attempt to capture an early picture of farmer adoption. A farmer was considered to have adopted the technology based not merely on whether seeds were obtained but on whether the farmer had actually planted the seed in an effort to obtain fruit production. Other pertinent issues were also addressed regarding farmers' opinions on the quality of the transgenic fruit and resistance to PRSV, production costs, and attitudes toward the technology.

Fig. 3. (**A**) Reclaimed transgenic papaya field in 1999. (**B**) Transgenic papaya field with maturing fruit in 1999.

In an effort to become familiar with the workings of the papaya industry, farmers and personnel from packing houses, the PAC, the University of Hawaii, HDOA, Hawaii County, the Hawaii Agricultural Statistics Service, and others were consulted one or more times. Many farms in the Puna area were abandoned with infected trees left in the fields. Unfortunately, these trees provided a virus source that aphids could feed on and transmit the virus to other trees in the area. In some places, the environment had changed drastically, such that tall weeds had grown in the fields and the HDOA personnel were not able to locate where some of the papaya fields had been. Some farms had been converted to pasture land. It was nearly impossible to find farmers working in their fields because many farms were abandoned, and as mentioned earlier, many farms ceased to exist. Furthermore, most of the papaya farms were established on leased land, thus there was little chance of finding a farmer's home next to his papaya field.

The interviews took place over numerous weeks from June through September 1999, concluding at 16 months following the release of transgenic seeds in May 1998. Because the growers had decided to organize themselves as a Marketing Order of the USDA, all of the commercial papaya farmers were registered with the PAC. A total of 262 farmers were located on the island of Hawaii. Of these, 171 growers farmed in the Puna area. Attempts were made to contact all of the Puna growers, either by telephone or by letter, and 93 farmers agreed to do interviews. This represented a response rate of 54%. Ninety-two of these registered farmers were qualified to receive transgenic seeds because they had also watched a mandatory training video or educational session and signed a sublicensing contract.

Farmer adoption of the transgenic papayas was astounding: 90% of the farmers had obtained seed, 76% adopted by planting their seed, and 19% were already harvesting fruit. Of a group of 93 respondents, 57 grew papayas in Kapoho, an area that had been hit early and completely by PRSV. The Kapoho farmers who were interviewed had a higher adoption rate, with 88% adopting the technology by planting the seed and 29% far enough along to be harvesting fruit. Not only was there a high rate of adoption but it also occurred rapidly. Overall, 61% of the Puna farmers who received seeds had planted their seeds within 3 mo of obtaining them.

Demographic data showed that most of the farmers were married males, average age 47 yr, primarily of Filipino ethnic heritage. Their education ranged from elementary school through college in the United States or the Philippines. The PAC was instrumental in assisting farmers in the qualification process and later in seed production and distribution to the farmers. Farmers gave most of the credit to the PAC for disseminating information on "Rainbow." But it is important to acknowledge that the University of Hawaii provided much support to the PAC by providing educational materials, research data, and field-testing demonstrations for farmers. Personnel from Cornell University and the University of Hawaii were involved in the development and initial testing of the transgenic materials required for deregulation from the various governing bodies. They also provided a tremendous support to the PAC at all stages of information dissemination and production of the transgenic materials that were released.

In general, the farmers appeared to be satisfied with the performance of the transgenic product. Farmers ranked resistance to PRSV as the greatest reason why they planted the transgenic "Rainbow" variety. When asked to rate the quality of "Rainbow" papaya (based on what they learned from others or experienced firsthand), farmers who commented noted that fruits possessed the desired sweetness and firmness and trees exhibited excellent resistance to PRSV. Although the transgenic seeds were given free of charge to farmers, 86% of the farmers expressed a willingness to purchase transgenic seeds in the future provided that the cost is reasonable. In addition, if new virus-resistant transgenic varieties were introduced in the future, 88% said they would be interested in trying them, and 12% were not sure whether they would try them or not. Concerning a question on labeling, farmer attitudes were optimistic, with a twist. In response to the question, "Do you think it's important to label "Rainbow" or "SunUp" so that the consumer knows that these are transgenic fruit?", 77% said yes, 12% said no, and 11% were not sure. The twist was that farmers were primarily thinking of whether to label their fruit "Rainbow" or "SunUp," and not whether they should label them according to whether they were GMO or not. Farmers felt that labeling by variety name would distinguish their fruit as a superior variety and thus would be a good marketing tool.

Apart from one change in establishing the crop, there were no major changes to their farming practice. Farmers began to establish and transfer transgenic seedlings to the field rather than using their routine practice of direct sowing. In the past, direct seeding of nontransgenic papayas was practiced by most farmers because they either produced the seed themselves, or purchased their seed at a nominal cost (in this survey, an average of $27 per acre when purchased). In direct sowing, 15 or more seeds, sometimes a handful, are tossed into the planting hole to insure that one of them could be maintained as the healthy plant of the hermaphrodite sex, which produces the typical papaya shape desired in the marketplace. Although the transgenic seeds were distributed free of charge, farmers needed to be judicious in planting because the seed supply did not allow for direct seeding. This change from direct sowing to transplanting seedlings allowed farmers to plant fewer seedlings per hole and to avoid skips and the need to replant in case there was a problem with seed quality or low germination rates.

Production costs of growing nontransgenic during the first year varied widely from farmer to farmer with the average cost during the first year at $2515 per acre. Many people are curious about whether it costs more to grow transgenic papayas than the nontransgenic ones. Most respondents (66%) were not sure that there was a difference or made no comment on the matter, 25% felt there was no cost difference, whereas 2% stated that growing transgenics costs less and 8% said that it costs more.

On a global scale, and unlike large-scale farmers who benefit from genetically engineered corn, soy, or cotton, the papaya growers in Puna and hence, in Hawaii as a whole, were small-scale farmers. Fully 95% of those who were farming in this study were growing from one to nearly 50 acres of papayas. Thus, this study shows that small-scale farmers can reap benefits from the fruits of agricultural biotechnology.

Taken together, the transgenic papayas were well accepted in Hawaii because there was PRSV destruction in the papaya fields, and there was a solution: virus-resistant "Rainbow" and "SunUp" varieties. It is undeniable that agricultural biotechnology provided an excellent and useful transgenic product that performed well in the hands of Hawaiian farmers.

3.5. Impact of the Transgenic Papaya in Hawaii

Hawaii Agricultural Statistics Service published Hawaiian papaya production data showing a steady decline in Puna papaya production from 1992 when the PRSV virus was first found in the Puna growing region. At the time, Puna's production was 53 million pounds (**Table 1**). By 1998, the year the transgenic varieties were available to the farmers, production had dropped nearly 50%, to 26.750 million pounds. Production increased since then to 40.290 million pounds in 2001. Remarkably, in 2000, HASS reported that the varietal distribution of papaya-bearing acreage was 56% "Rainbow" on Hawaii Island and 54% statewide *(27)*. These figures are in concert with the high rates of farmer adoption reported in the survey. One can visualize the impact of the transgenic papayas, which had never been grown before, as they help to revive an industry where plants that were once entirely susceptible to a killer virus are now resistant to the effects of the virus. Another testimony to the beneficial effect of agricultural biotechnology is the report by HASS in 1998 compared to the one in 2001. Papaya has been able to maintain its rank as Hawaii's eighth most important farm product in a listing of the state's top 20 products. Farm level revenue generated from papayas was $12.6 million in 1998 and grew to $14.6 million in 2001 *(27)*. Also, the papaya has maintained its spot as Hawaii's second most important fruit crop next to pineapple.

3.6. Some Challenges Facing Hawaii's Papaya Industry

Hawaii's papaya industry faces two challenges pertaining to its future: maintaining its share of the Japanese market and maintaining durability of resistance of the current transgenic papayas. A significant amount of Hawaii's papaya is shipped to Japan, where Hawaii's transgenic papaya is not yet permitted to be imported. Thus, Hawaii needs to grow nontransgenic papaya to retain its share of the papaya market in Japan. Furthermore, measures must be taken to minimize the accidental contamination of transgenic papaya in papaya shipments that are exported to Japan. The transgenic papaya actually has helped growers to economically produce nontransgenic papaya in Puna *(21,28)*. Currently, a number of growers use transgenic papaya as a buffer to cut down virus pressure in their nontransgenic blocks *(28)*. To accomplish this, nontransgenic blocks of papaya are surrounded by large plantings of transgenic papaya, and the nontransgenic blocks are constantly surveyed for virus-infected plants that are immediately removed. Presumably, this practice works because the transgenic papaya provides physical isolation by increasing the distance that viruliferous aphids need to travel to contact nontransgenic plants, and viruliferous aphids that do feed on transgenic papaya before migrating to the nontransgenic papaya would lose their virus dosage. Naturally, the constant vigilance to identify and remove infected plants cuts down on the secondary spread of the virus in nontransgenic blocks. A company that focuses almost exclusively on exporting papaya to Japan has successfully followed this practice for several years (K. Okamura, personal communication).

The issue of developing practices that minimize the introduction of transgenic papaya into Japan has also been addressed cooperatively by Japan and Hawaii. At the request of Japanese importers, HDOA adopted an Identity Preservation Protocol *(29)* that growers and to which shippers must adhere to receive an Identification Preservation Protocol (IPP) certification letter from HDOA that accompanies the papaya shipment. This is a voluntary program. Papaya shipments with this certification are allowed to be distributed in Japan without delay, while Japanese officials perform spot testing to detect contaminating transgenic papaya in the shipment. However, although papaya shipments without this certificate can be shipped to Japan, the shipments are not allowed to be distributed until the Japanese officials have completed their tests, which may take several days to a week. During this time, the quality of the papaya will decrease even before it is distributed for sale. Some significant features of the Identity Preservation Protocol are that the nontransgenic papaya must be harvested from papaya orchards that have been approved by HDOA. To get approval, every tree in the proposed field must be initially tested by the GUS test and found negative, the trees must be separated by at least a 15-ft nonpapaya buffer zone, and new fields must be planted with non-genetically modified organism (GMO) papaya seeds that have been produced in approved non-GMO fields. These tests are monitored by the HDOA and conducted by the applicant who must submit detail records to the HDOA. Before final approval of field, the HDOA will randomly test one fruit from 1% of papaya trees in the field. If approved by the HDOA, fruit from these fields can be harvested. In addition, the applicant must submit the detailed protocols that the applicant will follow to minimize the chances of contamination of non-GMO papaya by GMO papaya. This includes a protocol by the applicant on the random testing of papaya before they are packed for shipment. If the procedures are followed and tests are negative, then a letter from the HDOA will accompany the shipment stating that the shipment is in compliance with a properly conducted Identity Preservation Protocol.

The above procedure represents a good faith effort by the HDOA and applicants to prevent the contamination of transgenic papaya in nontransgenic papaya shipments being exported to Japan. It also illustrates meaningful collaboration between Japan and the HDOA to continue shipment of nontransgenic papaya to Japan with the least delay once it arrives in Japan, and yet adhere to the policy that transgenic papaya will not commercially enter Japan until it is deregulated by the Japanese government. These efforts, along with the effectiveness of the transgenic papaya in helping the economic production of nontransgenic papaya, have allowed Hawaii to maintain significant shipment of the latter to Japan.

Obviously, deregulation of transgenic papaya in Japan will circumvent much of the concern of accidental introduction of transgenic papaya into Japan. To this end, efforts to get the transgenic papaya in Japan were initiated by PAC soon after the transgenic papaya was commercialized in Hawaii. Again, the researchers took the lead in developing the petition. Approval of the transgenic papaya in Japan requires the approval of the Ministry of Agriculture Fisheries and Forestry (MAFF) and the Ministry of Health Labor and Welfare (MHLW). The petition to the MAFF was approved in December of 2000. The petition process for approval by MHLW is still in progress. An initial petition was submitted to MHLW in April 2003. MHLW requested more information and this information are currently being generated by the researchers. Canada approved the importation of Hawaii's transgenic papaya in January 2003.

Finally, the issue on durability of resistance has to be considered. Studies have shown that "SunUp" papaya has broader resistance than "Rainbow" (**18**), but the reality is that "Rainbow" is the dominant transgenic papaya grown in Hawaii ***(21,26)***. So far, we have not observed breakdown of resistance of "Rainbow" in Puna, and on Oahu. However, we need to be on guard for this possibility. The occurrence of new strains due to recombination of PRSV strains in Puna with the *CP* transgene of "Rainbow" is remote. A more realistic danger is through the introduction of PRSV strains resulting from outside of Hawaii. We have shown that "Rainbow" or hemizygous 55-1 is susceptible to many strains of PRSV from outside of Hawaii ***(18,30)***. These include strains from Guam and Taiwan, which are quite close to Hawaii. Technically, "SunUp" should be resistant to many strains of PRSV that might be introduced into Hawaii. However, as noted previously, the red-fleshed "SunUp" is not the preferred cultivar in Hawaii.

A potential solution is to develop transgenic "Kapoho" that is resistant to a wide range of strains. This could be used as a stand-alone cultivar, or it could serve as a transgenic parent for creating a new type of "Rainbow" by crossing the transgenic "Kapoho" with "SunUp." This F1 hybrid should have the horticultural characteristics of current "Rainbow" but very likely have much wider resistance than the current "Rainbow," owing to increase in *CP* gene dosage. We indeed have developed transgenic "Kapoho" that is resistant to range of PRSV strains (D. Gonsalves, *unpublished data*). However, the time frame to commercialize this transgenic "Kapoho" may be longer than the time it took to commercialize line 55-1. These circumstances point to the fact that we need to carefully guard against the introduction of PRSV strains into Hawaii and seek ways to develop new broader resistant cultivars by using the existing germplasm of line 55-1.

4. Transgenic Papayas in Jamaica

Jamaica is one of the few countries in the Caribbean that has maintained a consistent supply of papaya for its domestic and international markets. "Sunrise" solo selections originally from Hawaii are exported as fresh produce to the European Community, Canada, and the United States, whereas large fruited varieties, from Floridian and South American selections, are cultivated mainly for use in the local processing industries. Early accounts document a virus disease of papaya, presumably PRSV, in Jamaica in 1929 *(31)*; however, the first epidemic occurred in 1989 in the traditional papaya growing areas of St. Thomas and St. Catherine *(32)*. At this time, the papaya industry of the country was expanding and had grown from 0.1 ha, in 1986, to 60 ha *(33)*. In an attempt to eradicate the disease, a reported 50,000 trees on 5000 holdings were destroyed between July and November 1990 *(32)*. Recultivation was halted for about 6 mo. Production and expansion of the industry continued thereafter until produce inspectors in 1994 observed the characteristic ring-spotting pattern on fruit destined for export. Field visits confirmed that the disease was prevalent in St. Mary, Clarendon, Manchester, and once again in the parishes of the initial outbreak, St. Thomas and St. Catherine *(32)*. Forty percent of papaya orchards have been devastated by PRSV (Jamaica Papaya Growers Association, *personal communication*).

Shortly after the major outbreak of PRSV in 1994, a papaya-breeding program with collaborators at the University of the West Indies (Mona), Cornell University, and the private sector organization, the Jamaica Agricultural Development (JADF), was initiated. The major goal of the program was to develop new papaya varieties with durable resistance to PRSV using *CP* genes of the virus. Somatic papaya embryos (derived from "Sunrise" solo hybrids) were transformed with the *CP* gene of the virus collected from one of the traditional papaya growing regions of the island Caymans *(2)*. Two versions of the viral *CP* gene were cloned and bombarded into papaya tissues at the New York State Agricultural Experiment Station (NYSAES), Geneva, NY; a translatable version (CP_T) and an untranslatable version (CP_{NT}), based on the status of literature on transgenic virus resistance at the time. The literature then differentiated between CP-mediated and RNA-mediated protection elicited by CP_T and CP_{NT} viral genes, respectively. It was speculated that the mechanism of resistance in the former case was conferred by the *CP*, is broad spectrum against distantly related viruses *(34–36)*, and is effective at both low and high levels of virus inoculum *(37,38)*. Different lines of evidence also suggested that resistance exhibited by transgenic plants not carrying a CP_T version of the viral *CP* gene, is RNA mediated and results from the induction of a specific cellular RNA-degradation mechanism that may be induced by prior infection *(39–42)*. Because the diversity of PRSV in the country was not known, both *CP* versions were utilized in the transformation experiments to assure field resistance.

Within 1 yr after bombardment, some 39 R_0 transformants were acclimatized under greenhouse conditions at Cornell University. *CP* transcript was detected in Northern blot analyses with total RNA from established plants; the signal intensities were higher for some transgenic line designations than others *(2)*. High levels of resistance (78%) to manual inoculations with the homologous virus were observed under greenhouse conditions with plants carrying the CP_T gene. Lower levels of resistance (10%) were obtained with transgenic plants carrying the CP_{NT} and a recovery phenotype (15%) that was characterized by the development of chlorotic spots over the leaf lamina 20–58 d following inoculations with all subsequent new leaves free of symptom expression. Those R_0 clones that were not used in the infectivity assays were subsequently shipped to Jamaica for field testing in the same region of the island from which the homo-logous isolate was collected *(2)*. It should be noted that Jamaica did not have the regulatory framework in place for overseeing the importation and testing of genetically modified organisms when the project started in 1994. In 1997, the National Biosafety Committee (NBC) was set up as a subcommittee of the National Commission of Science, and Technology and the Plants (Importation) Control Regulations was passed in Parliament under Section 38 of the Plants (Quarantine) Act to allow for the importation and controlled field testing of the transgenic papaya. Therefore, the transgenic plants remained under greenhouse conditions at Cornell University for 2 yr while the appropriate biosafety regulations were put in place and a permit issued.

On transfer to Jamaica (1998), the primary transformants (R_0) were acclimatized for 2–3 wk under shadehouse conditions and subsequently transferred to the field. A plot of about 0.2 hectares was enclosed by fencing on the commercial Brampton Farm in St. Catherine. Thirteen rows of transgenic (184

Fig. 4. Transgenic papaya field in Jamaica (1999).

plants) and nontransformed "Sunrise" solo papaya (33 plants) were randomly established in the plot. The test plot was surrounded by a border row consisting of nontransformed "Sunrise" solo plants and some 60 hectares of mangoes. No mechanical inoculations were done and papaya fields in the nearby regions served as the primary source of PRSV inoculum (**Fig 4**).

Within 5 mo after establishment, PRSV was detected in the plot. Even though the permit issued allowed for field evaluation of the materials, the researchers were instructed by the NBC to remove the infected trees. The major concern was whether infected plants in the plot would act as a reservoir for the virus and contribute to the spread of the virus in the area. It was only after a series of discussions with members of the NBC, papaya growers, and a review of the literature on the field-testing of transgenic papaya in Hawaii that the NBC revoked the injunction and assessment of resistance was allowed.

By 17 mo, all nontransformed trees showed symptoms typical of PRSV infection. As for the transgenic trees, a similar trend in the levels of resistance conferred by the *CP* genes under greenhouse conditions was observed in the field. Transgenic lines with the CP_T gene showed strong resistance (80%) against field infections while those with the CP_{NT} gene were not as resistant (44%), showing tolerance, that is, attenuated symptoms, or severe symptoms after a delay (18–24 mo) in the onset of disease. In a second trial (1999), R_1 offspring exhibited similar levels of resistance as the parental lines (58%), improved resistance (26%), or lowered resistance (16%). Together, these observations suggested that even though the protein expression product is not involved in the resistance mechanism, resistance was only achieved by transgene messenger RNA (mRNA) from a CP_T gene. Other studies have similarly shown that the CP_{NT} gene is not as successful as the CP_T gene in conferring transgenic resistance. For instance, all 13 transgenic oilseed rape lines carrying CP_{NT} were as susceptible to *Turnip mosaic virus* as control plants; only transgenic lines carrying the CP_T gene effectively conferred resistance against the virus ***(43)***. Likewise, *Potato mop-top virus* CP_{NT} transgenic potato exhibited mild symptom expression whereas CP_T lines were completely resistant and did not develop symptoms ***(44)***.

The horticultural characteristics of the Jamaican transgenic papaya were also assessed relative to the nontransformed trees. Numerous lines were found to exhibit agronomic traits suitable for local and export markets. There were gynodioecious trees that produced fruit of red flesh (average weights between 260 and 535 g) and also gynodioecious trees with larger fruits (>535 g) with red or yellow flesh ***(45)***. Nutritional components (vitamin C, total sugars, fats, brixes) and the antinutritional component, benzyl isothiocyanate (BITC) were also examined ***(45,46)***. Statistical analyses conducted at the 95% confidence interval indicate that these components of the transgenic lines were not significantly different from the nontransformed controls. Although total sugars were lower in fruits from line 52.3 ($p < 0.05$) and fats slightly higher in fruits from line 52.24 ($p = 0.037$), the values were within the range documented for papaya.

Another examined characteristic of the transgenic lines of practical benefit related to biosafety. One of the concerns with foods derived through genetic engineering is whether transgenic products are safe for human consumption and there is no potential of introducing allergens into the food supply ***(47)***. Although the safety of NPTII, GUS, and viral CP gene products is documented ***(47–49)***, it is reassuring that neither CP nor NPTII protein was detected in the edible pulp of fruits from the Jamaican transgenic papaya trees. Surprisingly, however, GUS activity was consistently detected in only one transgenic line.

Encouraging data were also obtained in a 12-wk dietary study with adult Wistar rats. The rats were divided into three groups; (a) rats fed normal diet (a marketed laboratory rodent diet recommended for rats, mice, and hamsters), (b) normal diet supplemented with commercial papaya pulp, and (c) normal diet and transgenic papaya pulp. At the end of the study, blood, liver, and kidney were collected and total plasma protein determined as well as the activities of transaminases and phosphatases. Plasma proteins are made in the liver and are important in maintaining the pressure between fluids of the vascular system and surrounding cells. Although an insensitive indicator, measurement of the total plasma protein can point to liver dysfunction. On the other hand, levels of the enzyme transaminases (aspartate and alanine transaminases) and phosphatases (acid and alkaline phosphatases) in the plasma are a measure of liver damage; high levels are suggestive of disease or injury of the organ. In the analyses, there were no significant differences ($p < 0.05$) in the levels of acid and alkaline phosphatases in the liver of rats fed normal rat diet compared to those fed the different preparations of the transgenic and commercial papaya. The same trend was observed for the phosphatases in the kidney and plasma. Moreover, the protein content of the plasma was not statistically different among the groups. Together, data suggesting no negative effects of the papaya supplements (commercial or transgenic) on tissue and organ integrity has been obtained.

The study shown above represents the results of 6 yr of research focused on the development and transfer to the field in Jamaica of PRSV-resistant transgenic papaya lines. The resistance exhibited by the Jamaican transgenic papaya lines (four in particular, 52.2, 52.3, 52.22, 52.24) should be useful for commercial production and efforts are presently focused on stabilizing interesting lines by continued self-pollination. Moreover, Southern blot data recently obtained in the lab (Tennant et al., unpublished data) suggest a single insertion of the *CP* gene in some of the resistant transgenic lines (e.g., line 52.3). Invariably, this will be of benefit in transferring the resistant trait to local papaya varieties, namely the large fruited "Santa Cruz giant" and "Cedro" varieties. These varieties are as susceptible to PRSV as the commercial "Sunrise" solo variety and are grown for the local market, particularly for the hotel and baking industries. Given these results, the collaborators put together a timeline for moving the transgenic papaya toward deregulation and commercial release. In 2000, a field trial aimed at further field evaluation and building the seed supply would be set up at Bramptom Farm, in 2001 data would be submitted to the NBC for a decision on deregulation and setting up trials on farmers' orchards, and in 2002, pending a positive ruling from the NBC and the establishment of the necessary regulatory framework, large-scale field release of transgenic seeds to papaya growers.

However, the project is 2 yr behind the proposed schedule. The third field trial has been set up as planned but the question at hand is whether researchers will be allowed to conduct field-tests on growers' orchards and whether the transgenic papaya will be deregulated and released commercially in Jamaica.

Although Jamaica was ahead of its CARICOM counterparts and Small Island Developing States (SID) in having established a National Biosafety Committee in 1997 and initiated field-testing genetically modified organisms in 1998, the country now appears to be lagging behind. The required regulatory guidelines for the release of genetically modified organisms have not progressed despite formulation in 2000. Discussions on how to facilitate the extension of the field trials to growers' orchards (provided satisfactory review of the data generated by the previous field tests) were initiated last year. Meetings with Parliamentary Counsel representatives advised the NBC on examining the Plant Quarantine Act to accommodate this Post Quarantine stage or confined testing of the transgenic papaya on growers' orchards. No final decisions have been communicated to date.

Moreover, one of the collaborators financing the project (JADF) has requested that the next phase of the project involving field-testing on growers' orchards across the island, be postponed until the NBC has deregulated the transgenic papaya. The organization foresees legal repercussions with some growers should the transgenic materials be released under temporary provisions by the NBC.

However, there is interest among papaya farmers for the completion of the research and release of materials. Growers have visited the field trials and identified transgenic trees exhibiting acceptable commercial traits. Moreover, a recent survey of papaya growers producing on farms for the local (26%) or export market (74%), report that 80% are anxious to receive the transgenic papaya seeds and to participate in setting up experimental plots with the transgenic papaya on their orchards. The other 20% are hesitant of the introduction the material into commerce (Dawkins et al., unpublished data). This is because their major markets are in Europe and they fear genetic contamination of their nontransformed materials. Some of these growers have said that their European buyers explicitly stated that fruits from farms with transgenic crops (papaya or otherwise) would not be taken. Another reason given by these growers against adopting the transgenic papaya, is that they may not be safe for consumption and that the transgenic trees actually carry the virus (i.e., they are "immunized") and will spread the virus to other trees. Interestingly, these farmers are situated in the Western end of the island, where the virus only recently moved in and the disease pressure is very low. They are the ones presently exporting papaya fruit and maintaining the industry.

In a wider survey of the Jamaican public in the corporate area and outskirts, 40% are willing to try genetically modified products, 49% will not try these products, and 11% were not able to say whether they would purchase these products or not (Pinnock et al., unpublished data). The respondents felt that genetic engineering could probably have a positive effect on the quality of life (73%) and were in favor of the application of the technology to medicine (73%), the improvement of ornamentals (52%), and plant defenses (69%), rather than the improvement in the taste of foods (39%) and improvement of livestock (32%).

Thus, there is interest in the project from some papaya farmers and consumers, so the researchers will continue developing a transgenic product with acceptable commercial performance while the regulatory processes "catch up." The research will focus into developing homozygous transgenic lines and in transferring the transgenes to other local varieties, as well as developing new varieties using portions of viral genes to elicit resistance by chimeric transgenes (Gonsalves et al., unpublished results) and safer selection markers. Most importantly, the NBC has mounted an education program to inform the public on genetic engineering; activities at schools, as well as radio and television broadcasts, have been initiated to clear up some of the misconceptions on the technology. But as the old adage goes, only time will tell whether transgenic papaya will eventually be deregulated in Jamaica.

5. Transgenic Papayas In Venezuela

Among the three countries discussed in this chapter, Venezuela is the one producing the most but exporting the least of papaya. In 2002 Venezuela produced 120,000 Mt of papayas, and the United States (mainly Hawaii) and Jamaica produced 20,684 and 8,637 Mt, respectively. However, yields were higher for the United States and Jamaica (300,000 and 230,000 kg/ha) as compared to Venezuela (176,000 kg/ha), which exported papayas for an added value of only \$472,000 (1000 Mt), whereas Jamaica and the United States obtained gains as high as \$3.3 million (2200 Mt) and \$17.2 million (8300 Mt), respectively. The United States also imported papayas valued \$62 million in the year 2001 ***(5,50)***.

Venezuela is a country with an economy based on the exploitation of oil and where agriculture represents just 5% of the GDP. Papaya is a very popular fruit and its production high but, as shown before, almost entirely devoted to local consumption. Commercial and domestic orchards are severely attacked by PRSV, and the most efficient control measure is eradication. Big farmers use to switch to other crops when roguing does not suffice as a control measure of PRSV, but small orchards are kept until the virus decimate the trees. In Mérida, state papayas are mainly produced in the southwest, close to Lake Maracaibo. Although reports on early infection by PRSV in the area are lacking, in the beginning of the 1990s the main papaya orchards were severely attacked by the virus, and industrial papaya production declined dramatically. In an effort aimed at developing the infrastructure required to help advance projects in Biotechnology in Venezuela the Inter American Development Bank issued Universidad de Los Andes (ULA) with a grant for the creation of transgenic papayas resistant to local varieties of PRSV. Transgenic papayas were obtained in Venezuela (Universidad de Los Andes, ULA) thanks in part to a collaborative effort with Cornell University that started in 1992 ***(21)***.

In 1993, the *CP* gene of two different Venezuelan geographical isolates from Mérida were cloned at the New York State Agricultural Experiment Station (NYSAES) in Geneva, NY ***(51)***. The two isolates were collected in lowland areas where papaya is grown domestically (PRSV LA in Lagunillas,

Mérida) or commercially (PRSV EV in El Vigía, Mérida). Both isolates were the donors of *CP* genes that were cloned in their translatable, antisense or sense/untranslatable forms ***(51)***. Although the actual role of the protein was not clear when the transformations were done (1994–1995), only the translatable constructs were used for further experiments. Contrary to the Jamaica's case, plant material was transformed in Venezuela instead of NYSAES, after a transfer agreement of biological material was signed between ULA and Cornell University. The embryo donors for transformation via *Agrobacterium tumefaciens* were open-pollinated papayas grown commercially in El Vigía and known locally as "Tailandia Roja" ("Thailand Red"). Plant material was donated by an industrial grower willing to assay the transgenic papayas once they were developed. Plant material was transformed following a protocol designed in ULA for this purpose ***(52)***, and few transformants were recovered (1995–1996) and subsequently crossed (1997–1998). The progenies from several of these crosses were analyzed for three generations ***(3,19)***.

Venezuela's R_0 transgenic papayas consisted of few independent lines with single-copy insertions of either the PRSV LA *CP* or PRSV EV *CP* gene. Female, male and hermaphrodite R_0 plants were intercrossed or selfed in Mérida to obtain the first seed-derived generation; R_1 and R_2 generations were obtained and analyzed by one of us at NYSAES ***(3,19)***. Resistance seemed to be RNA mediated, and R_1 and R_2 plants showed promising levels of resistance not only to the homologous strain of the virus but also to distant geographical isolates of PRSV, such as one from Thailand with only 87–89% similarity to the Venezuelan transgenes ***(3)***. Among the R_2 plants, two hermaphrodite plants showing high levels and ranges of resistance were identified and kept for future multiplication and breeding programs. A replicate set of experiments was performed under greenhouse conditions in Mérida, and resistance to the homologous strains of the virus was high. Similar to the Jamaica case, but not the Hawaii case, neither the *NPTII* nor the viral *CP* are detectable in the transformed, PRSV-LA and -EV resistant transgenic papaya plants ***(3)***. Because the transformed tissue was donated by an open pollinated plant, segregation of important horticultural traits was observed. For instance, differences in color of the flesh, and size and shape of the fruits are common. However, all fruits tested were sweet and pleasant (G. Fermin, personal observations). Further characterization of the transgenic fruits is stalled, for reasons that will be explained below. However, at this point it is worth recalling that Venezuelan papayas were transformed with PRSV *CP* genes isolated from viruses in areas where papayas are grown for domestic or industrial purposes (i.e., the transgenes are homologous to the virus present in the area of the potential release of the engineered plants), that they harbor a single copy of the transgenes *npt*II and *CP*, that these two proteins are not detectable in the transformed plants, that all of the transgenic plants lack reporter marker genes and that the safety concerns of their release and future consumption were diminished to its minimal potential. Although transgene insertions by *Agrobacterium*-mediated transformation are "cleaner" than the ones produced by biolistics ***(53)***, we still lack data supporting the idea than the single insertions harbored by the Venezuelan transgenic papayas are composed only of transgenes and T-DNA sequences with or without rearrangements.

The purpose of creating PRSV-resistant papayas was to offer a clean and reliable solution to the problem associated with the presence of a virulent strain of the virus in the papaya growing area of Mérida. The next natural step was to release some experimental plants in the field and assess their degree of resistance under the local pressure of the virus. It is worth recalling that in the year 2004, Venezuela still does not have a set of rules and regulations regarding the manipulation, release, import, and commercialization of GMOs. There is a Biosafety National Committee in charge of presenting some rules for the discussion of future national laws, but the work is being drastically slowed by some anti-GMO activists. Curiously, the same kind of fierce opposition to transgenic papayas has not been heard of in Hawaii and Jamaica to the level we have suffered in Venezuelan. Yet, even in the absence of a national mandatory rule, the Venezuela transgenic papayas were the subject of a study aimed at analyzing all potential sources of risk related to their release, and a methodology to liberate them in the field at a small scale was proposed ***(54)***, years before the transgenic plants were developed.

A small plot of transgenic papayas was set at Lagunillas, Mérida. The area of Lagunillas was chosen for this set of tests for various reasons: one of the virus samples was isolated in the area, papaya is being grown only domestically far from any commercial orchard (and from the experimental field), and the plot is close to the researchers in charge of the field tests. The lack of a national legislation pertaining to the management of GMOs obliged some researchers to look for creative solutions. Laboratory work and field experiments with GMOs were not expressly prohibited in the country, but a special permit from the Ministry of Health of Venezuela (MHV) was granted to perform the field testing (1999–2000) in the small plot described above (**Fig. 5A**). The plot was planted with R_1 individuals previously selected in the greenhouse as PRSV-resistant. Later on, the plants continued to show very good performance in the field under the local pressure of the virus. When the transgenic, PRSV-resistant papayas were set to flower unexpected problems started to emerge.

Parallel to the development and characterization of the transgenic papayas for Venezuela, open opposition to GMOs from nongovernmental organizations (NGOs) flourished in Mérida by the end of the 1990s. The appearance of misinformed but well-organized opposition groups, the poor preparedness of

Fig. 5. (**A**) Transgenic papaya field in Venezuela (May 2000). (**B**) Transgenic papaya field after being set in fire (December 2000).

the public opinion, and the lack of supporting legislation on GMOs all conspired against the acceptance of the transgenic papayas, engineered in the cleanest way possible to solve a practical problem. The main arguments against the transgenic papayas were that the experiments were being performed in secrecy, the risks associated with the creation and field release of the transgenic papayas were unknown, and the country lacked the appropriate regulations to deal with GMOs in a responsible and organized manner. Only the third claim had certain truth. The news on the grant obtained to develop transgenic papayas in Mérida was highly publicized nationally. Besides, all the data was publicly available to anyone interested, mainly as monographs from one of the authors of this chapter (G. Fermín) but also through articles in local journals and seminars across the country. The claim of secrecy is probably the most unfair of all. Later, we understood that the accusation of secrecy was just another element of a very well-orchestrated campaign to discredit the scientists involved in the experiments and to predispose the public against the transgenic technology as such. The few transgenic papayas tested in the field (**Fig. 5A**) were held responsible for the increase in mutation rates, abortions, "contamination" of food and water, and so on; scientists were held as criminals and their names painted on the city walls with degrading epithets. The people of Lagunillas suffered the most, being constantly bombarded with a collection of "information" aimed at creating panic. Opponents did not accept the legality of the permit issued by MHV and the matter was discussed and solved by the State Legislature. The scientists were allowed to continue the experiments until the plants set fruits, which could be collected and transferred to the laboratory. However, the plants should be incinerated after fruits were collected. A serious discussion about GMOs was lacking, the fierce battle ended up exhausting all reasoning, and legal decisions were not accepted nor respected by the infuriated people who were convinced by few NGO representatives that transgenic papayas would trigger serious health problems. The most fanatical opponents of transgenic plants violently attacked the plot's guardian and set the small plot on fire in December 2001 (**Fig. 5B**).

At the time this chapter was written, transgenic papayas were still considered a "bad thing" and the legislation on GMOs was moving forward slower than necessity dictates. Some NGOs call for a total ban on GMOs, and opinions in the government are divided. When the political uncertainty of the country dissipates, Venezuelans will probably recover the time to think about more pressing and fundamental issues, like how to speed the process of developing a sound Plant Biotechnology program that might help modernize Venezuelan agriculture. In the meantime, a well-thought plan is under way to mitigate the damage done by few anti-GMO activists. The media, mainly the local media, is surprisingly better informed than the anti-GMO activist and crave for more reliable information on transgenic food. Several interviews in newspapers and radio and local TV stations are helping dissipate the mystified negative aura that surrounds the issue. A more proactive approach to interact with State legislators is currently underway, and workshops to educate qualified people are being organized in the University on the potential risks associated with the release of transgenic plants. Once the papaya growers feel enamored of the many benefits this technology might bring, we will probably reach a stage of more conscious and informed acceptance of this precious biotechnological good.

6. Concluding Remarks

We have presented three examples in the adoption of a transgenic crop: the engineered papayas resistant to PRSV. They differ in the degree of acceptance, despite the unequivocal success the transgenic papayas from Hawaii, Jamaica, and Venezuela have shown in resisting PRSV attack. The Hawaiian transgenic papaya story often is viewed as a rather unique case because the Hawaiian papaya industry had no alternative solution. However, what is often missed is that the papaya case represents efforts by researchers who were simply trying to do their job of addressing and developing solutions to a defined problem. The fact that the problem became of crisis proportion and that the solution was rather dramatic should not be interpreted to mean that the transgenic approach should only be used as a last resort. Instead, it should be viewed as a component of a system for developing effective, safe, and sustainable efforts to address agricultural problems. The Jamaican case is potentially a good example of a successful story on technology transfer, if it can overcome the serious challenges on the deregulatory stage to bring the papayas to the market. However, Venezuela faces more fundamental problems: a society still in search of a political identity spends little time solving problems of longlasting effects (e.g., food production and safety). From the three countries discussed here, only Venezuela has a very involved and aggressive group of anti-GMO activists. If the success of the transgenic papayas in other latitudes has not convinced them on the many virtues of the biotechnological approach to solve agriculture-related problems, then the potential success of the Venezuelan transgenic papayas will probably do so.

References

1. Gonsalves, D. (1998) Control of *Papaya ringspot virus* in papaya: a case study. *Ann. Rev. Phytopathol.* **36,** 415–437.
2. Tennant, P., Ahmad, M., and Gonsalves, D. (2002) Transformation of *Carica papaya* L. with virus coat protein genes for studies on resistance to *Papaya ringspot virus* from Jamaica. *Trop. Agr.* **79,** 105–113.
3. Fermín, G., Inglessis, V., Garboza, C., Rangel, S., Dagert, M., and Gonsalves, D. (2004) Engineered resistance against *Papaya ringspot virus* in Venezuelan transgenic papayas. *Plant Dis.* **88,** 516–522.
4. Manshardt, R. (1992) Papaya, in *Biotechnology of Perennial Fruit Crops.* (Hammerschlag, F. and Litz, R., eds.), CAB International Press, Cambridge, UK, Chapter 21, pp. 489–511.

5. Food and Agriculture Organization of the United Nations. (2002) The state of food and agriculture. Agriculture and global public goods ten years after the Earth Summit. FAO Agriculture Series. Editorial Group—FAO Information Division: 212, Rome, Italy.
6. Powell-Abel, P., Nelson, R., De, B., et al. (1986). Delay of disease development in transgenic plants that express the tobacco mosaic virus coat protein gene. *Science* **232,** 738–743.
7. Sanford, J. and Johnston, S. (1985) The concept of pathogen derived resistance: deriving resistance genes from the parasite's own genome. *J. Theor. Biol.* **113,** 395–405.
8. Lomonossoff, G. (1995) Pathogen-derived resistance to plant viruses. *Ann. Rev. Phytopathol.* **33,** 323–343.
9. De la Rosa, M. and Lastra, R. (1983) Purification and partial characterization of papaya ringspot virus. *Phytopathol. Z.* **106,** 329–336.
10. Purcifull, D., Edwardson, J., Hiebert, E., and Gonsalves, D. (1984). Papaya ringspot virus. *CMI/AAB Descriptions of Plant Viruses* **292** (84 Rev.), p. 8.
11. Shukla, D. (1994) *The Potyviridae.* Wallingford, CAB International.
12. Teakle, D. and Pares, R. (1977) Potyvirus (potato virus Y) group. Ultrastructure of biological systems, in *The Atlas of Insect and Plant Viruses.* Maramorosch, K., ed.), Academic Press, New York, NY.
13. Shukla, D., Frenkel, M., and Ward, C. (1991) Structure and function of the potyvirus genome with special reference to the coat protein coding region. *Can. J. Plant Pathol.* **13,** 178–191.
14. Urcuqui-Inchima, S., Haenni, A.-L., and Bernardi, F. (2001) Potyvirus proteins: a wealth of functions. *Virus Res.* **74,** 157–175.
15. Yeh, S.-D. and Gonsalves, C. (1984). Evaluation of induced mutants of papaya ringspot virus for control by cross protection. *Phytopathology* **74,** 1086–1091.
16. Manshardt, R. (1998) *'UH Rainbow' Papaya.* College of Tropical Agriculture and Human Resources Germplasm, University of Hawaii, Honolulu, HI, pp. G-1:2.
17. Fitch, M., Manshardt, R., Gonsalves, D., Slightom, J., and Sanford, J. (1990) Stable transformation of papaya via microprojectile bombardment. *Plant Cell Rep.* **9,** 189–194.
18. Tennant, P., Fermin, G., Fitch, M., Manshardt, R., Slightom, J., and Gonsalves, D. (2001) Papaya ringspot virus resistance of transgenic "Rainbow" and "SunUp" is affected by gene dosage, plant development, and coat protein homology. *Eur. J. Plant Pathol.* **107,** 645–653.
19. Fermín-Munoz, G. (2002). Use, application, and technology transfer of native and synthetic genes to engineering single and multiple transgenic viral resistance. Department of Plant Pathology. Cornell University, Ithaca, NY, p. 293.
20. Gonsalves, C., Cai, W., Tennant, P., and Gonsalves, D. (1998). Effective development of papaya ringspot virus resistant papaya with untranslatable coat protein gene using a modified microprojectile transformation method. *Acta Horticult.* **461,** 311–314.

21. Gonsalves, D. and Fermin, G. (2004) Transgenic papaya to control papaya ringspot virus in Hawaii and technology transfer to other countries, in *Handbook of Plant Biotechnology*, Christou, P. and Klee, H., eds.), John Wiley & Sons, Chichester, West Sussex, UK.
22. Yeh, S.-D., Gonsalves, D., Wang, H.-L., Namba, R., and Chiu, R.-J. (1988) Control of papaya ringspot virus by cross-protection. *Plant Dis.* **72,** 375–380.
23. Gonsalves, D. and Ishii, M. (1980). Purification and serology of papaya ringspot virus. *Phytopathology* **70,** 1028–1032.
24. Fitch, M., Manshardt, R., Gonsalves, D., Slightom, J., and Sanford, J. (1992) Virus resistant papaya derived from tissues bombarded with the coat protein gene of *Papaya ringspot virus. Bio/Technol.* **10,** 1466–1472.
25. Lius, S., Manshardt, R., Fitch, M., Slightom, J., Sanford, J., and Gonsalves, D. (1997) Pathogen-derived resistance provides papaya with effective protection against papaya ringspot virus. *Mol. Breeding* **3,** 161–168.
26. Ferreira, S., Pitz, K., Manshardt, R., Zee, F., Fitch, M., and Gonsalves, D. (2002) Virus coat protein transgenic papaya provides practical control of papaya ringspot virus in Hawaii. *Plant Dis.* **86,** 101–105.
27. Gonsalves, C. (2001) Transgenic virus-resistant papaya: farmer adoption and impact in the Puna area of Hawaii. Graduate School, Empire State College, State University of New York, NY 170.
28. Gonsalves, D. and Ferreira, S. (2003) Transgenic papaya: a case for managing risks of papaya ringspot virus in Hawaii. *Plant Health*, doi:10.1094/PHP-2003-113-03-RV.
29. Anonymous (2003) *Identity Preservation Protocol for Non-GMO Papayas*, Revised April 7, 2003. Hawaii Department of Agriculture, Quality Assurance Division, Commodities Branch, Honolulu, HI.
30. Tennant, P., Gonsalves, C., Ling, K., et al. (1994) Differential protection against papaya ringspot virus isolates in coat protein transgenic papaya and classically cross-protected papaya. *Phytopathology* **84,** 1359–1366.
31. Jensen, D. (1949) Papaya diseases with special reference to papaya ringspot virus. *Phytopathology* **39,** 191–211.
32. Young, F. (1994) Diseases, pests and other field problems: the papaya ringspot virus in Jamaica. Proceedings of the Papaya Industry Seminar.
33. Thomas, O. (1993) The current status of papaya production in Jamaica. Proceedings the Papaya and Mango Seminar, Kingston, Jamaica.
34. Namba, S., Ling, K., Gonsalves, C., Slightom, J., and Gonsalves, D. (1992) Protection of transgenic plants expressing the coat protein gene of watermelon mosaic virus II or Zucchini yellow mosaic virus against six potyviruses. *Phytopathology* **82,** 940–946.
35. Pang, S.-Z., Nagpala, P., Wang, M.-B., Slightom, J., and Gonsalves, D. (1992) Resistance to heterologous isolates of tomato spotted wilt virus in transgenic plants expressing its nucleocapsid protein gene. *Phytopathology* **82,** 1223–1229.
36. Maiti, I., Murphy, J., Shaw, J., and Hunt, A. (1993) Plants that express a potyvirus proteinase gene are resistant to virus infection. *Proc. Natl. Acad. Sci. USA* **90,** 61,110–61,114.

37. Nelson, R., Powell-Abel, P., and Beachy, R. (1987) Lesion and virus accumulation in inoculated transgenic tobacco plants expressing the coat protein gene of tobacco mosaic virus. *Virology* **158,** 126–132.
38. Providenti, R. and Gonsalves, D. (1995). Inheritance of resistance to cucumber mosaic virus in transgenic tomato lines expressing the coat protein gene of the white leaf strain. *J. Hered.* **86,** 85–88.
39. Farinelli, L. and Malnoe, P. (1993) Coat protein gene-mediated resistance to potato virus Y in tobacco: examination of the resistance mechanisms—Is the transgenic coat protein required for protection? *Mol. Plant Microbe Int.* **6,** 284–292.
40. Kollar, A., Thole, V., Dalmay, T., Salamon, P., and Balazs, E. (1993) Efficient pathogen-derived resistance induced by integrated potato virus Y coat protein gene in tobacco. *Biochimie* **75,** 625–629.
41. Lindbo, J., Silva-Rosales, L., Proebsting, W., and Dougherty, W. (1993) Induction of a highly specific antiviral state in transgenic plants—implications for regulation of gene-expression and virus resistance. *Plant Cell* **5,** 1749–1759.
42. Smith, H., Swaney, S., Parks, T., Wernsman, E., and Dougherty, W. (1994) Transgenic plant virus resistance mediated by untranslatable sense RNAs: expression, regulation, and fate of nonessential RNAs. *Plant Cell* **6,** 1441–1453.
43. Lehmann, P., Jenner, C., Kozubek, E., Greenland, A., and Walsh, J. (2003) Coat protein-mediated resistance to Turnip mosaic virus in oilseed rape (*Brassica napus*). *Mol. Breeding* **11,** 83–94.
44. Barker, H., Reavy, B., McGeachy, K., and Dawson, S. (1998) Transformation of *Nicotiana benthamiana* with the potato mop-top virus coat protein gene produces a novel resistance phenotype mediated by the coat protein. *Mol. Plant Microbe Int.* **11,** 626–633.
45. Chin, M., Pinnock, S., Roberts, M., et al. (2003) Field resistance and nutritional analysis of papaya ringspot virus coat protein transgenic papaya. Seventeenth Annual National Conference on Science and Technology, Science and Technology for Economic Development, Technology Driven Agriculture and Agro-Processing, Kingston, Jamaica.
46. Roberts, M., Minot, D., Tennant, P., Jackson-Malete, J., and Gonsalves, D. (2003) Comparative Evaluation of transgenic and nontransgenic papaya (*Carica papaya* L.) fruit developed for the control of Papaya ringspot virus in Jamaica. Proceedings of the Sixth Conference of the Faculty of Pure and Applied Sciences, Mona, Kingston, Jamaica.
47. Kaeppler, H. (2000) Food safety assessment of genetically modified crops. *Agron J.* **92,** 793–797.
48. Kuiper, H., Kleter, G., Noteborn, H., and Kok, E. (2001) Assessment of the food safety issues related to genetically modified foods. *Plant J.* **27,** 503–528.
49. Chassy, B. (2002) Food safety evaluation of crops produced through Biotechnology. *J. Am. Coll. Nutr.* **21,** 166S–173S.
50. World Bank (2003) World development indicators 2003.
51. Fermín, G. (1996) Análisis molecular de variantes geográficos del virus de la mancha anular de la lechosa (PRV) y cloneo de diversos constructos (Molecular

analysis of geographical isolates of Papaya ringspot virus (PRSV) and cloning of diverse constructs). Graduate School in Fundamental Medical Sciences, Faculty of Medicine. Universidad de Los Andes, Mérida, p. 106.
52. Inglessis, M., Vielma, M., Dagert, M., Contreras, I., and Herrera-Estrella, L. (1997). Preliminary studies for an efficient regeneration and transformation of *Carica papaya* L. var. Thailand red. *Acta Cient. Venez.* **48,** 154–159.
53. Smith, N., Kilpatrick, J., and Whitelam, G. (2001) Superfluous transgene integration in plants. *Crit. Rev. Plant Sci.* **20,** 215–249.
54. Fermín, G. (1996). Elementos para la pre-evaluación del riesgo biológico de plantas transgénicas de lechosa resistentes al virus de la mancha anular (PRV) ensayadas en campo a pequeña escala. Graduate School in Fundamental Medical Sciences, Faculty of Medicine. Universidad de Los Andes, Mérida, p. 86

Index

E

F

G

H

I

K

L

M

N

O

Kolibri 360 GmbH
Ferdinand-Jühlke-Straße 7
99095 Erfurt, Deutschland
produktsicherheit@kolibri360.de